Major General Sir C.W. Pasley, K.C.B.
(courtesy of Royal Engineers Barracks, Chatham)

PRACTICAL ARCHITECTURE

C. W. PASLEY

With an introduction by Lawrance Hurst

DONHEAD

First published as lithographed notes in 1826

Reprinted in 1862 by the Royal Engineer Establishment, Chatham

Donhead Publishing Ltd
Lower Coombe. Donhead St Mary, Shaftesbury, Dorset SP7 9LY
Tel. 01747 828422
www.donhead.com

ISBN 1 873394 47 0

A CIP catalogue for this book is available from the British Library

Printed in Great Britain by J. H. Haynes & Co. Ltd., Sparkford, Somerset

Donhead Publishing would like to thank Professor Sir Alec Skempton FRS of Imperial College, London, for making the original of this work available from the Historical Collections in Imperial College, Department of Civil Engineering Library, and Mr Michael Chrimes, Head Librarian, Institution of Civil Engineers, for drawing it to the attention of the publisher.

Donhead Publishing would like to acknowledge the Royal Engineers Library, Chatham for the loan of an additional copy of the original work.

British Library Cataloguing in Publication Data

Pasley, C. W.
Practical architecture – 2nd ed.
1. Building
I. Title
690
ISBN 1873394470

Introduction to the 2001 Edition

Although the title of this book is misleading, its importance as a textbook should not be underestimated. It is all about how brick buildings were actually built and the materials that were being used at the start of Queen Victoria's reign. A brief note of how and why it was written follows.

In 1825 Lieut-Colonel Charles William Pasley was ordered to begin an architectural course at the Royal Engineers' school at Chatham which he had established in 1812. This was not to teach his students to design buildings, as architecture is now generally understood, but to teach them building construction, or practical architecture. He enlisted the assistance of Robert Howe, their experienced Clerk of Works, who was appointed as Professor of Practical Architecture, and looked for suitable textbooks. He found that Nicholson and Tredgold concentrated mainly on carpentry and joinery, assuming that details of foundations and walling generally were already known to their readers, so Charles Pasley filled this gap for his students by compiling and issuing lithographed notes, which were reprinted without alteration or revision, as a book, published by the Royal Engineer Establishment in 1862.

This scarce volume that Donhead have now produced in facsimile therefore sets out best practice in 1826 as researched and reported by Pasley using his extensive network of contacts. His fellowship of the Royal Society and membership of the newly formed Institution of Civil Engineers will have provided access to authorities on most scientific subjects, and references in the text to what such notable figures as Soane, Robert and Sidney Smirke, Nash and Rennie were actually doing on their current projects provide details of contemporary practice in 1826 which are simply not available elsewhere. There are more than six references to the details employed by Robert Smirke at the London Custom House, then in the course of underpinning, alteration and reconstruction following the collapse on 26 January 1825, and also reports of other of Robert Smirke's then current projects, as well as references to what Nash was doing at Buckingham Palace.

Since this text was written in 1826, references to "cement" refer to Roman cement, because the Portland variety had only just been patented by Aspdin and had not yet been manufactured. This book shows us how important Roman cement was in the nineteenth century as a quick setting and waterproof binder for renders and mortars. It also describes the use of trass and pozzulana, as hydraulic additives to lime, and there are several references to the beneficial use of brick dust in mortars and renders as a substitute for trass or pozzulana.

Although Pasley concentrates on all aspects of brickwork and mortars, from footings to bonds to vaults to flues – did you know that 9" x 13½" (225 x 340 mm.) was big enough to provide access for a climbing boy? – to all the details needed for ordinary buildings, he also provides valuable information on such ancillary subjects as the use of hollow pots for fireproof flooring and on the use of timbers in walls. It is the only book I know which not only describes the use of chainbond timbers in the centres of walls, then common practice, but also sets out why they were used and why they were superior to common bond timbers and wall plates flush with the inside of the brickwork. It is interesting to note that twelve years later, in his book on limes etc., Pasley reports chain bond timbers suffering from dry rot, and advocates substituting four or five courses of bricks laid in pure cement mortar strengthened by hoop iron in the joints.

C. W. Pasley's best known work is his *Observations on Limes, Calcareous Cements, Mortars, Stuccos and Concrete etc.*, of 1838, and Donhead reprinted in facsimile the Appendices in 1997. That is the standard text on those subjects current for most of the nineteenth century. His experiments on the manufacture of artificial cements had started in 1826, and are reported in this earlier text. *Practical Architecture* also includes other gems in connection with the interaction of different materials of which this is probably the earliest mention, but which are still being regularly re-learnt today, such as spalling of soft bricks pointed with cement (then Roman, now Portland) mortar, and differential movement resulting in separation between brickwork built in cement mortar and brickwork built in common lime mortar. This last observation is sensibly extended to a recommendation that alterations to existing building be built in cement mortar, so that it continues to align with old matured walling and does not slump or settle as it sets.

Pasley, in this 'introduction', reviews the up to date use of limes and cements, including their incorporation in concrete foundations as then being pioneered by Robert Smirke. Hollow walls are mentioned for the control of dampness and ventilation. Descriptions of Hiort's patent and improved chimneys, recently introduced and used by Smirke and Nash, leads on to a discussion of a variety of new heating and ventilation systems in course of development and use.

The purpose of this *Course* was to enable young officers to construct, extend, maintain and repair the carcase of ordinary buildings for barracks, hospitals, store-houses and other military purposes, and which could also be applied to civil buildings in times of peace. The intent was to explain all the details in the art of building in brick that could not be learned except by attending the construction of one or more buildings from beginning to end.

The fact that Pasley's original notes of 1826, of which only a few copies were lithographed, were deemed suitable for reprinting and more general issue in 1862, implies that they were still current and useful 36 years later. Today they are particularly useful as an illustrated guide to how ordinary buildings were actually being built a hundred and fifty years ago, and to the limes and cements that were used then and would still be eminently suitable for their repair. This book will therefore be welcomed by anyone interested in mid-nineteenth century buildings or concerned with work to them, as it provides practical guidance to the details and materials, particularly mortars, they can expect to find.

Lawrance Hurst
January 2001

Lawrance Hurst, BSc, FCGI, CEng, FICE, FIStructE, FConsE, FBEng
Hurst, Peirce & Malcolm, founded by his father in 1910, is a small practice of consulting engineers for structural work connected with buildings, new and old, in most materials. Lawrance joined Hurst, Peirce & Malcolm in 1968 where he was a partner and is now retained as a Consultant. He prefers old buildings and has been fortunate to have been concerned with the Royal Albert Hall, the Nelson Monument, the London Custom House and Finsbury Barracks, as well as a number of less prominent, but equally important and interesting old buildings. He has written papers and lectured on nineteenth-century concrete and cements, fireproof flooring, early iron and steel in buildings and the history of party wall legislation. He is convenor of the Institution of Structural Engineers History Study Group and chairs the Institution of Civil Engineers Archives Panel.

Biography

PASLEY, Charles William (1780–1861), general, royal engineers, was born on 8 September 1780 at Eskdalemuir, Dumfriesshire, Scotland, the son of a London merchant. He was educated in the school of Andrew Little of Langholm and later at Selkirk. Pasley entered the Royal Military Academy at Woolwich in 1796 and graduated the next year, receiving a commission in the royal artillery. In 1798 he was transferred to the corps of royal engineers and posted to Portsmouth. During the period 1799 to 1809, Pasley served on the east coast of England as well as in Minorca, Malta, Naples, Sicily and various European theatres of the Napoleonic Wars. He was severely wounded at the siege of Flushing during the Walcheren expedition and this incapacitated him for further combat duty. Thereafter, Pasley's career in the corps was to focus on the education of military engineers and building technology research.

In 1810 Pasley joined a group of six royal engineers in forming a society for procuring useful military information. The next year, while in command of a company of royal military artificers at Plymouth, he instituted at his own expense a course of instruction for the non-commissioned officers and men in order to improve their knowledge of fortifications and fieldworks. On 23 April 1812 a Royal Warrant established Pasley's school permanently at Chatham, making him director, and opened instruction to junior officers of the royal engineers as well as the military artificers, now renamed the royal sappers and miners. This new military educational institution was called the Royal Engineer Establishment. Pasley began with a teach-yourself system of training in which non-commissioned officers led the lessons. This method was influenced by Pasley's visits to the schools of Joseph Lancaster and the Reverend Andrew Bell, the two dominant figures in the primary education field in Britain. Pasley produced a three volume textbook, *Course of Instruction Originally Composed for the Use of the Royal Engineer Department,* and began a professional library at the Establishment.

Pasley married Harriet Cooper in 1814 but she died after a only a few months. Five years later, he was wed to Martha Matilda Roberts, by whom he had six children. His son, Charles Pasley (1824–1890), would also serve in the royal engineers, including as Commissioner of Public Works for Victoria, Australia and as Director of the Admiralty Works Department. As early as 1816,

Pasley was elected a fellow of the Royal Society. Three years later, he was appointed to a committee of experts to consider the Thames tunnel scheme of Sir Marc Isambard Brunel. Well recognised as a man of practical science and possessed of high social standing as an army officer, Pasley was asked by Thomas Telford, a life-long friend, to become a member of the recently founded Institution of Civil Engineers in 1820. For the rest of his career, Pasley participated regularly in discussions and presented papers at meetings of the Institution.

In 1825, following upon orders from the Master General of the Ordnance, the Duke of Wellington, Pasley began an architectural course at the Establishment. He developed another textbook, *Outline of a Course of Practical Architecture,* and appointed a civilian clerk of the works as instructor. Pasley was to serve for twenty-nine years as director of the Royal Engineer Establishment, and from 1839 to 1855 acted as a public examiner at Addiscome, the military college that trained engineer officers for service in India. He thereby became a major formative influence on the education of British military engineers in the nineteenth century.

Pasley's position as director of the Establishment induced him to pursue research in building technology. Initially, he undertook experiments with wooden models to determine the stability and most efficient form of retaining walls. Pasley published the results in the textbook for the architectural course, *Course of Elementary Fortification,* the century's first fully comprehensive English work on the subject. In 1826 he began experiments in an attempt to make an artificial Roman cement and four years later succeeded. Pasley tried his cement in structures at the Brompton Barracks, Chatham. He also printed an essay on his research and distributed it to all royal engineer stations at home and in the colonies, and to engineer officers of the East India Company. By 1836 his cement was equal in quality to the best of its kind. Pasley's achievement was founded on a prodigious programme of testing in which he used a variety of standard methods of the time for determining the tensile strength of mortar. In 1838 Pasley published his master work, *Observations on Limes, Calcareous Cements, Mortars, Stuccos and Concrete etc.,* wherein he described his experimental findings and gave an historical account of cementitious materials unequalled in the nineteenth century. The book was highly commended by the technical press and reprinted in 1847. Pasley's principal influence on the manufacture and use of cementitious materials was in the promotion of Roman cement, both natural and artificial, establishing its superiority over ordinary lime and supporting an English preference for its use over hydraulic limes in civil engineering work.

As an adjunct to his cement experiments, Pasley was involved in the testing and promotion of hoop iron reinforced cement bond brickwork as a substitute for bond and chain timbers in brick walls and in door and window lintels. His experiments, which began in 1837, were modelled on the testing to destruction of brick beams undertaken respectively by Sir Marc Isambard Brunel and Messrs. Francis and Sons, cement manufacturers. Pasley was credited with resolving the debate between the two over the effectiveness of hoop iron in reinforcing the test beams; he concurred with Brunel that the material greatly increased strength. Pasley's 1838 book became the recognised authority on the subject.

Another achievement of Pasley's experimental work at the Establishment was with the use of military mines in blasting under water. Between 1838 and 1844, he carried-out the removal of shipwrecks that obstructed navigation at Gravesend, Spithead and St. Helen's. In 1841 Pasley was appointed inspector-general of railways. This government post had been created only the year before and he became an easy target for criticism by civil engineers who objected to such state regulation. Nevertheless, Pasley kept a diary during his five-year term and it demonstrates that he attempted to overcome his lack of practical knowledge of railways by great industry and conscientiousness.

Pasley was made a KCB in 1846. He became colonel-commandant of the royal engineers in 1853 and a general in the army in 1860. Pasley died on 19 April 1861 at his home, 12 Norfolk Crescent, Hyde Park, London, from an ailment of the lungs. A portrait of Pasley, by Eddis, hangs in the Royal Engineers Headquarters Officers Mess at Brompton Barracks, Chatham.

Reproduced from: *Biographical Dictionary of Civil Engineers of the British Isles, 1500–1830* (Institution of Civil Engineers, 2001) by kind permission of the Institution. Article by John Weiler.

OUTLINE

OF A

COURSE

OF

PRACTICAL ARCHITECTURE,

COMPILED FOR THE USE

OF THE

JUNIOR OFFICERS OF ROYAL ENGINEERS.

BY

C. W. PASLEY,

LIEUT.-COLONEL IN THE CORPS, F.R.S., AND HONORARY MEMBER OF THE INSTITUTION OF CIVIL ENGINEERS.

REPRINTED 1862.

CHATHAM:
ROYAL ENGINEER ESTABLISHMENT, BROMPTON BARRACKS.

Errata.

Page 54, last line, for *there* read *other*.

" 89, first line of note, for *ingenuous* read *ingenious*.

" 244, line 43, for *loose* read *lose*.

CONTENTS.

PREFACE.

His Grace the Duke of Wellington, Master General of the Ordnance, having been pleased to direct that Practical Architecture should in future form part of the course of study of the junior officers of Engineers, and having approved of General Mann's proposition that it should be conducted at this establishment, and that Mr. Robert Howe, Clerk of Works of the establishment, whose mechanical skill and ingenuity, as well as his knowledge of the practical details of architecture are well known to the Corps, should be employed to assist in carrying on this branch of duty, as Professor of Practical Architecture,* I have been induced, as Director of the Establishment, and responsible for the general execution of all the duties, to give my attention to this subject, and to endeavour, by the able assistance of Mr. Howe, to bring the proposed course of Practical Architecture to a system. I have, accordingly, commenced the following work, which I have entitled an Outline, &c., because it is not my intention to write a complete course, even if I were qualified for so doing. The most difficult branches of Practical Architecture have for some years past been treated of in this country by very able writers, amongst whom Mr. Peter Nicholson deserves the praise of having been the first to show so useful an example, which has since been followed by Mr. Tredgold and others. But there are several parts of this important subject which those writers have scarcely noticed, probably from the circumstance of their being generally known to all persons regularly brought up to the profession of architecture or to the mechanical trades connected therewith.

* The order of the Master General and Board directing this measure to be carried into effect is dated the 15th September, 1825.

My object has been to endeavour to fill up those deficiencies, an attempt which, if successful, may be useful to the junior officers of the Corps, who are often sent to the British Colonies soon after they enter His Majesty's Service, and are there required to perform duties analogous to those of architects or civil engineers without having had any previous opportunity of acquiring a practical knowledge of the details of those duties; and where, although they will derive considerable assistance from a good collection of the books before alluded to, they will still find themselves at a loss in respect to many particulars of considerable importance, which are not sufficiently explained.

In the course of this undertaking I have already derived considerable assistance from the liberal communications and information afforded me by some of the most eminent civil engineers and builders of this country and the persons in their employment, and I hope to derive equal benefit from the same source in the remainder of my work. A series of plates connected with it has been drawn by Mr. Howe, but, according to the custom adopted in my former elementary writings, I have generally preferred introducing figures extracted from those plates into the body of the work, opposite to those parts of the subject which they are intended to illustrate.

Having, from circumstances, had less experience in those matters which form the subject of this work than most officers of the same standing in the Corps, if my brother officers will favour me with any remarks which they may consider useful I will be much obliged to them, as a few copies only have now been lithographed, under an expectation of improving the work in a second impression, after more information shall gradually have been collected.

I have heard it remarked by some gentlemen in civil life, that it seems almost too much to expect the same individuals to combine the habits and feelings of the military life with the laborious pursuits of men of business. In exacting both from the corps of Royal Engineers the Government of this country appear, however, to have acted more wisely than if they had confined the Corps to military duties solely. The same kind of talent and knowledge that render an Engineer officer capable of forming the project of military works, and of arranging and directing great masses of workmen in a siege, or in fortifying the position of an army in time of war, by a very little more study will enable him to superintend the construction of public works of equal importance in time of peace; and if the Government did not provide employment of that description, such as the superintendence of public buildings, barracks, &c., and the execution of other public works for the officers of the Corps of Royal Engineers and the soldiers under their orders, it would be necessary in time of peace either to employ them in the common routine of infantry duties or to disband them. There is no other alternative, for their being allowed to remain idle is out of the question.

By combining science and industry with the activity, zeal, and spirit of the military character, and by considering none of the multifarious duties that he may be required to perform as a drudgery, since they are all equally useful to his country,* the young officer of Engineers has it in his power not not only to establish a reputation for himself, but to contribute towards maintaining the fame of his corps as one no less useful in peace than distinguished in war.

* In respect to those details of duty alluded to, which are the most uninviting, such as the superintendence and measurement of barrack repairs, the receipt of materials, &c., the young officers of Engineers will do well to recollect that upon the strict and faithful execution of their orders when so employed the saving of thousands to the public may depend.

PRELIMINARY OBSERVATIONS.

——:o:——

The buildings about London, and in most parts of England, are of brick; and if a person understands the construction of brick buildings, he may easily make himself master of the method of building with stone.

The bricklayer's work and carpenter's work of a building are intimately combined, and these two form the principal part, or carcase, of the edifice, to which the work afterwards performed by the other mechanics employed —such as the joiner, plasterer, &c.,—is secondary.

In the present mode of finishing dwelling-houses in England, everything, excepting the outside walls, is cased up by the joiner or plasterer in such a way, that the whole of the carpenter's work, as connected with the walls, (being, as before mentioned, the most important part of every building,) is completely hid from view.

The object of this course of Practical Architecture is, therefore, chiefly to explain all those details in the art of building, which cannot otherwise be understood without attending the execution of one or more buildings, whilst in progress, from the beginning of the foundation to the completion of the edifice. Thus, without pretending to lay down those rules for proportioning the various parts of an edifice, and for designing the decorations, which form the study of the professed architect, this little treatise will draw the attention of the student to every point that requires consideration, and to each several branch of work that is to be executed in a common military building, such as a barrack, hospital, or storehouse.

This will be done in an elementary manner, so as to explain the nature and general principle of every part of a building, adding, occasionally, an example or two, but without enlarging upon each subject to that extent, which would be proper in a professed treatise on the subject of architecture. To supply this deficiency, the student will be referred to certain articles in the most useful works on architecture, carpentry, &c. Verbal explanations will also be given; explanatory models will be provided; and it is hoped, in course of time, to obtain complete drawings and specifications of approved buildings, which will form the groundwork of useful exercises.

The measuring of artificer's work will be taught according to the London method, as it is presumed that an officer who understands that, will easily be able to make himself master of the varieties which occur in other parts of Her Majesty's dominions, which every person employed in building ought to inquire into, on his arrival at a new station.

The practice of this art will be learned by attending those measurements which take place from time to time, in the department of the Commanding Royal Engineer at this station.

After copying a sufficient number of architectural drawings, and learning the rules, and attending the practice of measurement, the student will be expected to draw up an estimate of the expense of a given building, from the drawings and specifications only, according to certain supposed prices, which may be taken from one of the London price books for the current year.

Having developed the plan of the proposed course, and of this treatise, we will proceed to particulars, explaining more especially, as was promised, the nature of brick buildings.

Of Bricks.

The dimensions of the bricks used for building in this country are from $8\frac{1}{2}$ to 9 inches long, from 4 to $4\frac{1}{2}$ inches broad, and from $2\frac{1}{2}$ to $2\frac{3}{4}$ inches high. By Act of Parliament, it is directed that they shall not be lsss than the smallest dimensions above stated; but it is usual to average the length of a brick at 9 inches, and, in the walls of buildings, the thickness is always computed by so many bricks and half bricks, according to the above dimensions, which also makes allowance for the mortar laid in the vertical joint between two bricks.

Thus, for example:—

A 1 brick	wall	implies	one	9 inches thick.
A $1\frac{1}{2}$	,,	,,	,,	$13\frac{1}{2}$,,
A 2	,,	,,	,,	18 ,,
A $2\frac{1}{2}$	,,	,,	,,	$22\frac{1}{2}$,,
A 3	,,	,,	,,	2 feet 3 inches thick.
A $3\frac{1}{2}$	,,	,,	,,	2 ,, $7\frac{1}{2}$,,
A 4	,,	,,	,,	3 feet thick.

Whilst this method is followed in describing the walls of common buildings, the dimensions of the revetments of fortresses, or other more substantial masses of brickwork, are always stated in feet.

The different kinds of bricks made near London are principally *Place*, *Grey*, and *Red Stocks*, *Marle Stocks*, and *Red Rubbers*.

They were formerly burnt in kilns, but are now almost always burnt in large masses, called *clamps*.

The bricks burnt in clamps are of various qualities, as they are liable to be occasionally injured by weather and other causes.

The well burnt bricks are called *Best Greystocks;* next to these in quality are the *Second Greystocks*. With these bricks the fronts of most of the houses in London are faced. The least burnt, or worst description of bricks, are called *Place Bricks*.

The *Marles* are prepared with more care, and greater attention is paid to burning them. They are of two kinds—*Firsts* and *Seconds*. The former are selected for arches, being of a superior colour and quality; the latter are used in facing the fronts of principal buildings, being superior to Greystocks.

At a village near Ipswich, in Suffolk, a very good kind of facing brick is made of a much paler colour than the Marle or Greystocks.

At Hedgerly, a village near Windsor, an excellent brick is made, capable of resisting the greatest violence of fire. These are called *Windsor Bricks*, or *Fire Bricks*.

Besides the bricks above mentioned, there are a variety of others, such as Paving Bricks, Coping Bricks, Concave Bricks, used for small drains; Dutch Clinkers, for paving stables, &c.; Welsh Fire Lumps, &c., &c.

The art of brickmaking is explained in "Nicholson's Architectural Dictionary," and in other books; but the best method of forming a just notion of this process is to visit the brick-works in this neighbourhood, of which there are several, one at Gillingham, one near Upnor, and another near Hoo; the two last are the most considerable, and bricks of an excellent quality are made there.

Specimens of various bricks, with their names, will be kept in the Architectural Model Room.

The specimens of buildings in this neighbourhood, executed with the various kinds of bricks enumerated, are as follows:—

The Ordnance Barracks and Hospital, and several of the new buildings in the Dockyard, are built with Greystocks.

The Garrison Library is built with the best Greystocks, of a quality nearly equal to second Marles.

The Royal Marine Barracks, and most of the buildings within the Infantry Barracks, are built with Redstocks.

Place bricks, whether red or not, being the refuse of the kiln or clamp, are of a dull colour and soft perishable quality, of which specimens may be observed in some small buildings in this neighbourhood, but none in Government buildings.

Marle bricks (by workmen styled Malms), of the second quality, have been used in some new buildings at Ordnance Place, in Chatham.

Of Lime.

Lime is obtained by breaking any calcareous stone, such as marble or chalk, into fragments of a convenient size, and burning it in a kiln, and afterwards pouring water upon it, which causes it to effervesce and fall into a fine white powder, in which state it is said to be slaked. The powder should then be sifted, in order to remove the stony parts, being those not perfectly burnt, which will not slake properly, and are unfit for making mortar.

The celebrated Mr. Smeaton made a great number of valuable researches into the properties of lime and cements in general, and he found, contrary to the then received opinions, that the goodness of lime did not depend upon the hardness of the stone.

He divides limestone into two qualities, one of which he calls *water lime*, as it forms a mortar more capable of resisting the action of water than the other, although the latter is of a much purer calcareous substance, when chemically analyzed; this he calls *common lime*.

All the water limestones examined by Mr. Smeaton were found to contain, besides calcareous matter, a proportion of brick earth or clay.

The best water limestone is considered the Blue Lias stone of Glamorganshire, Somersetshire, and Dorsetshire. Some kinds of chalk have likewise the same properties; and these have usually a yellowish colour, as being mixed with clay, and are rather harder than common white chalk of a purer quality. The Sussex clunch lime, obtained in the neighbourhood of Lewes; the grey lime of Berryton, near Petersfield, in Hampshire; and the Surrey lime, obtained in the neighbourhood of Dorking, are all of this quality. The latter, from its proximity to the metropolis, is most used in the neighbourhood of London, where it is improperly called stone lime, to distinguish it from that made of the more common and purer qualities of chalk.

Mr. Smeaton found that all the water limes, although the stone or chalk from which they were produced were of different colours—blue, white, or yellowish—invariably assumed a dun or buff colour, after being burned and slaked.

The Blue Lias limestones contained about one-eighth of clay; the Dorking lime one-seventeenth; that of Barrow, in Leicestershire, which was rather of an inferior quality, contained no less than three-fourteenths of clay.

Limestone is said to lose about four-ninths of its weight in burning. On being slaked, it increases considerably in bulk, and again increases in weight by the absorption of water. Mr. Smeaton states that lime powder occupies double the space by measure that it does before slaking. When mixed with just sufficient water to form it into a paste, it again decreases in bulk so much, that one measure of paste will require two measures of powder.

In respect to the other qualities of limestone before alluded to, Mr. Smeaton found that Plymouth marble and common chalk, which, when analyzed, proved to be both composed of very pure calcareous matter, yielded lime, as nearly as possible, of the same quality, but not at all adapted for water cements, although good enough for the common purposes of building, and for inside work.

Mr. Smeaton also tried lime made of the shells of cockles, &c., which, contrary to the common opinion, he found to be the worst of all. It sets, or hardens, rapidly; but after a certain time, it decomposes under water.

It is essential that burned limestone should be carefully preserved from the air, for which purpose it should either be used as soon as it comes from the kiln, or it should be kept in water-proof casks, which precaution is necessary in moving it to any distance. The same rule should be followed in regard to slaked lime, which should be used the moment it is slaked. If that is not convenient the workmen cover it over with sand, to save it from the air as much as possible.

It is essential that lime should be well burned, and capable of slaking the moment that water is applied to it. It is true, that those stoney particles of good limestone which are imperfectly burned, will slake in course of time by exposure to the air, but that destroys the virtue of the mass of lime in general, and, therefore, such particles should always be rejected.

Of Mortar.

The common mortar used for the walls of buildings is composed of slaked lime and sand, which are beaten up together, and intimately mixed, with no more water than is necessary to give them a proper consistency.

The goodness of the mortar depends partly upon the quality of the lime and sand, being the principal ingredients, and partly on the care with which it is mixed. Too much pains can scarcely be taken in this operation, which, in great works, is often done by a mortar mill, in preference to manual labour.

The sand used for mortar should be sharp (not round or smooth), and free from earthy particles, for which reason it should be washed in fresh water if necessary, and for this purpose exposure to a running stream, when practicable, is the best expedient.

This should always be done in using sand from the sea shore, in order to separate the saline particles, for mortar containing salt will never dry properly. For this reason, sea water should never be used for the walls of buildings, nor even sea sand, unless no other can be had; but there is no objection to the use of salt water for wharf walls, or other revetments exposed to the action of the tides.

In revetments, and other walls of considerable thickness, finer and more expensive mortar is often used for the outside work than for the remainder of it. The two kinds used in this case are called *Face Mortar* and *Backing Mortar*.

The sand for making mortar, when it cannot be procured of the proper degree of fineness, should be passed through a screen, to remove the coarser particles.

According to Dr. Higgins, the author of a treatise on calcareous cements, published in 1780, in which he laid down many useful rules, since acted upon by the most eminent architects and engineers, but contrary to the received opinions of the period at which he wrote, it is better to use two kinds of sand mixed. One, which he calls coarse sand, should pass through a sieve, having openings one-sixteenth of an inch square; the remainder, which he calls fine sand, should pass through a sieve, having openings one-thirtieth of an inch square; but of the last, he rejects all that in fineness equals the Lynn sand, used in glass works.

The sifting should always take place when the sand is in a wet state. He considers the best proportion for mortar to be four parts, by weight, of the coarse sand, three parts, by weight, of the fine sand, and one part, by weight, of lime powder of the purest quality, that is of common lime, according to Mr. Smeaton's definition, as stated in the foregoing article.

Dr. Higgins recommends that, in making mortar, the sand shall always be wetted first, and then the lime added, and mixed up with it, without any further addition of water; and he was the first who recommended that mortar should be used as soon as made, contrary to the former practice of the times.

It had been known before that limestone is converted into lime by depriving it of carbonic acid gas. Dr. Higgins discovered that, by exposure to the air, burnt lime will absorb the same gas by degrees, and becomes no better, as an ingredient for cement, than the dust of the same quality of limestone in its original state. If, on the contrary, it be intimately incorporated with sand, so as to form mortar, and used in building as soon as it is made, in joints of moderate thickness, the absorption of carbonic acid gas, by gradually restoring to the lime its original chemical properties, produces an intimate union of the whole mass of solid substances with which it is incorporated, or to which it is attached, and the mortar hardening by degrees will become stronger than brickwork.

Mr. Smeaton agrees that a mixture of coarse and fine sand is better for mortar than if it were all of the same quality, but he does not attach much importance to the precise proportion between the two. His rule is to mix the water and lime first.

In respect to the proper proportion of lime and sand which ought to be used in mortar, it is known that when the lime is in excess the mortar may be plastic, and convenient for use, but that it never hardens properly. If, on the contrary, the sand be in excess, the mortar becomes too short, as the workmen style it—that is to say, it is not sufficiently plastic at first, and may eventually crumble to pieces. It is agreed, however, that all the particles of sand should be just separated, and, consequently, cemented together, by the smallest quantity of lime capable of effecting this object, which can only be judged of by experience.

Having explained, in the foregoing article, that limestone alters in weight by burning, and in bulk, first by slaking, and secondly by mixing up into a paste with water, and it being known also that dry sand will compress into a smaller space after being wetted, it follows that, in order to judge of the proper proportions of lime and sand, the state of the lime, and of the sand, at the period of mixture, must be accurately defined. It is unfortunate that Dr. Higgins should have assumed weight as his criterion, contrary to the usual custom, for the worst burnt lime of the same quality is the heaviest, although the weakest, and nothing is more variable in weight than the different kinds of sand that may be used for making mortar.

The prejudices of common workmen are all in favour of using an excess of lime, upon which they consider the whole essence of good mortar to depend. In public works, it therefore becomes necessary for the Engineer to guard against this propensity, as they imagine that they are doing a service to Government by wasting the more expensive material; and, if they are not strictly looked after, they will make the mortar according to their own judgment, in spite of general directions to the contrary.

It must, therefore, be understood that, in estimating the comparative proportions of lime and sand for making mortar, unslaked lime—as it comes from the kiln—and dry sand, are generally implied, unless the contrary should be expressed; and, in making a contract, or in giving directions, in respect to mortar, these particulars should always be stated beyond the possibility of mistake.

In making inquiry into this subject, whether personally or by reference to books, equal care must be taken to obtain, if possible, a correct specification of the state of the lime at the period of its being mixed.

When lime is of inferior quality, or there are not means for mixing the mortar well, it is considered that one part of unslaked lime to two parts of sand is the proportion necessary for obtaining a compound of proper tenacity.

If the lime be of good quality, and dependence can be placed upon the diligence of the persons employed in making the mortar, one part of unslaked lime to three of sand, has been held out as a better proportion than the above, and has very often been used in Government works.

It appears to me that, for common mortar for the walls of buildings, the former may be considered the maximum, the latter the minimum, proportion of lime that ought to be used. The latter will probably agree with Dr. Higgins' rule; for, considering the specific gravities of lime powder and Thames sand, one part to seven, by weight, will be nearly the same as one part to three, by measure.

When mortar cannot be all expended as soon as it is made, the remainder of it should be carefully covered from exposure to the air, which is sure to deteriorate its quality; and it should be beat up again when about to be used, but without adding water, unless absolutely necessary.

In Government works in the Ordnance department, it has always been the custom to measure lime and sand by the cubic yard, which is the simplest method.

In London, sand is sold by the load of one cubic yard; but lime is sold by the hundred, which implies 100 pecks or 25 bushels; and this quantity, even according to stricken measure of the old standard, ought to have filled a space of about 3 feet $2\frac{1}{8}$ inches cube; but, as it has been the custom to serve out lime in baskets at the kiln, the purchasers have usually been satisfied if the quantity tendered as a hundred of lime proved capable of filling a cubic yard measure well, that is by standing about one inch higher than the top.*

It ought to have been observed that Dr. Higgins recommends mortar to be made with lime water, not with common water; but I am not aware that this nicety has ever been attended to by practical builders.

Recently, a custom has been introduced of grinding, or pounding, the lime about to be made into mortar or grout, previously to slaking it; it has been stated that this practice is chiefly advantageous when the water limes are used, which do not slake so quickly as common lime.

Remarks. The mortar of many old Roman buildings has been justly celebrated for its superior strength; and age certainly improves the quality of mortar when made of good ingredients, and in just proportions, but not otherwise, for many ancient buildings are very defective in this respect. I have observed, in particular, that the mortar of the walls of Alexandria, supposed to have been built by the Arabs after their conquest of Egypt, is exceedingly bad, and crumbled in the hand like sand.

It appears from Vitruvius that the Romans considered one part of lime mixed with three parts of sand as the best proportion for common mortar; and this agrees nearly with the practice of the most eminent engineers and architects of our own times; and if that quality of lime, termed by Smeaton water lime, be mixed with Thames sand, which is considered the best sand that can be procured in the south of England, I believe that no better proportion, or ingredients, can be adopted for buildings of any importance.

The objection to using common lime in such buildings is, that the surface more immediately exposed to the air is apt to decay, which, after some time, requires all the joints to be pointed with fresh mortar.

At the new British Museum, where the walls are much more massy than is common in buildings of equal height, the brickwork was laid in mortar composed of one part of Dorking lime, mixed with three parts of Thames sand. At the end of about one year it was resolved to substitute some granite pilasters, in place of brick ones, in the interior of the building. This change having taken place very recently, which required the brickwork to be cut away, I had an opportunity of observing the great tenacity of the mortar, which, after so short a period, rendered the operation exceedingly difficult, even with pointed tools. At the same time, I observed that the mortar of some buildings which were pulled down to make room for the proposed Parliament Mews, near Westminster Abbey, was quite rotten, so that it might be broken with ease by the hand, although these buildings were probably at least 100 years old.

I ascribe the comparative inferiority of the older mortar, in this case, to a practice which seems to have prevailed in London in the last century, of building with mortar composed of about equal parts of lime and sand for inside work, but of no less than two parts of lime to one part of sand for outside work. (See a book on Bricklayer's work, published by Mr. Batty Langley, architect, in 1729.)

* The cubic yard measures, used in great works, both public and private, are of wood, open at top and bottom, and usually in two pieces, one of which is placed upon the other to make up the height.

In the new works in Sheerness and in Chatham Dockyards, Mr. Rennie directed the fine mortar to be made with one part of Dorking lime to three parts of Thames sand; but the backing mortar to be made with one part of Dorking lime to four parts of sand. The lime was ground in a mortar mill previously to its being used.

Of Grout.

Grout is mortar or cement in a liquid state, which being poured into the joints of rough masonry, or brickwork, such as foundations, &c., fills up all the crevices, and hardens by degrees, though not so speedily as the same quality of mortar or cement when worked up stiff in the usual manner.

In the new docks in Her Majesty's yards at Sheerness and at Chatham, grout was used for all the rough parts of the foundations and walls. It was composed of Dorking lime ground in the mortar mill, but not slaked, and of Thames sand. One measure of the former to five measures of the latter was the proportion used for the lower part of the foundation, built with Kentish ragstone, about the heads of the piles and between the sleepers. Above the planking, where bricks were used, the proportion was one part of lime to four parts of sand. These ingredients, mixed together dry, were wheeled to the walls, and made into grout, by stirring up the composition with water and pouring it instantly into all the vertical joints of each course of the brickwork.

It is said, that on recently cutting away part of the grouted brickwork for a trifling alteration in the new dock at Chatham, the excellence of this kind of cement was proved by the very intimate cohesion of the parts, which was such, that a small horizontal gallery was driven into the mass of brick-work, the roof of which remained perfectly firm without support.

Mr. Smirke, the architect, has recently made great use of grouted gravel for the sub-foundation of considerable buildings. One measure of pounded Dorking lime, not slaked, and five measures (in some cases, but more usually seven measures) of clean gravel are mixed together with water, and poured into the trench, or excavation, prepared for the foundation of the building. In the gravel with which the lime is to be incorporated, no pieces larger than an egg ought to be used; but there should be no limit in regard to the fineness of the minuter particles of the gravel used.* This compound, which will be more particularly noticed hereafter, forms, by degrees, a sort of artificial rock.

Hot lime grout has also frequently been used in rough work. This is made by stirring up slaked lime in water, without any other ingredient, and pouring it into the joints of each successive course; and, in so doing, the courses have sometimes been laid in a horizontal bed of common mortar, and the face of the wall secured with the same, so that the grout enters into the vertical joints of the body or mass of the work only.

Another method of obtaining grout for the same purposes, is to dilute common mortar, or other cement, previously made in the usual manner, by adding a sufficient quantity of water, and beating and stirring the whole up

---

* In fact, if only large gravel could be procured, it would be proper to mix a proportion of sand with it.

together, until an intimate mixture of the same ingredients is again obtained, but in a more liquid state.

Of Lime Putty.

The term putty, better known as the cement for fixing glass in windows, is applied in brickwork to a very different substance, which is nearly the same as hot lime grout. It is made by dissolving, in a small quantity of water, as much hot lime as when slaked, and continually stirred up with a stick, will assume the consistency of mud, so that, when the stick is taken out, it will just drop. It is then sifted, in order to remove the unburnt parts of the lime, and should be used without delay.

It is only proper for gauged brickwork, or for the ornamental outside work of brick walls, which will afterwards be described.

Of Pargetting.

Pargetting is a mixture of lime with horse dung, or cow dung, of which the last is preferred. The common proportion is one part of lime to three of dung.

This composition is used for plastering the inside of the flues of chimneys, whilst the brickwork is in progress.

Of Cements in general.

The term cement, although sometimes applied to common mortar, generally implies some compound, useful in building, either of a finer quality, or possessed of some superior property.

The most useful cements are of two kinds—viz., gypsum or plaster, and water cements.

Of Gypsum.

Gypsum, which is the sulphate of lime of the chemists, is found in great abundance in the neighbourhood of Paris, from whence it is often styled plaster of Paris. It is also found in Staffordshire, Derbyshire, and other parts of England.

It is used by the plasterers for all projecting ornaments in the interior of buildings, such as cornices, &c.; and at Paris, where it most abounds, it is also frequently used like mortar, for the common purposes of building. This practice is reprobated by M. Rondelet, an elaborate modern French writer on the art of building who observes, that plaster cement, when used in walls, decays by age; whilst mortar cement, of a good quality, continually improves. Hence, he states, that frequent failures occur in buildings in and near Paris, such as cracks, bulges, &c., which are not to be seen in the provincial towns of France, where lime mortar only is used.

Gypsum in its natural state is not acted upon by acids like limestone.

After being burnt, it is ground or pounded, and mixed with water, when about to be used as a plaster. It sets or hardens very quickly, but its durability depends entirely upon its being kept dry, which is, of course, only practicable in inside work. When exposed to moisture, it gradually loses its strength.

Of Gueesh.

In Minorca, a species of gypsum called gueesh is used, which indurates so very quickiy that, in turning arches with it, the use of regular centering is dispensed with, each set of archstones being supported by reeds or wooden rods, until the keystone is introduced, after which those temporary props, extending throughout the whole width of the arch, are removed, and used for supporting another set of arch stones. Thus, the arch, which is commenced at one end, is continued in portions until the whole is completed.

In using this cement, the outside joints of adjacent stones being secured with common mortar, the gueesh, in powder, is stirred up in a bucket with water, until it becomes well mixed, but in a fluid state; after which it is poured into all the other joints, which are previously scored for the purpose, in the manner represented in the annexed figure.

Fig. 1.

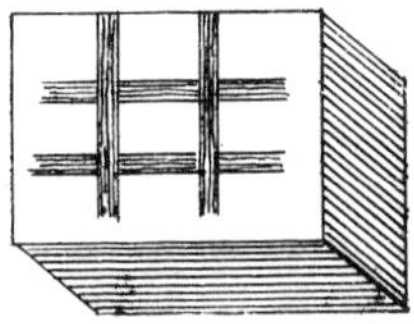

This cement sets almost instantaneously, and eventually becomes harder than the stone itself;* but it must be carefully secured from wet, which renders it useless by reducing it to a pulp. For arches and partition walls, protected by a roof, nothing can be better. The latter are often built in that island, with stones only three or four inches thick, set on edge, which are sufficiently strong when cemented in the manner that has been described.

Of Water Cements in general.

Mr. Smeaton ascertained that a ball of mortar, made of sand and common lime alone, will never harden or set, if immersed in water, but that it is even liable to be entirely decomposed in process of time. Hence, he recommends that such mortar shall never be used under water, although, from a very curious fact recorded by him, it would appear that common mortar, found in a damp state under a large stone, at the bottom of a malt cistern, at the end of probably more than a century after it had been placed there, so far retained its virtue that, when beat up again, and used in the open air, it appeared to be as good mortar, for dry work, as any which that part of the country produced.

It therefore appears proper, in buildings exposed to the action of water, either to use water lime, as defined by Mr. Smeaton, which is in itself a water cement, if common sand only be employed; or if common lime be used at all, it must be mixed with some efficient kind of water cement, such as Puzzolana, Tarras, and other substances which will afterwards be

* In 1800, a 24 pounder was dragged up by a tackle attached to a ring fixed in Gueesh, in one of the walls of Fort George, in Minorca, in a hole cut for the purpose only a couple of hours before, for the sake of experiment.

treated of; it being understood, however, that the best is not in all cases the most suitable, for economy must always have its weight, so that the expense of the material must be regulated by the importance of the work.

As a proof of the efficacy of good water lime, may be mentioned a fact communicated to me by Mr. Telford, the eminent civil engineer. That gentleman observed a pier building at Watchet, by the common masons of the country, who used no other material than bolder stones, or round pebbles, found on the beach, which they cemented together with hot lime grout, of the blue lias stone of that country. Thus, a wall was formed, which resisted the tides of that coast, which rise to a much greater height than in the south of England.

A curious incident, in corroboration of the above, occurred at Waterloo Bridge. In driving piles for one of the coffer dams, the workmen, most unexpectedly, encountered what appeared to be a stratum of rock. This, on examination, was found to be the common gravel at the bottom of the river, which had been indurated by a large load of Dorking lime, previously sunk over that spot by accident. This circumstance being mentioned by Mr. Rennie to Mr. Smirke, induced the latter to adopt the grouted gravel before alluded to, for the foundations of several important edifices.

Of Puzzolana.

Puzzolana is a volcanoic substance found near Naples, and in other parts of Italy, which has the invaluable property of hardening under water, and has, accordingly, been used for cisterns, drains, &c., since the time of the Romans. It is said to contain iron, but in what proportions I do dot know, never having seen a chemical analysis of it. Mr. Smeaton used Puzzolana from Civita Vecchia, in building the Edystone Lighthouse, which he mixed with an equal quantity of lias lime powder, using it in stiff mortar for the horizontal courses and facings, and in grout for the vertical joints. He considered this Puzzolana to be the strongest of all water cements that he experimented upon.

In other works, not so much exposed to the violence of the elements, he used smaller proportions of Puzzolana, and added sand instead of employing nothing but cement and lime; and in the backing, in particular, he used less Puzzolana than in his face mortar.

M. Rondelet states that the Neapolitan is inferior to the Roman Puzzolana, and that the former will not make into a proper cement with lime alone, but requires an addition of sand to render it fit for use. He also states that the Neapolitan Puzzolana effervesces slightly with acids, whilst the Roman does not; hence it is probable that the former contains a greater proportion of carbonate of lime, which, together with silica, alumina, and iron, are understood to be the component parts of this excellent cement.

In the piers of new London Bridge, built under the direction of Messrs. John and George Rennie, which are composed of large blocks of Scotch granite and Yorkshire stone, the cement consists of two parts of Neapolitan Puzzolana, one part of Aberthaw (or Welsh lias) lime, and two parts of Thames sand.

Of Tarras.

Tarras (called Trass by the Dutch and Germans) is a kind of Puzzolana found in the neighbourhood of Cologne, and much used in Holland, from whence it is imported into this country, and from which circumstance it is usually called Dutch Tarras. Mr. Smeaton states that this is an excellent cement, when always under water, but that he found it liable to decay when exposed to air, or alternately wet and dry. He also mentions that it sometimes grows in a very inconvenient manner, forming a hard excrescence in every joint, projecting beyond the face of the wall, which is exceedingly troublesome, and may even require to be cut off where there are flood gates, mill-wheel floats, &c. M. Rondelet, on the contrary, states that the Tarras commonly used in Holland, Germany, and the north of France, is equally proof against damp, dryness, and all the inclemencies of the weather, and that it is, in those countries, considered equal to the best Italian Puzzolana. Belidor gave it an equally high character about a century before, but with this proviso—if the Tarras could be obtained of a pure quality, which he states to have been very difficult in his time. Hence it may be inferred, that the Tarras experimented upon by Smeaton may have been adulterated, or otherwise injured in its passage from the Rhine to England.

Mr. Smeaton found, in using common lime with Tarras, that one measure of slaked lime, in powder, and one of Tarras, was a better proportion than two measures of the former ingredient, which was the more common proportion then used in England.

By employing water lime, half the quantity of Tarras might suffice that was proper in using common lime. He also found that, to a mixture of two measures of water lime and one of Tarras, three measures of sand might be added, without injury to the virtue of the mixture as a water cement.

Of Roman Cement.

Roman cement is the name given to a powerful water cement, discovered in this country subsequently to the publication of Mr. Smeaton's work, and which has recently, in a great measure, superseded the foreign water cements, unless in works of very great importance. It is used not only as a water cement, but for pointing, and plastering the walls of brick buildings, which, in the new streets of London, have thereby been rendered susceptible of those architectural decorations hitherto confined to buildings of fine stone.

Recently, it has also been employed for forming artificial stones of considerable size for drains, copings, chimney pieces, &c.

The stone from which the Roman cement is made is procured near the Great Nore by dredging vessels, and also on the coast near Harwich.

Roman cement, made at Sheerness under the direction of the Resident Engineer superintending the execution of the great naval works there, is used in the most important parts of the new basin and docks, and those most exposed to the action of the water, and especially in all the joints of the granite work, whilst the remainder of the same walls has been completed with the kind of mortar or grouting before described. The same system is followed in the new docks at Chatham.

Roman cement is an article, of the purity of which it is obviously much more difficult to judge than of most other materials used in building. In Ordnance works it will, therefore, generally be more advisable to procure it from the Government manufactory at Harwich, than to purchase it of private dealers.

*Of the Harwich Cement.**

The stone from which this cement is made, on being analysed in 1812, by Sir Alexander Bryce, then Commanding Royal Engineer in the Eastern District, was found to contain—

Carbonate of lime	70	per cent.
Aluminous earth	24	
Iron	6	

Hence it agrees, in the two first of its component parts, with the blue lias or other water lime stone, and in the last with Puzzolana; and thus all efficient water cements known appear to be partly composed of clay, or of iron, or of both combined.

In preparing the Harwich cement iu the kiln, a proportion of one part of coals to eight parts of stone is found sufficient to burn the latter thoroughly. In burning it loses nearly a third of its weight, and its gravity in different states is as follows:—

A bushel of stone, unburn, broken small	118lb
A bushel of stone, when burnt	82lb
A bushel of cement, when ground	88lb

Every 10 bushels of stone put into the kiln produce about 11 bushels of cement, exclusive of clinkers.†

The Harwich cement appears to be somewhat darker in colour, and stronger than any other with which it has been compared.

The cement may be used either by itself, or with various proportions of lime and sand. When used alone it has been for tanks and ornamental work, and for the covering of walls or casemates; in this case it is diluted with water, and well mixed with a wooden beater till it assumes the consistency of plaster. Care must be taken not to mix more at a time than can be used in 10 minutes, or before it begins to set, the process of remixing being very injurious to its quality.

In building or pointing, an excellent mortar is made by using one part of cement, one part of ground lime (unslaked), and one part of sand; these ingredients should be intimately mixed by passing them through a sieve, after which water is added, and the mortar carefully mixed and beat by the hand, or, still better, by a mortar mill, continuing to add water until the whole is thoroughly incorporated, and of the consistency of fine plaster.

This mortar is much longer in setting than the cement alone, and may, therefore, be used almost with as much ease as common mortar, for building or pointing. It becomes very hard in a few days; and, at the end of a fortnight or three weeks, is harder and stronger than the bricks themselves.

The quantity of sand has, in some cases, been increased in the proportion of a bushel and a-half to a bushel of each of the other two materials, and it seems difficult to determine which of the two proportions above mentioned is the best for building; in fact, it is understood, that the private manu-

* This article is chiefly extracted from a report of Sir Alexander Bryce, dated the 20th November, 1812.

† The expense of the cement manufactured at Harwich in the summer of 1812, appears to have been about 1*s*. 6*d*. per bushel, exclusive of cooperage; but the experiments which have been made show satisfactorily, that when the demand shall be sufficient to keep the establishment regularly at work, so as to manufacture 200 bushels per diem, the expense of a bushel of cement would not exceed 1*s*.

facturers of Roman cement are in the habit of adding a proportion of fine sand to the powder before it leaves the manufactory.

The mortar above described sets thoroughly in a fortnight, or three weeks at farthest, by which time it becomes as hard as freestone, and there is every reason to believe that it will last as long.

As exposure to the air is found to injure the quality of the cement when in the state of powder, it is recommended, on opening a cask, if the whole contents be not immediately required for use, that the remainder shall be repacked in a smaller cask till wanted, and that the greatest care be taken to exclude moisture from the cement powder.

Here, it may be remarked, that the same precaution is of indispensible importance in using Puzzolana, Tarras, or any other good water cement.

Lieutenant-Colonel Fanshawe observes, that Harwich cement, unmixed with sand, will not adhere to any substance that is sufficiently solid to prevent evaporation passing through it; hence, as one of its properties, after being set, is its rejection of water, he recommends that a second coat of it should never be used, for, in this case, the air will harden the outer crust of the second coat before the body of it, and the evaporation taking place inwards, will be rejected by the first coat of cement, and thus the two will never combine properly.

Of Basalt Cement.

The Basaltic rocks in various parts of the world, of which the most celebrated is the Giant's Causeway, on the coast of Antrim, having been found by chemists to contain the same component parts as the most celebrated water cements, there can be no doubt of calcined basalt proving a useful cement of that description when mixed with lime; but I am not aware that the trial has, as yet, been made in any work of importance.

Of the Cement obtained from Forge Scales, or Calcined Ironstone.

A water cement, of a tolerably good quality, may be made by mixing lime with a proportion of common forge scales pounded.

Mr. Smeaton, who used this ingredient in the same proportion and manner precisely as Puzzolana, considered it an excellent expedient for a small job; but, as he observes, it would be more expensive to collect forge scales in quantity than to purchase Puzzolana at once. He therefore tried ironstone, which had been calcined or burned with lime, being the first process in making iron; and this material, which he calls Minion, a local name given to it in some parts of England, proved to be a useful water cement, though not equal to Puzzolana, nor even to the pounded forge scales; for, he states, that both of these will answer with common lime, whereas he fouud that the Minion was not efficacious, unless mixed with lias or water lime. Mr. Smeaton further states, that this cement was in general use in the iron districts of England.

Of Coal-Ash Mortar.

By Mr. Langley's book, before quoted, it appears to have been customary in England to use not only forge scales, but the ashes of coals, and even coal dust, as ingredients for water cement, instead of excluding the latter entirely, as was done by Mr. Smeaton. Mr. Langley recommends a mixture of three pecks of ashes from a smith's forge, including iron flakes; one peck of coal dust, not burnt; and two bushels of lime, as a good composition to be used in lieu of Tarras. The former ingredients—namely, the forge ashes —are still used in the Royal Engineer Department at Devonport, it having previously been found, by the experience of the inhabitants of that humid climate, to be the only cheap cement that could secure the slated walls of dwelling-houses from being penetrated by beating rains.

In the neighbourhood of Tournay it has long been the custom to use the contents of the ash pits of lime kilns, consisting of coal ashes intermixed with small fragments of half burnt limestone. This ingredient has, for more than a century, been considered an excellent water cement in France, where it is known by the name of Cendrée de Tournay.*

Of Tile-Dust Cement.

The practice of using pounded tiles or bricks as an ingredient for water cement is very ancient, and has been general in almost every part of the world; and, from the circustance of argillaceous limestone only yielding a good water lime, as stated by Mr. Smeaton, it would seem to be founded on just principles. Dr. Higgins, in his work before quoted, gives, however, a strong opinion against it; but, in this particular, he does not appear to have been borne out in his opposition to received opinions.

Mr. Batty Langley asserts, that a mixture of powdered brickdust and lime yields an excellent mortar for tile pavements in situations generally damp, and that it is superior to Tarras in situations alternately wet and dry. M. Rondelet, the French writer before quoted, speaks in favour of it as making an efficient water cement, when mixed with lime; and he states that tiles well burnt should always be selected, and that old tiles which have been used on roofs are better than new tiles or bricks. He also states, that when these cannot conveniently be procured, it may be worth while to burn small fragments of brick earth for the purpose.

In the "Quarterly Journal of Science" this practice is recommended in a communication from John White, Esq., on the subject of calcareous cements.† He states, that by burning a chalky clay to such a degree as just to assume the appearance of commencing vitrification, a material will be obtained, which, if broken into small particles, but not absolutely pulverised, will, when mixed with Dorking lime, produce a cement, to which he has thought proper to give the name of British Puzzolana, a title which, considering that the same cement has been used from the earliest ages, and in all countries, seems to me rather inappropriate.

On mixing the brickdust that has been described with various proportions of pounded Dorking lime—sometimes with, and sometimes without, an addition of sand—Mr. White tried therewith a number of interesting experiments, recorded in the above-mentioned work, which yielded very satisfactory results as to the strength of this kind of cement, the usual proportion of lime being one-fourth part of the whole compound. Mr. White infers,

---

* See Belidor's "Architecture Hydraulique," and Rondelet's "Art de Batir."
† See No. XXXIX., Art. 8.

that if a brick wall be built with that cement, 23 tons weight will be required to crack, and 50 tons to crush, a superficial foot of such brickwork; but that a bedding, or joint of the cement itself, is not destructible by a weight of 173½ tons, which is just sufficient to split the best Portland stone.

Mr. Smirke, the architect, who was present at some of those experiments, has recently ordered the brickwork in the foundations and walls of those parts of the new London Custom-house, reconstructed by him, to be built with a mortar consisting of one part of pounded Dorking lime, one and a-half part of pounded clinkers, or bricks nearly in a state of vitrification, and one and a-half part of Thames sand.

Here it may be remarked, that, for the common walls of a brick building, it seems quite unnecessary to use any other cement than common mortar, made with water lime and sand, which, in process of time, will become equally strong with the bricks themselves, or perhaps rather stronger, for, according to the common but obvious adage, no compound substance can possibly be stronger than its weakest part.

There is one circumstance, however, which may sometimes render it advisable to use cement of a more expensive quality in a common building. It is known that walls built with mortar always settle, more or less, for a certain period, in consequence of the weight above compressing the lower joints; whereas, Puzzolana, Roman cement, and even what Mr. White terms his British Puzzolana, are said to be incompressible. Hence, in adding a new wall, pilaster, &c., to an old building which has already settled to the utmost, it may be proper to use cement for the new work. This is actually the case at the new Custom-house, where a part of the old walls have been left standing.

Of Mastic.

Mastic is a kind of cement which, for the external plastering of houses, yields a smoother surface than Roman cement. I am informed that it is made of a mixture of oil and pounded Portland stone. The front of the Union Club, in Pall-Mall East, is coated with this species of cement. Whether it will prove so durable as Roman cement is as yet considered doubtful; time will show.

Remarks on Obtaining Water Cements by Artificial Means.

Mr. Brande observes, in a brief notice of mortar, in his "Manual of Chemistry," that "the addition of calcined ferruginous clay, or of calcined basalt, " or of black oxide of iron, gives mortar the property of becoming hard " under water;"* and, as the component parts of which the best water cements are composed, are now known, or may be subjected to chemical analysis, it has been justly observed, by the author of the article Stone Masonry, in the supplement to the "Enclycopædia Britannica," that "there " are very few situations where the means of forming an artificial cement " are wanting." He states that, "in general, it will be sufficient to mix

---

* Mr. Brande quotes a useful French work on this subject by M. Vicat, which I have not yet been able to procure.

" slaked lime* with the substance containing silica, alumina, and oxide of " iron, and calcine them together, and afterwards grind them. The iron " should be in the state of protoxide to the amount of from 13 to 18 per " cent., by weight; the quantity of lime may be about 30 per cent., or 53 " per cent. of carbonate of lime, the silica and alumina being in equal pro- " portions; hence, in the raw materials, 100 parts by weight may consist " of—

" Carbonate of lime	53
" Protoxide of iron	18
" Silica and Alumina	29
	100 "

Of Mortar for Ovens, Furnaces, &c.

As common bricks and mortar do not resist an intense heat well, it is usual in building ovens and furnaces, not only to use fire-bricks but to lay the joints which are immediately exposed to the flame, with the same Windsor loam, or peculiar clay, of which such bricks are made, moistened into the consistency of cement by water. This clay cement extends about 2 inches into each joint, after which the remainder of the joint is finished with common mortar, which is also used in every other part not exposed to the flame.

Of the Foundation of Buildings in Good Soil.

The ground on which a building is to be erected should be carefully examined to a considerable depth, and if it proves to be good throughout -- such as rock, gravel, or soil of a sound uniform quality, not liable to be altered from time to time, entirely or in part, by the effects of water acting below the surface—a trench is cut all round, and carefully levelled, on which the walls are commenced, giving, however, to each wall a base somewhat thicker than that part of it which is to appear above ground.

If the ground to be built upon should happen to be very irregular, such as the side of a hill, which often occurs in practice, instead of cutting the whole trench intended for the foundation of a building to one general horizontal level, it is usual to cut it in several levels, in steps or terraces. It should not on any account be cut according to an inclined plane, although this method is sometimes used in low boundary walls, as in the north inclosure of the Infantry Barracks at Chatham, where the courses of bricks are all laid sloping. In any other description of wall this arrangement would be perfectly inadmissible.

Of the Footings of Brick Walls.

The base of a brick wall is technically called the *footing*, and it should always exceed the thickness above ground. The custom is to diminish the

* This appears a mistake. I apprehend that slaked lime should be mixed with the other substances specified, after the latter have been calcined, not before.

base of the wall from the bottom upwards by small steps or offsets of a quarter of a brick on each side, making half a brick in all saved upon each projecting course. Sometimes smaller offsets have been used, but recently it has been very unusual to employ greater in the footings of brick walls.

The offsets that have been described are sometimes laid out in every successive course, which in brickwork does not exceed three inches in height for each step, until the diminution proposed is obtained. Sometimes they are laid out in every second or third course, which latter arrangement appears to me to be the best.

Fig. 2.

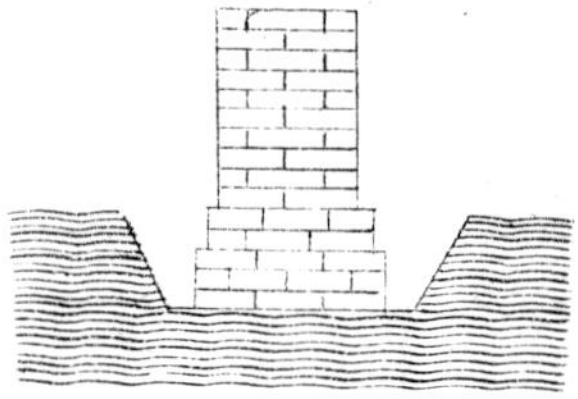

The annexed figure represents a wall of two bricks in thickness, having a base of three bricks thick, which is supposed to have been diminished in the manner last described.

The offsets at the base of a building are usually covered with earth by the filling in of the original trench excavated for the foundation; but sometimes there may be offsets on a higher level, in which case there is seldom more than one externally, which is usually at a moderate distance from the ground. The others are internal. These will afterwards be treated of.

The propriety of giving to the walls of buildings a footing such as has been dsscribed arises from the well-known circumstance that the stability of bodies depends, in a great measure, upon the magnitude of their base. To add further to the security of very extensive walls serving as inclosures, they are usually built not only with footings but also with buttresses, at moderate intervals apart, as may be observed in the walls of dockyards, &c. In common buildings the several walls serve as buttresses to one another; still, the footings are useful in good soil; in bad sail they are indispensibly necessary, as will afterwards be explained.

Of the Foundations of Buildings in Unfavourable Soil. Piled foundations explained.

It was before observed that the soil upon which a building of any importance is to be erected should always be well examined before the work is commenced, which may be done by striking it to ascertain if it sounds hollow, and then by boring it in various parts with an earth borer. Small shafts, cut at certain intervals, may also answer the purpose of discovering the nature and depth of the several strata, when the ground is not of uniform consistency.

When the soil is partly good and partly bad, the bad parts may be cut deeper until good soil is obtained, if it can be found at a moderate depth; and thus the footings of the wall may be built as in irregular ground, not on one continued level, but in steps or terraces.

When the soil is all of the same quality, but looser in some parts than in others, ramming it well in the loose parts has been considered a very useful, and sometimes a sufficient precaution, and this method is particularly recommended by M. Rondelet.

Even in buildings erected of brick it has been common to commence the foundation with flat stones, placing the largest as regularly as possible, and filling in the intervals with smaller ones, and ramming them down.

When the soil is of a middling quality, not absolutely good nor yet altogether unsound and treacherous, a grating, or planking of continued woodwork, has been recommended, and very frequently used, as an expedient for insuring the stability of an edifice.

In this case sleepers of about five or six inches square are laid transversely, at intervals of three or four feet apart, under the whole base of the proposed wall. Over these are fixed longitudinal planks forming a continued platform under the whole foundation of the walls of the proposed building, as represented in section in the annexed figure. The planks must be carefully pinned down upon the sleepers, and the intervals between the latter are usually filled with bricks or stones, not with earth.

Fig. 3.

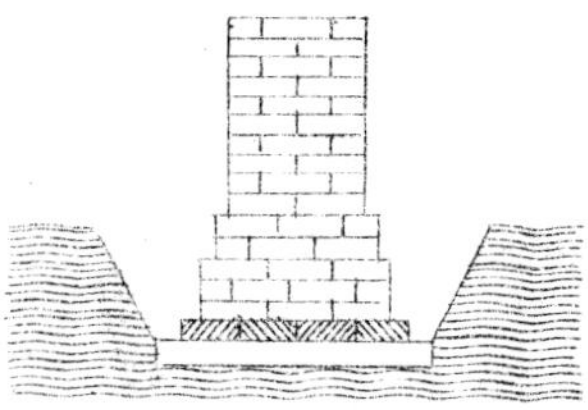

As every heavy body, if placed upon soft soil, has a tendency to sink more or less, which tendency is directly in proportion to its weight, but inversely as its base, the footing given to the walls of brick buildings has, in such soil, the effect of diminishing, or altogether rendering null, the said tendency, which might operate in a manner unfavourable to the stability of the edifice if the walls had no such increase of base, but were of uniform thickness throughout, from the bottom upwards. In respect to the wooden platform that has been described, its use is of a temporary nature, as it merely serves to prevent the brickwork, if it sinks at all, from sinking unequally whilst the mortar is yet soft, or the work green as it is technically termed. After a certain number of years, when the mortar and bricks shall have become completely consolidated into one compact mass, they are no longer liable to separate, even if the woodwork beneath the footings of the walls could be removed.

It will perhaps be scarcely necessary to remark that the base of every brick wall of the same height and thickness, or in other words, of the same weight, should be greater in direct proportion to the softness of the soil on which it is built; also, that the projecting parts of the footing of a wall should have a certain depth to give them proper solidity, otherwise the greater weight upon the central part of the base might fracture the bottom of the wall, and thereby render the footing of no use; and for this reason I prefer not having the offsets, when there are several, in every succeeding course, but at greater intervals apart, as was mentioned in the former article.

Moreover, it appears to be advisable, in erecting a building in middling soil upon a wooden platform, such as has been described, either to make all the walls of equal thickness, and consequently of equal weight throughout, or if that arrangement would not be convenient, to give to the thicker wall the greater base, in direct proportion to its weight.

When the soil on which a building is to be erected is very soft or swampy for a considerable depth below the surface, which is the worst kind of foundation that can occur, it has been usual to secure the base of the walls by a wooden grating or platform of the nature above described, supported upon strong wooden piles of a length sufficient to penetrate into harder soil, and acquire a fixture there, which are driven until they will go no deeper.

Such piles are usually shod, or pointed, with iron, and are driven at intervals of three or four feet apart, and there ought to be at least two rows of piles under the foundation of every wall, as in the annexed section, so that every pair of piles supports a transverse sleeper.

Fig. 4.

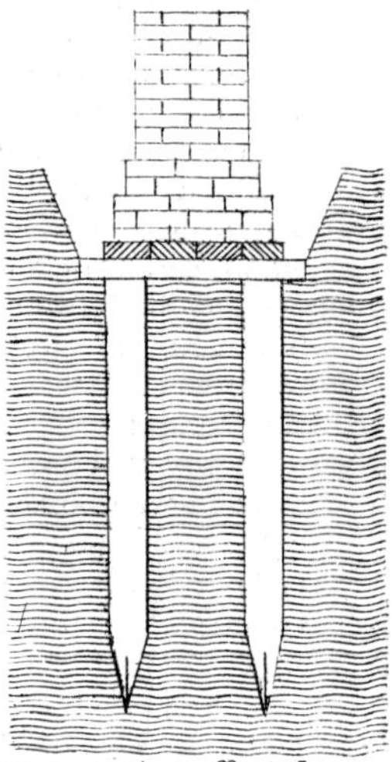

Less than two rows of piles would not give sufficient stability to the wall even of a common building. More than two rows, or three at the utmost, can only be required for the foundation of very substantial masses of masonry, such as revetments, wharf walls, &c.*

The principle of the stability derived from a piled foundation is this:—By the laws of mechanics the force of percussion greatly exceeds that of simple pressure by dead weight; and in the pile engines used in this country, the ram (sometimes termed the monkey)—that is the weight by which the pile is driven—falls upon the head of it from a considerable height, which greatly increases the effect, as compared with the old system of pile driving still used abroad, in which the ram is not raised more than four or five feet. Now the momentum of 10 or 12 cwt., a common weight for the ram of a pile engine falling from the height of 15 or 20 to 30 feet, produces a greater effect than the dead weight in any common mass of building; hence, if a pile be driven by such an engine until it can go no further, there can be little risk of the foundation giving way afterwards, in consequence of the mere pressure of the walls.

Foundations secured by piles are, however, liable to failure:—first, from the piles losing their original position, in consequence of the soft ground offering too little resistance against a lateral movement, so that the walls built upon them may have a tendency to overset.

In order to guard against this evil, the piles for the walls of buildings are sometimes driven obliquely, so as to diverge at bottom; and those especially for leaning revetments, or for the abutments of bridges, may be driven in such oblique direction as to oppose the greatest resistance to the unequal pressure, which acts chiefly on one side of the wall.

Secondly, a failure may arise from the woodwork of a piled foundation rotting, or being corroded by worms, in process of time; the former of which evils is almost sure to occur in foundations alternately wet and dry. Fortunately, the kind of soil where piles are most required is that where water is found a little below the surface, and if the woodwork be kept at such a level as always to remain wet, there is little or no risk of its decaying by rot.†

In the piled foundations of the fortifications of Portsmouth, of the new Docks of Chatham and Sheerness, and of the Waterloo, Southwark, and new London Bridges, this important circumstance has been attended to, and when such foundations are used for docks and bridges it can scarcely

---

* Practical examples of several foundations of this description are given in Lieut.-Colonel Pasley's "Elementary Forfortification."

† The perfect state in which trees have been found in bogs at the end of many centuries, is well known. A few years ago, on throwing two arches of Rochester Bridge into one, the piles of the intermediate pier, which must have been there several centuries, were taken up perfectly sound.

be otherwise, for the whole foundation, woodwork included, must almost of necessity be under low-water mark.

In respect to common buildings, a strong impression against using piled foundations has recently been excited in this country, for when the woodwork cannot be sunk low enough to remain always wet, such foundations are certainly very precarious. The cracks in the walls of the new storehouses at Woolwich are a proof of the danger of trusting to piles.

I shall not quote the recent failure of the new London Custom-house, although built on piles, in corroboration of the same opinion, because it was afterwards discovered that the foundation had not been executed as it ought to have been.

A remarkable circumstance, however, which bears upon the subject under consideration, occurred at Amsterdam in 1822. A very extensive brick building, four stories high, surmounted by an attic with dormer windows, which was probably at least a hundred years old, as it had been the warehouse of the old Dutch East India Company when in their prosperity, after having been empty for some years, had been converted into a granary, and was entirely filled with corn. In this state about two-thirds of the building fell down, and almost entirely disappeared in the mud, excepting the roof. This happened about 5 o'clock in the afternoon, without any previous warning, except a crackling noise that had been heard at times in the course of the day. The general opinion at Amsterdam was, that the piles being decayed had given way under the unusual weight with which the building had been suddenly loaded.* As a proof of the inconveniency attending piled foundations, it has been found necessary to prohibit the use of wheeled carriages both at Amsterdam, and at Bilboa, in Spain; in the former city both goods and passengers are conveyed on sledges.

M. Rondelet, the French writer, whom I have already frequently quoted, also expresses himself very strongly against the use of piles for the foundations of buildings.

It is proper, however, to remark that the system of piling has been generally adopted for all the foundations in the new naval works at Sheerness, including not only the revetments of the basin and docks, but also the whole of the storehouses, and other great public buildings.

To advert once more to the first expedient mentioned in this article, it may be observed that a wooden platform, formed of continued planking, under the foundation of a building, is, of course, liable to decay in process of time, under most circumstances. Now, since the uniform thickness of such a platform cannot exceed a few inches, this kind of failure taking place after the building has become completely consolidated, will not be attended with those injurious consequences which would arise from the decay of a piled foundation. Still, it is desirable to avoid it; and, therefore, instead of using continued planking, it will be best to employ a couple of parallel longitudinal timbers for the same purpose, laid at a sufficient interval apart which, when surrounded by brickwork or masonry, are called chain timbers.

The annexed figure (*Fig.* 5) represents a foundation secured in this manner. Two parallel chain timbers, seen in section, which may be about five inches square or thereabouts, are laid upon transverse sleepers, placed also parallel to each other at intervals of about feet apart. If this arrangement

* Captain Harry D. Jones, who informed me of this occurrence, saw the building in ruins about a fortnight after its downfall.

be adopted, it will be evident that in the event of the woodwork decaying, the base of the wall will have a sufficient and solid footing on the natural ground; for the intersecting grooves of about five inches square that will be occasioned by the rotting of the timber are too small in themselves and at too great intervals apart to injure the stability of the brickwork. It will be understood that the chain timbers, which extend all round the building, and which, therefore, must be in several pieces, are firmly connected together at the ends by scarfing or otherwise, and they are pinned down to the sleepers upon which they lie. A third chain timber may also be used, but not on the same level with the others, in which case it might weaken the wall too much, but in the centre of the wall two or three feet higher; as, for instance, at the part where the uppermost offset of the footings terminates. This third chain timber is also represented in the figure, and it will be evident that the three combined will nearly produce the same favourable effect in the first instance as a continued planking, but without being liable to any of its inconveniences.

Fig. 5.

Of Arches, and more especially of Inverted Arches, as applied to Foundations.

When the soil is soft near the surface, but proves to be of rock or of a hard incompressible nature, after excavating to a certain depth, it is obvious that a saving of materials may be effected by sinking shafts at certain intervals down to the solid, and building therein a set of piers along the whole line of the proposed foundation, and connecting these by arches at the proper level. It is to be observed, however, that this is only a saving of materials; and, therefore, all reasons for and against it ought to be fully considered before the expedient of a foundation, arched throughout its whole extent, is adopted under the circumstances above supposed.

If, however, the foundation of a proposed building should prove to be excellent in general, with the exception of one or two unsound parts of moderate extent, not occurring at any of the angles—in this case there can be no doubt of the propriety, and, indeed, of the necessity, of turning an arch over each of those unsound parts.

The above remarks apply to the use of common arches. I shall now treat of the inverted arch, which is a far more important means of securing the foundation of a building; for the former, as I have just explained, only applies to the foundations of walls built wholly or principally on rock or other excellent soil.

The tendency of all new walls to settle to a certain degree was before alluded to. This will happen even on the best foundations, for the weight of the superincumbent mass will compress the mortar in the joints of the lower courses; but, in such foundations, no evil can result from it if the building be carried up gradually and equally all round, as it ought to be. But in building on bad soil, this circumstance of itself may cause the heaviest parts of the walls to settle more than the others, by sinking deeper into the ground itself, and hence unseemly cracks may arise to disfigure, if not absolutely to endanger, the edifice. Now, it will be evident, on inspecting a dwelling-house or other common building, that those portions of the wall in which the doors and windows are situated must be much lighter than the solid parts of the same walls. Hence, if a building be erected on soft soil, having the foundation laid on sleepers and a planking of the nature explained in the foregoing article, it will be obvious that if inverted arches be turned a little above the woodwork and immediately under the position of every tier of doors and windows throughout the whole extent of building, this expedient will equalise the weight of the mass of brickwork and diffuse it generally over the whole foundation, so that, although it may still have a tendency to sink to a certain degree in the soft soil supposed, it will be compelled to sink equally, and without any risk of a fracture taking place.

Fig. 6.

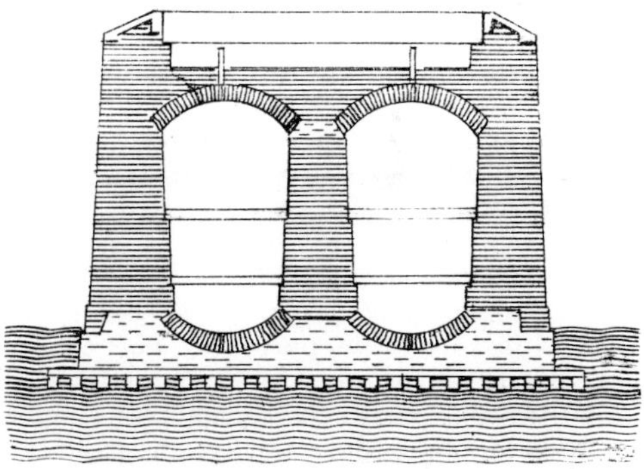

The annexed figure being a rough section of one of the Martello Towers on the coast of Suffolk, explains this expedient. First was laid a grating, composed of two rows of long elm planks, crossing one another at right angles, each being 12 or 13 inches wide, and 6 inches thick, placed at clear intervals of about 1 foot apart, and firmly pinned together.* Over these were laid some courses of large flat Purbeck stones, upon which the brickwork was commenced. This consisted of an inverted annular arch, immediately under the great annular arch of the tower, and pressing laterally on the same abutments and central pier.† It will be evident that if there were no inverted arch in this section, there would be a very great pressure upon the foundation immediately under the said abutments and pier, but nowhere else. Hence the elm grating being acted upon unequally, and not deriving proper support from the soft soil below, might bend or break under the walls of the towers, which might consequently crack or go to pieces.

Although the inverted arches that have been described, by diffusing the weight of a building equally over the whole of the wooden grating or platform, will render it impossible for the brickwork to separate, even in the softest soil, these precautions alone are not sufficient. Care must also be taken that the proportions of the building shall be so regulated, that the centre of gravity of the whole mass shall stand as nearly as possible over the centre of the grating or platform, otherwise the building, without any tendency to go to pieces, may sink deeper on one side than on the other. Hence, in erecting Martello towers in very soft soil the wall should be of uniform thickness all round, contrary to the usual construction. The necessity of this precaution was proved, not only in one or two of the towers

* The interstices of the grating, or spaces between the elm planks, were filled up with grouted brickwork.

† A model, explaining the interior construction of one of those towers, may be seen in the Model Rooms of the Establishment.

on the coast of Essex, built on marshy ground, where the thickest side sunk deeper in the mud than the others,* but also at Gillingham Tower in this neighbourhood, which, from the same cause, has, by degrees, assumed an inclined position, like the celebrated leaning tower of Pisa, at the same time that the brickwork remains in a perfect state.

All inverted arches should be executed with the greatest care; for, if they should not be perfect in every respect, it will be evident that the desired object of strengthening the building in the manner that has been supposed, will not be obtained.

Of the use of Grouted Gravel, or other small materials in Building, and especially in Foundations.

It appears to have been a common practice with the ancient Romans to build massy walls, consisting chiefly of small fragments of stone not larger than a man's fist, which having been thrown, with seeming irregularity, into a wooden case, or into the intervals between regular masonry of large squared stones, have evidently been cemented with hot lime grout.†

That distinguished people used the same expedient for building several piers or sea walls in the Mediterranean, and this practice still prevails in the south of France, where it is called Beton masonry. The method of proceeding has been particularly described by Belidor in his "Architecture Hydraulique:"—He states that in forming a pier or other maritime work of small rubble in the manner alluded to, an open case must be made with two lines of planks, parallel or nearly so, fixed to a sufficient number of piles driven for the purpose. This case marks the profile of the intended wall, which is formed by throwing down into the intermediate space, first a number of small stones, and then a sufficient quantity of Beton mortar or grout to fill all the interstices between them, and thus the work proceeds, using the stones and mortar alternately, so that the masonry shall rise in courses of moderate height. When the water is of any considerable depth, Belidor observes, that the grout must be lowered in a close cubical box, until it very nearly reaches the rubble work, when it must be emptied by a contrivance which it would be needless to describe. This precaution was adopted under the supposition that the grout might be partially decomposed before it reached the bottom, if thrown down from a height.

The grout for such foundations was always a good water cement, usually composed in a great measure of Puzzolana.

At Carlscrona, in Sweden, the same method has been resorted to for forming the foundation of a fine dock, on a spot which was originally composed entirely of mud in so very fluid a state that a rod could be pushed down to a depth of 40 feet below the surface. This was done by clearing away the mud and throwing down small fragments of granite, and grouting them with Puzzolana and lime.‡

---

* At the first tower, when this defect was discovered, small trenches were cut behind the wall on the reverse side to bring it back to its original perpendicularity. This measure caused it to return, but, unfortunately, rather too far, so that it leaned the contrary way, the attempts to rectify which were attended with much trouble.

† The same practice may be frequently traced upon the works of the Arabs, who probably borrowed it from the Romans.

‡ This circumstance was mentioned to me by Captain Harry D. Jones, who saw the dock when finished in perfect good order in 1823, with a line-of-battle ship on the stocks, and who was assured that no piling whatever had been used in the foundation. He received his information from the Swedish Admiral Nordenshiould. Before I was acquainted with this fact, Mr. Smirke, in reply to a question from me, stated that he would not scruple to use a grouted foundation for a naval dock, or other massy work of equal importance, in the softest soil.

In preparing the foundation of one of the Martello Towers in the Eastern District, in 1810, which happened partly to consist of stiff soil, and partly of soft mud, extending to a great depth, the latter substance was cut out to the depth of from about 10 to 12 feet, and the excavation was filled in with small gravel grouted with hot lime, until a general level was obtained, upon which the usual elm grating was laid, so that it rested partly upon this artificial substance, and partly upon the sound portion of the natural soil. This precaution was found perfectly sufficient to secure the stability of the Tower, as no failure or alteration occurred afterwards in the slightest degree.*

But the use of grouted foundations, although a very old expedient, has recently been more systematically adopted by the eminent architect, Mr. Smirke, than was, perhaps, ever before attempted. I have already related the incident which drew his attention to this measure. He first made use of it at the Penitentiary, at Milbank, where the soil was of the most unfavourable nature possible, consisting of the softest description of mud, mixed with peat moss, to a very great depth. There he cleared out the whole space, intended to be covered by the proposed building, with some feet more on each side, beyond the base of the external walls, which great excavation he filled with grouted gravel to the depth of from 12 to 18 feet. The quality and proportion of the lime used having been before described, it is only necessary to state that the rule, in executing the work, is to spread the composition regularly throughout the whole foundation, and to raise it everywhere to the same general horizontal level, in courses of about six inches, and in so doing labourers puddle it down with their feet. Unless the trench itself be full of water, care must be taken to wet the gravel well in mixing it, because dry sand or gravel will shrink; whereas, after being well wetted, they naturally subside into the smallest possible space, and have no tendency to shrink or settle afterwards in drying, like masonry or brickwork built with common mortar.

When we consider the great extent of this foundation, and the nature of the substance employed, which becomes in process of time a kind of artificial rock, it will appear evident that the edifice itself rests on much too large a base to render it possible that the weight of the walls should ever press the foundation deeper into the mud. But Mr. Smirke does not in general recommend the grouting to be laid under the whole area of a building, as he has done at the Penitentiary. He is usually satisfied if it extends some feet each way beyond the footing of every wall; but at that edifice the apartments were so very small, that it would have been more troublesome and expensive to cut separate trenches for the foundations of the several walls than to excavate the whole space.

In the reconstruction of the New Custom House, after completely removing the piled foundation of the former architect, which had failed, he also grouted the whole area of the building, including not only the foundation of the new walls, but also of such of the original walls as were allowed to stand, which was done with great care by small portions at a time, and by underpinning the old wall whilst this operation was in progress.

---

* This fact was communicated by Mr. Howe, who was then employed in the Eastern District under Captain Whitmore. The tower alluded to was built upon the marsh near Walton Ferry, opposite to Harwich.

In respect to the foundation of the new walls in particular, he caused a course of large flat Yorkshire stones, each six inches thick, and averaging 25 superficial feet,* to be laid over the grouted gravel, and upon these two chain timbers of scantling 9 inches wide and 12 inches deep, which were firmly bedded in the brick footing of the walls, without resting on sleepers in the usual manner. This judicious arrangement possesses the advantages of a continued wooden platform without its disadvantages.

Mr. Poel's house, near Whitehall, which was built by Mr. Smirke, affords another instance of a grouted foundation in very bad soil, and as all the other buildings near it are piled, time will show which of these two expedients is the most effectual.

In the New Parliamentary Mews, about to be built near Westminster Abbey, Mr. Smirke has directed a foundation of grouted gravel 8 feet deep and 8 feet wide to be prepared for each of the walls.

I shall conclude by giving one more example of this kind of foundation. The new buildings called Lancaster Place, adjoining to Waterloo Bridge, are founded on very bad soil, which proved on examination to have consisted originally of soft mud, but which was intersected in all directions by the remains of the foundations of old houses and wharves. Trenches having been cut to a sufficient depth and width along the proposed line of the new foundations, fragments of the old walls were laid therein as regularly as possible, in the same manner as large flat stones might have been used. These were well sluiced with water, and mixed with grouted gravel of the nature before described, and rammed down flat, and thus the work continued, until the fragments alluded to were expended. Then the rough foundation was continued to the proposed footing of the brickwork with grouted gravel alone, and at that level transverse sleepers were used at intervals of about 4 feet apart,† upon which were laid two longitudinal chain timbers 9 inches square, completely surrounded by brickwork, but immediately under and flush with what was intended to be the proposed surface of the principal wall of the building after the work should be raised to a certain height, and a few feet higher, opposite to the top or uppermost offset of the brick footing, a third chain timber of the same scantling was bedded in the middle of the wall. Thus those three chain timbers combined cover a space of no less than two feet three inches wide, being equal to the thickness of a three-brick wall, and this arrangement is continued throughout the whole extent both of the external and cross or party walls of that row of buildings.

The annexed figure represents the profile of the foundation which has just been described, and which is everywhere 8 feet wide, and of the same average depth. The brick footings have a base of $5\frac{1}{2}$ bricks, which, by four successive offsets on each side, at intervals of three courses apart, diminishes, at the height of 3 feet, to $3\frac{1}{2}$ bricks, which is the thickness of the front and back walls of the lowest stories of the building, which are vaulted and used for stables or stores. Over this the walls of the basement floor of the inhabited part of those buildings are three

---

* Those stones of the description usually called Yorkshire Landers originally composed the pavement of the long room, which was supported by a series of groined arches resting on square piers and piles, the greater part of which suddenly fell down, but, fortunately, before the room was filled with persons on business, so that no lives were lost.

† But all under piers, none under the position of doors or windows.

Fig. 7.

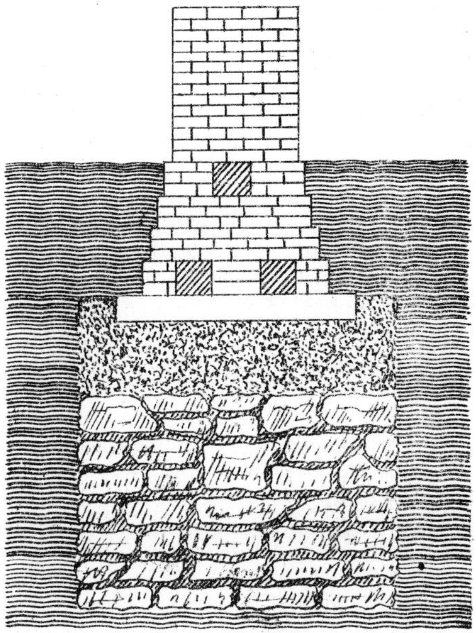

bricks thick, which dimension is gradually diminished by successive offsets to $1\frac{1}{2}$ bricks in the uppermost storey of the house.

The height from the top of the brick footing to the coping of the parapets in front and rear is not less than 88 feet, to which, adding the footings and foundation, it appears that there is a mass of brickwork and rough masonry and grouting nearly 100 feet in height, that is, on the side nearest to the river. If no failure should hereafter occur in this great building, which, as I said before, rests upon soft mud, and there seems no symptom of any such failure at present, it will afford a most convincing proof of the efficacy of the peculiar kind of foundation treated of in this article.*

I am also informed that Mr. Gwilt, the architect, and an author of several useful works, has adopted a similar principle. He forms his foundations of small broken stones, such as are now used for what have recently been termed Macadamized roads. These he cements with hot lime grout, and he states that he has never known any crack or failure in the walls of buildings having such foundations, even in very indifferent soil,† although he has frequently witnessed the failure of other methods, more commonly used.

On comparing together the two kinds of foundations that have been described in this and in the former article, it must be allowed that a grouted foundation, if equally efficacious, must be considered preferable to piling; first, as being a great deal cheaper; secondly, because piling, under many circumstances, may give way, after remaining good for a limited number of years; whereas, if a grouted foundation, once properly made, does not give way at first, there can be no doubt of its lasting for ever, as good gravel cemented by water lime is neither liable to decay nor decomposition.

By an experiment recently tried by Mr. Smirke at the New Custom House, in which he grouted two equal parts of gravel, the one with Dorking lime, the other with Roman cement, it was found that the latter

* The general design of those buildings was formed by Mr. Smirke, the architect, but the foundation that has been described and all the details of execution were determined by Mr. Baker, of Rochester, the proprietor, by whom they were also built, and whose skill and experience as a builder are well known.

† Mr. Gwilt's method was communicated to me by Lieutenant Gregory, to whom he described it in conversation.

would not answer at all, for, instead of becoming harder, the mixture fell to pieces.

Rules for judging of the Quality of Bricks.

In building, two things are evidently requisite: first, that the materials shall be of good quality; secondly, that they shall be skilfully put together.

In the foregoing articles the properties of mortar and cements have been sufficiently illustrated, and the unfavourable chemical changes to which they are liable, if improperly exposed to air or water before they are used, have also been explained.

In the article upon bricks an omission, however, will be observed, which it is proper to correct before we proceed farther. In stating the several kinds of bricks used, no rule was given for judging of their quality.

The value of bricks in the market, as an article of commerce, depends partly upon their colour; the light or buff-coloured bricks, such as grey stocks, being preferred to red stocks, although the latter may be equally good in every other respect; and as there will be several shades of colour, even amongst bricks obtained from the same clamp, it is usual for builders to purchase picked bricks of uniform colour for the fronts of buildings at a higher price.

There is also a quality of bricks made at Southampton, not mentioned before, which, being nearly white, are used in ornamental fronts, combined with Portland stone, of which a specimen may be seen in the new church at Windham Place, Bryanston Square.

The durability of bricks as a building material, which also has, of course, a great influence on their price, depends entirely upon their hardness and weight. The former quality may be judged of by taking a sound brick, without cracks, and striking it with another brick of the same description, or with a piece of metal. That which sounds the clearest is the best. In respect to weight, a good brick ought not to weigh less than about five pounds, but in all cases those which average the heaviest are the best.

Soft light bricks, of the quality called *place* or *samel* bricks, should invariably be rejected in every building of importance, for bricks of this description are liable to be crushed by the mere weight of the upper parts of a building in which they may be employed, and to moulder away by the action of the weather.

Having corrected the omissions in question, we will now proceed to treat of the art of bricklaying.

Of the manner in which Bricks ought to be laid. The term Bond in Masonry or Brickwork defined.

In building with bricks, as all the parts are equal and regular, they are laid in horizontal courses from the foundation or bottom of the wall upwards. The reason is obvious. Regular bodies, such as bricks, placed on a horizontal base, and with vertical sides, have no tendency to move laterally. If, on the contrary, bricks were laid on an oblique base, they would have a tendency to slide downwards in one direction. Hence the wall might eventually bulge or crack, or lose its perpendicularity.

If we suppose a wall of only half-a-brick thick, all the bricks used would of course be laid lengthwise, so as to show their whole length in the face of the wall, in which case they are called *stretchers.*

In laying a second course of bricks care would be taken to prevent any two vertical joints from coinciding. This would be effected by placing the joints or meeting of the ends of the second course of bricks over the middle of the bricks of the first course. The same arrangement would be attended to in the third course, and in all the succeeding courses.

Figs. 8 and 9 represent the plan and elevation of part of a brick building, in which the walls are supposed to be only half-a-brick or about four inches thick, and built in the manner that has been described.

Fig. 8.

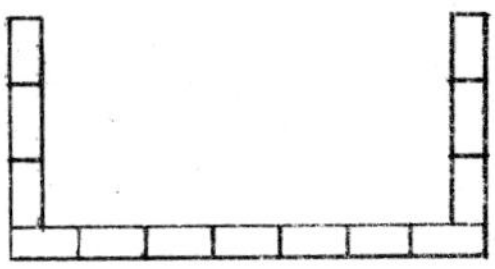

The arrangement of never allowing any two vertical joints to coincide is technically called *breaking joint*, and a wall thus built is said to have proper *bond*, a term which implies that the parts are well connected.

Fig. 9.

It will be observed, that all the bricks represented in the elevation are stretchers, showing their whole length in the surface of the wall except one brick at the angles of the wall in each alternate course, which shows its end in the elevation. A brick thus laid is called a *header.*

By this arrangement of having headers at the angles in each alternate course, bond is obtained. If the whole were laid stretchers throughout, as in *Fig.* 10, all the vertical joints would coincide, and the parts of the wall would evidently be very imperfectly connected.

Fig. 10.

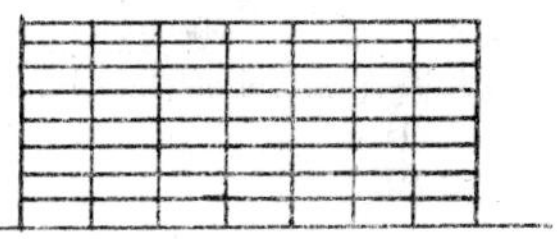

It will be remarked, that the above supposition of a building only half-a-brick thick was introduced merely to illustrate the principle of what is technically called *bond* in building. No brick walls so slight as the above are ever used, this dimension being much too weak to afford proper stability even to the smallest buildings, unless the brickwork be held together by a wooden frame, of which it fills the intervals. This style of building, used for economy, is altogether unsuitable for public buildings, unless of a temporary nature, and is called *brick nogging,* of which specimens are very common, especially in villages.

Brick nogging is also sometimes used for the partitions of dwelling-houses.

In houses built of regular squared stones the arrangement of laying the walls entirely with stretchers is, however, not uncommon. In the Island of Minorca, for instance, the stones generally used for dwelling houses are cut two feet long and one foot square. Stones of this description are called Cantoon stones, and are of a soft but handsome and durable description, like Bath stone. The walls are usually made one foot thick. Hence all the stones used are laid in the manner shown by the first elevation in this article; every stone of each several course throughout the whole external walls of the building being a stretcher, but so connected at the angles that the end of each corner stone, whilst it is a stretcher in one wall appears as a header in the adjoining one, and thus proper bond is obtained throughout.

Regular squared stones, such as have been described, are called *ashler*, whilst rough irregular stones are called *rubble;* but as it is not our object to enlarge upon masonry, we shall return to brickwork.

Having already stated that proper brick walls are always one brick thick or upwards, it is to be observed that in such walls bond cannot be obtained without introducing parts smaller than a regular brick near each angle and opening, that is to say, not only at the corners of the walls, but also at each side of every door and window, break, or recess, throughout the whole building.

For example, if we suppose a wall to be one brick thick, that is, about nine inches in thickness, the bricks may evidently be laid either as headers, in which case one brick occupies the whole thickness of the wall, or as stretchers, in which case two bricks placed alongside of each other occupy the whole thickness.

Fig. 11.

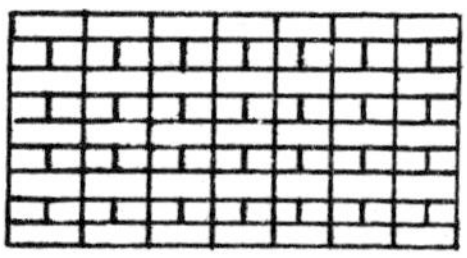

Now, if we suppose *Fig.* 11 to represent the elevation of a wall in which the bricks are all of the regular full size, and laid stretchers and headers alternately, in successive courses, it will be evident that one half of all the vertical joints must necessarily agree, because every two headers occupy, in the face of the wall, precisely the same space as one stretcher.

Fig. 12.

To obviate this inconvenience, which would render the bond of the wall very imperfect, small headers are used, only a quarter-of-a-brick in width, near each angle, opening, or break in the wall, as was before mentioned, the utility of which will be evident on comparing *Fig.* 12, in which this method is represented, with the preceding figure.

This arrangement is invariably followed in brick buildings, as may be observed by inspection, and thus a perfect bond is produced throughout the whole wall, it being observed that these narrow bricks appear in alternate courses only, and are the second brick from the angle of their respective courses, the first or corner brick of the same course always being a header.

It is to be observed that, instead of using smaller bricks made for the purpose, the bricklayer, in the act of building, breaks off pieces from a common brick in order to obtain the narrow bricks that have been mentioned, which, from the circumstance of their closing or connecting the work properly, are termed *closers*.

These closers, which show as headers in the elevation of a building, are not in reality such, for they do not extend the whole length of a brick, but are some shorter fragment, it being found difficult to break a whole brick longitudinally down the middle.

Broken bricks are generally called brickbats, but the term bat is by workmen specially applied to one half of a brick broken transversely across the middle.

A regular bat or half brick broken a second time produces two pieces, each a quarter of a brick in width and half a brick in length, which combine to form a closer, those pieces singly being half closers.

When a brick is broken transversely into two unequal parts the larger of the two is called a three-quarter bat.

The remarks hitherto made upon the bond of brickwork apply only to the external surface of the wall, as seen in elevation, which may be judged of by the eye. In walls of a certain thickness the internal bond, which cannot be seen or examined after the work is finished, is of equal importance.

Fig. 13 represents the section of a wall two bricks in thickness, in which the courses are supposed to be laid alternately all headers and all stretchers, so as to correspond with the foregoing elevation; and yet, owing to a faulty arrangement of the bricks used, all the vertical joints in the middle of the wall coincide with each other. It will be evident that such a wall, although it would appear strong and perfectly bonded, by merely inspecting the external surfaces of it, has no proper bond internally, for one vertical joint prevails from top to bottom, and from one end of the wall to the other, so that such a wall would be liable to split in two by a longitudinal fracture.

Fig. 13.

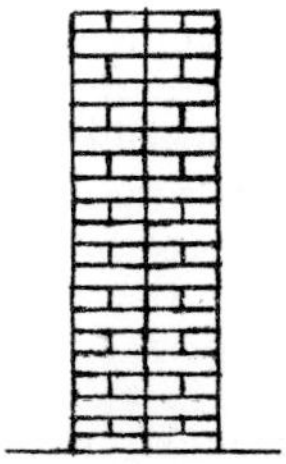

By adopting a different arrangement, as in *Fig.* 14, which represents the section of a wall of the same thickness, the above defect will be completely obviated, for it is evident that all the vertical joints, internal as well as external, are now properly broken, and that a sufficient bond is thereby obtained throughout every part of the wall.

Fig. 14.

At the same time it will be understood, on carefully comparing this section with the former one, that although the one is exceedingly weak, whilst the other is sufficiently strong, they may both have had precisely the same external appearance, or elevation, as viewed by the eye of a spectator, after the work was finished.

It will hereafter be seen that it is often convenient, if not absolutely necessary, for the purpose of obtaining the most perfect bond, to finish one side of a wall, not exactly, although nearly, in the same manner as the other; so that, although both elevations may appear alike to an inexperienced observer, the arrangement of the joints of the one will in reality be less regular than the other. In this case *the face of the wall*, which in a building implies the external surface, being that exposed to public view, must be finished in the most regular manner, in preference to *the back* or internal surface of the wall, which is eventually concealed from view by plastering or papering.

Having defined the bond of brickwork in general terms, the succeeding articles will treat fully of the different kinds of brickbond which either have been actually adopted in the practice of building, or which may be used. But, before these articles can be read to advantage, the student ought to provide himself with a number of wooden models of bricks, such as are usually made for the amusement of children. The whole bricks should be two inches long, one inch broad, and two-thirds of an inch thick. Bats, or half bricks, and closers, or quarter bricks, should also be provided, together with a small proportion of three-quarter bats. With these models the student should practice upon a table, building the representation of brick walls of different thicknesses, and in two or three courses, such as will afterwards be described. This will give him a much better notion of the subject, and in a much shorter time than any other method that can be suggested, not even excepting the inspection of bricklayers at work.

Of English Bond, as it has been defined in some Books of Architecture.

In some architectural books recently published, English bond is defined to be that in which the bricks in successive courses of the wall are laid alternately all headers and all stretchers from the bottom upwards; the former from this circumstance being called heading courses, and the latter stretching courses, and in which, whilst the bricks of successive courses break joint with each other, according to rule, as seen in elevation, those of the same course also break joint with each other, as seen in plan.

The elevation given in one of the figures (*Fig.* 12) of the foregoing article is therefore a correct representation of the external appearance of English bond, whilst *Figs.* 15 and 16 show the plan of two successive

Fig. 15.

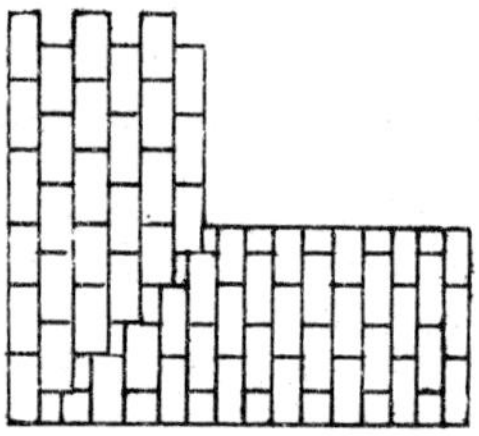

Fig. 16.

courses of it, supposed to be executed according to the above definition.

These represent the corner of a building, three bricks thick, where two walls meet at right angles, according to custom.

A sufficient portion of each wall is shown to illustrate this style of building. Closers or quarter bricks in each course are laid near the angle, without which, the external joints could not be broken in the manner that was shown in the elevation.

The various sections of this wall, for it is capable of several, are as follows:—

First, under the supposition that the bats and whole bricks of the successive heading courses happen to be so arranged that they shall regularly alternate with each other from the bottom of the wall upwards. In this case, the several vertical joints, after coinciding for three courses in height, will be broken by a header in the fourth course, as in *Figs.* 17 and 18.

Secondly, under the supposition that all the bats of the successive heading courses happen to be placed exactly above each other. In this case, all the vertical joints will partially coincide from the bottom to the very top of the wall, as in *Figs* 19 and 20.

Fig. 17.

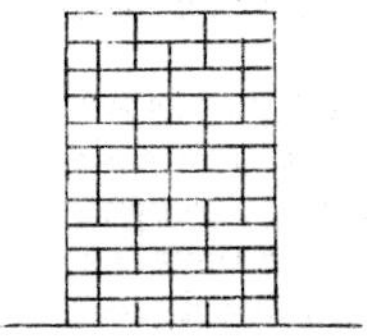

Fig. 18.

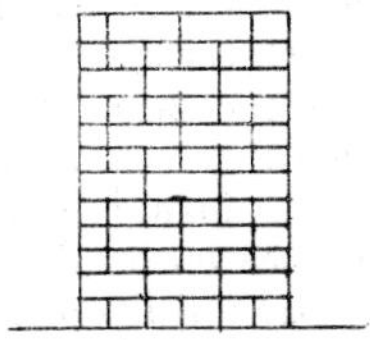

Fig. 19.

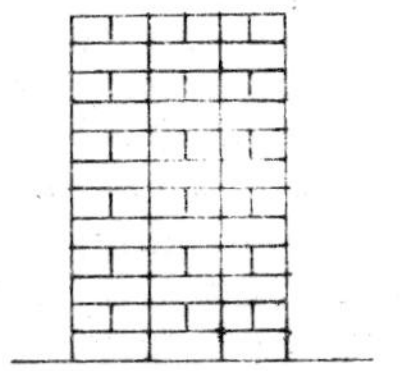

Fig. 20.

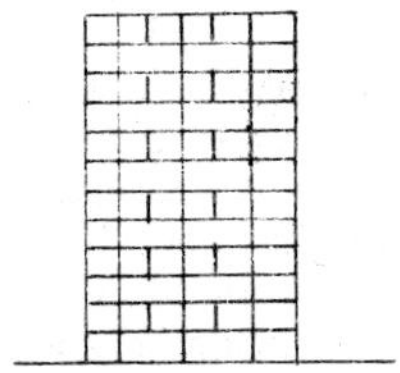

It is to be observed, that those joints which thus coincide or are not properly broken, form a series of fissures in parallel rows, which, according to the circumstances before supposed, may either be partial and broken at intervals, or which may be equal in depth to the whole height of the wall. They extend in a longitudinal direction from one end of the building to the other, not in one continued right line, but broken at intervals of about four inches or half-a-brick apart, to which dimension they are also equal in length. The number of rows is one less than the number of half bricks in the thickness of the wall, and the fissures in each row are not exactly opposite to each other, but arranged checkerwise, as in *Fig.* 21, which is a horizontal section of the wall, corresponding with the plan of one of the heading courses, whilst the work is in progress.

Fig. 21.

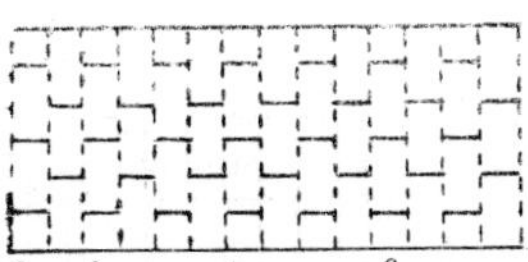

In this figure all the joints excepting the unbroken ones, or unfavourable fissures alluded to, are dotted.

Taking into consideration the unfavourable circumstance of so many unbroken joints occurring, even under the supposition of every brick being laid with the extreme regularity shown in the two first figures of this article (*Figs.* 15 and 16), it appears to me that this species of English bond ought to be rejected, as being a very weak and defective arrangement, for such extraordinary precision cannot be expected in practice, for many reasons which it would here be needless to enlarge upon, and the slightest deviation from it would increase the number of false joints, of which there are already too many.

I have never seen any wall built in this manner, excepting a part of the great mass of grouted brickwork in the foundation of the new dock now in progress in Chatham Yard, of which some courses were executed nearly according to the above arrangement, which did not, however, by any means prevail systematically throughout the whole work.

Indeed, I have great doubts whether this kind of English bond was ever used at all, for one cannot pull down ancient buildings to ascertain this point, and there is no judging of the inside of the walls of such buildings merely by their external appearance, and I should therefore have passed it over in silence, as an arrangement absolutely unworthy of notice, if it had not found a place in the architectural books alluded to, the authors of which have apparently copied their definition of, and remarks upon, English bond, from one another, without giving due attention to this part of their subject. There is, however, one book specially written on the bond of brickwork, from which more satisfactory information might be expected, but I have not yet been able to procure it.*

The next article will treat of a more rational kind of English bond, which in all probability accords with the practice of former times, before the peculiar style of brickwork called Flemish bond, which is now in more general use, was introduced into this country.

Of Common English Bond.

In the common English bond alternate courses of headers and stretchers appear on each side of the wall, so that the elevation of it is the same which has already been described (see *Fig.* 12). But, with the exception of the above-mentioned stretchers, which appear externally, no others whatever are used, so that the remainder of the wall consists exclusively of headers.

In looking at a wall thus built, those courses where headers appear to the eye are called *heading courses*, but those in which stretchers appear to the eye are called *stretching courses*, notwithstanding that the inside of the latter may be filled up with headers.

It is essential, in common English bond, that the horizontal joints in the same course shall never be broken, but that the bricks shall be so arranged that the whole of the opposite joints shall extend transversely in one and the same right line, from one side of the wall to the other, that is to say, the bricks of the same course must always be arranged as in *Fig.* 22, not as in *Fig.* 23.

Fig. 22.

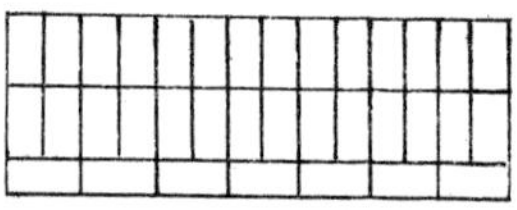

Fig. 23.

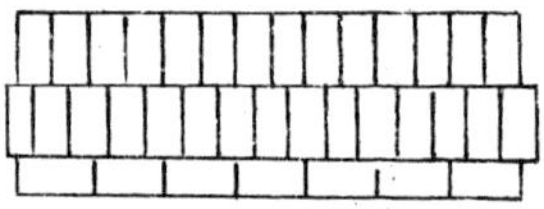

By attending to the above rules this style of building will be found extremely simple.

In a wall whose thickness is either one brick, two bricks, three bricks, four bricks, or any other dimension involving a whole number, the stretching courses agree on both sides of the wall, but if the thickness be one brick and a-half, two bricks and a-half, three bricks and a-half, or any other dimension involving half a brick, that which is a stretching course on one side of the wall is a heading course on the other side, and *vice versâ.*

In order to exemplify the common English bond thus described in general terms, we will suppose a portion of a brick wall equal to seven bricks in length, and terminating with both its ends at right angles to the general line of the wall. This portion of brickwork may represent part of a wall contained between the openings of two adjacent windows. It is true that the sides of windows are not now usually finished in

Fig. 24.

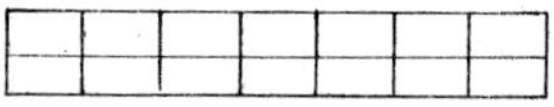

Fig. 25.

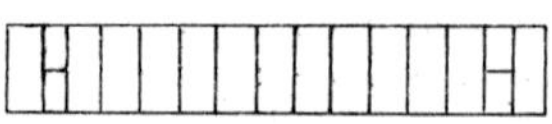

Fig. 26.

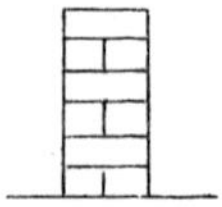

---

* Entitled "A Tract on Brick Bond," by Mr. G. Saunders.

such a simple manner,* but for the present we will suppose them to be so, for the sake of illustrating the peculiar kind of brickwork under discussion.

Fig. 24 represents the stretching course and *Fig.* 25 the heading course of a wall of this description, one brick thick. In the heading course it will be observed that closers are laid next to the extreme header at each end, without which the proper bond could not be obtained. *Fig.* 26 is the section or end elevation of the same wall.

Figs. 27 and 28 represent two successive courses of a wall of the same length, but one brick and a-half thick, and *Fig.* 29 is a section of the same wall.

Fig. 27.

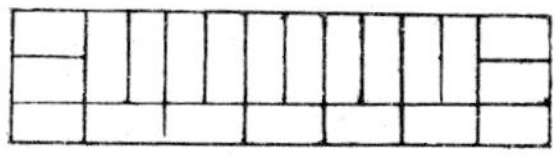

Fig. 28.

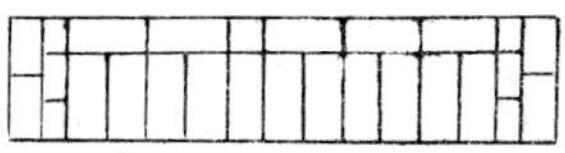

Fig. 29.

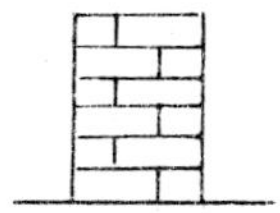

In this case it will appear by the figures that in consequence of there being a half brick in the thickness, that which is the heading course in front is the stretching course in rear of the wall, and *vice versâ.* At the ends of the wall three-quarter bats are used in alternate courses, without which proper bond could not be obtained there, in a wall of this thickness.

In this example it will also be observed that two sides of the wall are not exactly alike, and the same circumstance will occur in all walls built according to common English bond, in which a half brick is involved in the thickness of the wall. This variety has been already alluded to, and according to the rule stated, that side of the wall which is towards the bottom of the paper in all the figures representing the plan, being the most regular of the two, is supposed to be *the face* of the wall, and, as such, to coincide with the external surface of the building. In all walls, on the contrary, whose thickness is measured by an integer number of bricks, both sides are precisely alike.

Figs. 12 and 30 represent the respective elevations of the *face* and *back* of a brick wall, in which the difference alluded to occurs. *Fig.* 12 is the regular English bond, the representation of which is here repeated. *Fig.* 30 varies from it, inasmuch as the stretching courses have a header and closer at each end, a half brick in the middle, whilst the heading courses have a stretcher at each end.

Fig. 12.

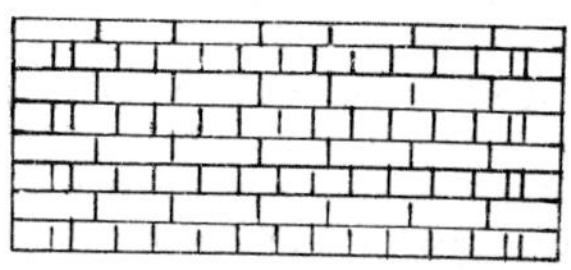

Fig. 30.

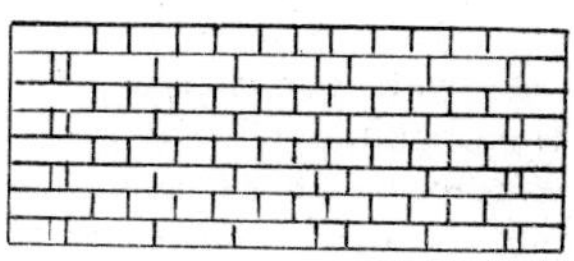

Fig. 31.

Fig. 31 is the end elevation of the same wall, in which the two three-quarter bats before alluded to appear in alternate courses.

Figs. 32 and 33 represent the plan of two successive courses of a wall of the same length, but

* It was, however, the method used formerly, and which is still retained in respect to the windows of inferior buildings.

two bricks thick. On attaining this thickness the ends of the wall present a more considerable surface than in the former suppositions, and therefore they require to have their joints broken by means of a couple of closers, in alternate courses, it being proper to treat the square ends of walls of a certain thickness precisely in the same manner as if they were short portions of the front of a building.

Fig. 34 is a section of the same wall.

Fig. 35 is an end elevation of it.

Figs. 36 and 37 represent the plan of two successive courses of a wall of the same description, but two bricks and a-half in thickness.

Fig. 38 is the section and *Fig.* 39 the end elevation of the same wall, in which a half brick, shown also in *Fig.* 37, is necessarily used in the centre of each alternate course, in order to obtain proper bond.

Figs. 40 and 41 represent in like manner the plan of two successive courses of a wall of the same description three bricks thick, whilst *Fig.* 42 is its transverse section, and *Fig.* 43 is its end elevation.

Having sufficiently explained the method of breaking joint in short portions of walls built according to the common English bond, it remains to treat of the method of effecting the same object, at the corners of buildings where two external walls meet at right angles.

In this case it is proper, commencing as usual from the corner header at the external angle, to introduce a set of closers, covering an extent equal to the whole thickness of the particu-

Fig. 34.

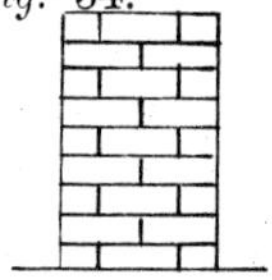

Fig. 32.

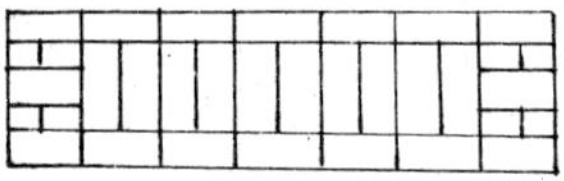

Fig. 35.

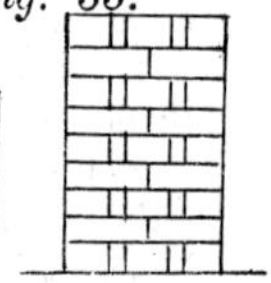

Fig. 33.

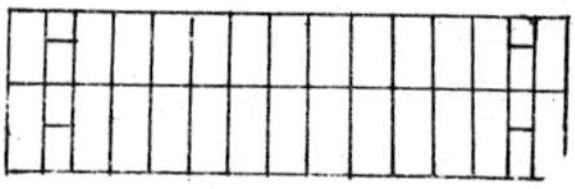

Fig. 36.

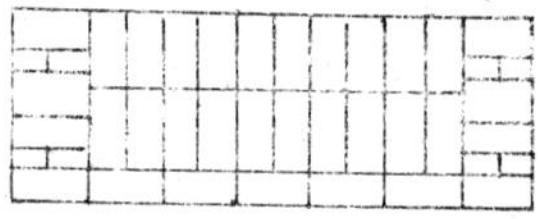

Fig. 38.

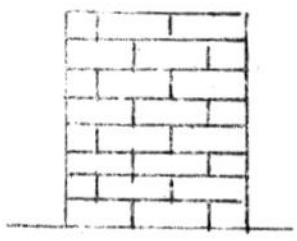

Fig. 37.

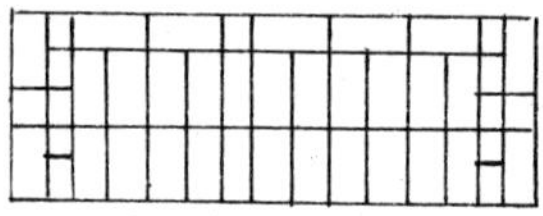

Fig. 39.

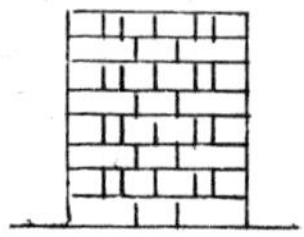

Fig. 42.

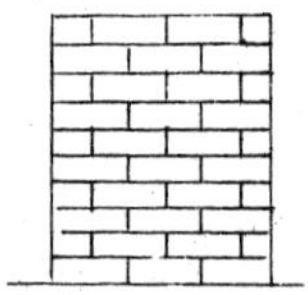

Fig. 40.

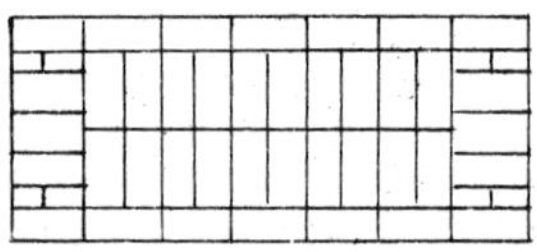

Fig. 43.

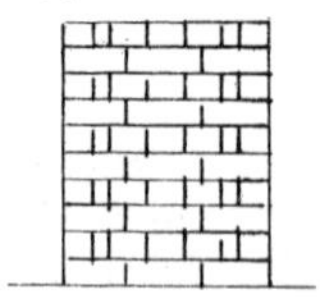

Fig. 41.

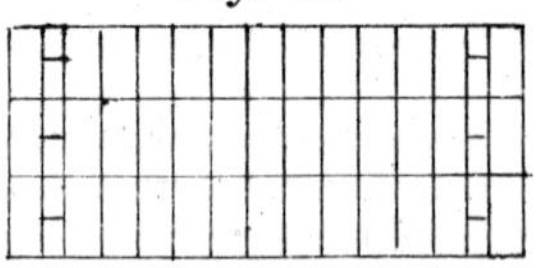

lar wall of the two* to which that header belongs, which closers need not, however, be disposed in a continued straight line, but checkerwise, so as to approach towards the re-entering angle, but without absolutely reaching it; and thus none of them will be seen in the inside of the building. They must all be laid parallel to each other, or, as it may be termed, in *échelon*.

Figs. 44 and 45 represent this arrangement, showing two successive

Fig. 44.

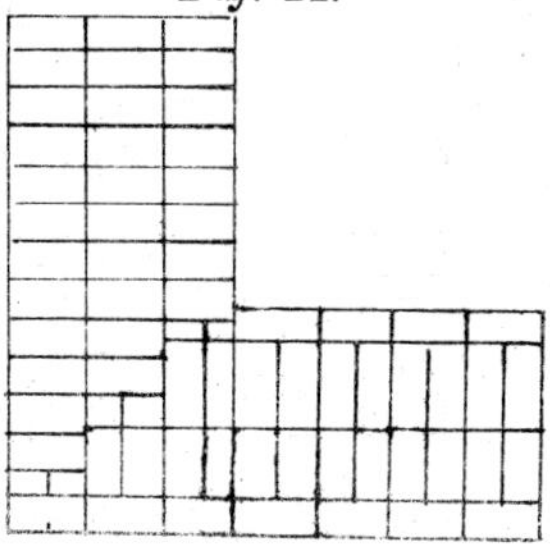

Fig. 45.

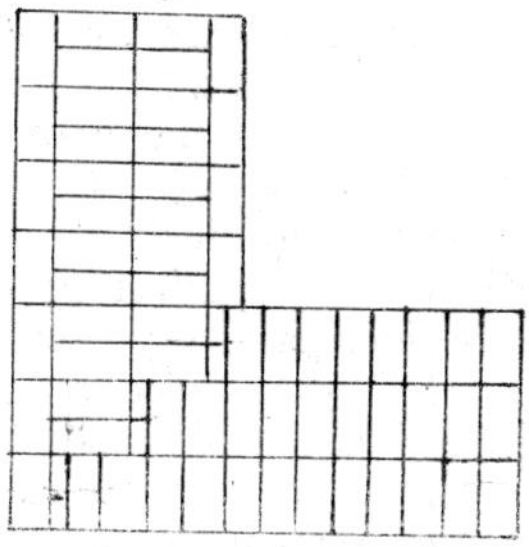

courses of the corner of a building, formed by the meeting of two walls, each supposed to be three bricks thick.

In this arrangement it will be observed that, of the two walls which meet each other at the angle, one shows a stretching course whilst the other shows a heading course on the same level, and *vice versâ*; and these two arrangements meet each other at those parts of the brickwork where the three closers, which are equal in all to the length of three bricks or to the thickness of the wall, extend checkerwise entirely across the wall.

It is to be observed that the same effect of breaking joint may also be obtained by extending the closers in a straight line from the external angle, as represented in *Figs.* 46 and 47; the former is, however, the more usual method.

Fig. 46.

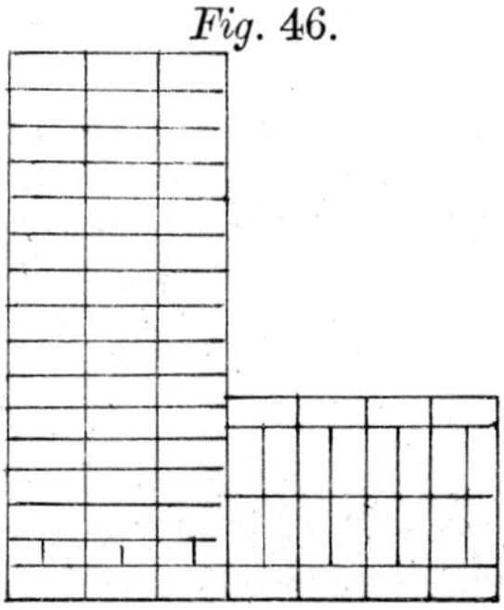

Fig. 47.

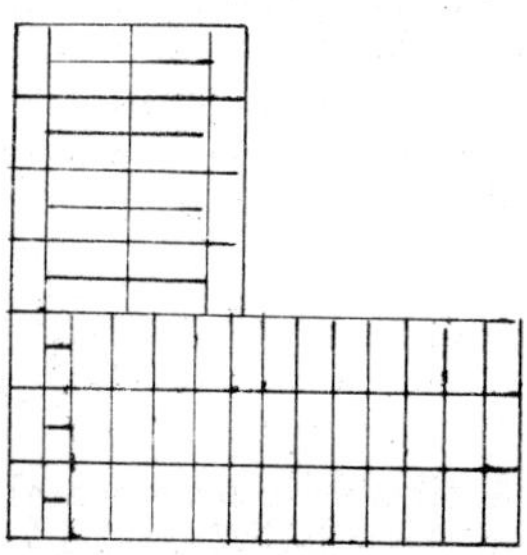

* That is if the two walls be of unequal thickness, which may often happen in practice.

I shall give another example of the corner of a wall, supposed to be three and a-half bricks in thickness, in which the closers are supposed to be laid checkerwise.

Figs. 48 and 49 represent the plan of two successive courses of the wall*

Fig. 48.

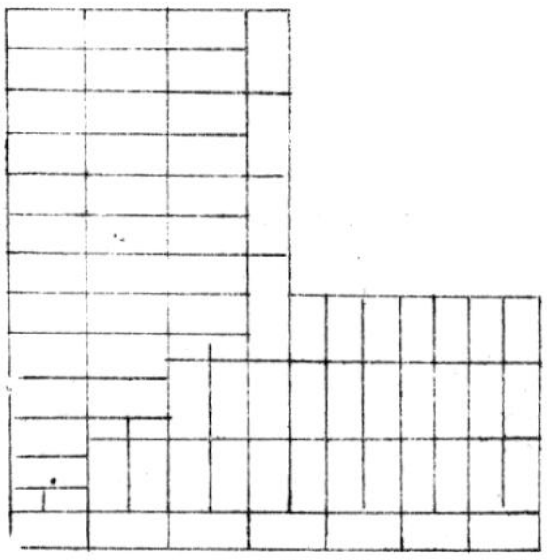

Fig. 49.

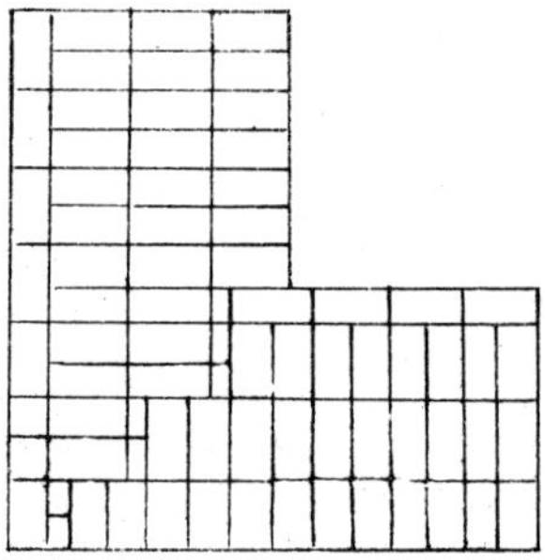

It would be superfluous to add any more examples of common English bond, as the principle of it has now been sufficiently illustrated. This arrangement has the peculiar advantage of not only being simpler than any other bond, but of breaking all the vertical joints from the bottom of the wall upwards. It was used in some of the counterscarp revetments of Portsmouth about 30 years ago, and recently in the lower part of the brick wall of the New British Museum, which is five and a-half bricks in thickness. It has also been used in the whole of the walls of some large extensive storehouses in the Commercial Docks at Rotherhithe; but I am not aware of its having been used in the external walls of any other new buildings of importance. It is, however, continually used for the cross walls, or party walls, of buildings, but as these are generally concealed from view by papering or plastering after the edifice is finished, this circumstance can seldom be remarked. There is an example of it in the Soldiers' Quarters in the Ordnance Barracks at this station, where all the inside walls, which are two bricks thick, are finished in this manner.

English bond may he remarked in many old buildings, as for instance in a part of Cobham Hall, the residence of the Earl of Darnley; in Restoration House,† opposite to the field called the Vines, in Rochester; in some of the storehouses in the Gun Wharf, and in the Poor House of Chatham.

Of a variety of Common English Bond, in which the use of closers is dispensed with.

In building English bond, a variety may be introduced by using so many three-quarter bats, instead of whole bricks, as corner headers at

* In a wall where the thickness involves a half brick it is necessary that three half closers shall be in the same line as in those figures, otherwise they would extend across the wall, and appear inside, contrary to the rule before laid down.

† A remarkable old building, said to have been so named from King Charles II. having slept there on his return from exile.

the extremities of every heading course, as will make up the thickness of the adjoining wall, and in this case no closers are required.

Figs. 50 and 51 represent the plan of the corner of a building three

Fig. 50.

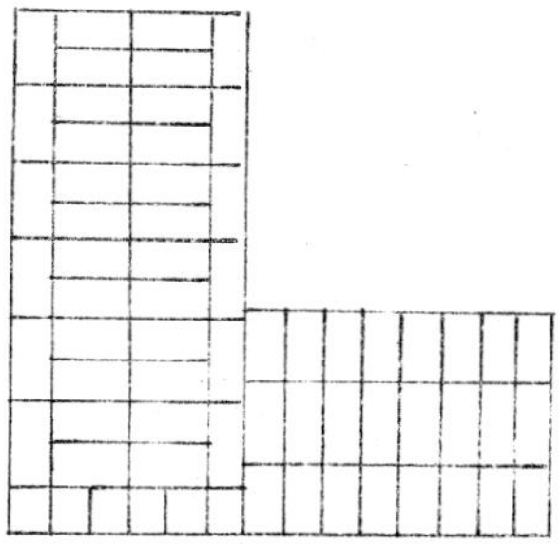

Fig. 51.

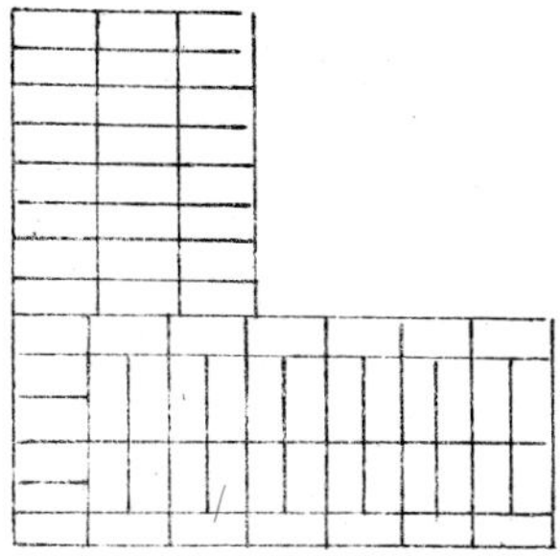

bricks thick, in which this arrangement has been followed. In each heading-course six three-quarter bats are used near the angle, which makes a space equal to three bricks, or to the thickness of the wall; and in this case it is evident that all the joints will be perfectly broken, without the aid of closers.

It will also be observed that the corner header of one wall shows as a short stretcher in the stretching course on the other side of the angle.

Fig. 52 is an elevation of the wall, which, combined with the two former figures, will fully elucidate all the particulars which have been explained.

Fig. 52.

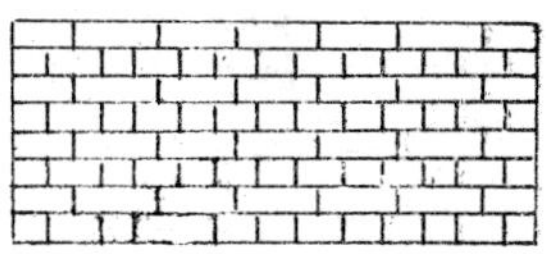

The variety of English bond that has just been described is equally good, perhaps better, than the common method, but it is not so economical of materials, because, by the exclusive use of three-quarter bats, a number of whole bricks would require to be broken in order to obtain those bats, and the smaller fragments would be entirely wasted; whereas, by the system of closers, all the bats or smaller fragments of broken bricks become available. For this reason the common practice of bricklayers may upon the whole be considered the most expedient.

Of Heading Bond.

A brick wall may be built all heading bond, that is, witbout using any stretchers at all, excepting in alternate courses, at the corners of the building or *coins*, as they are technically called, and at the various other angles or breaks occasioned by projections and recesses, or by the openings of doors and windows.

Figs. 53 and 54 represent the plan of two successive courses, and

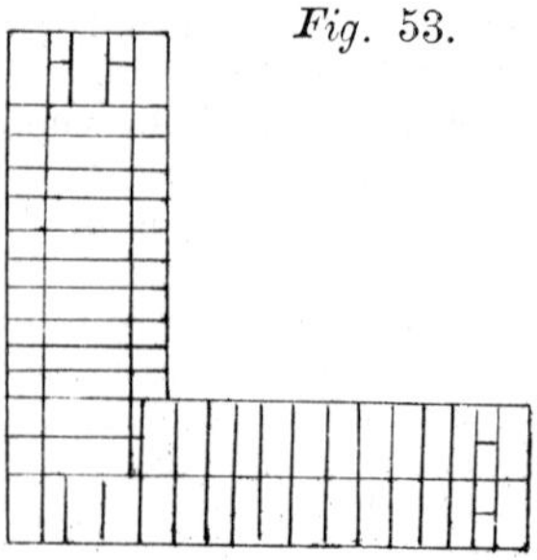

Fig. 53.

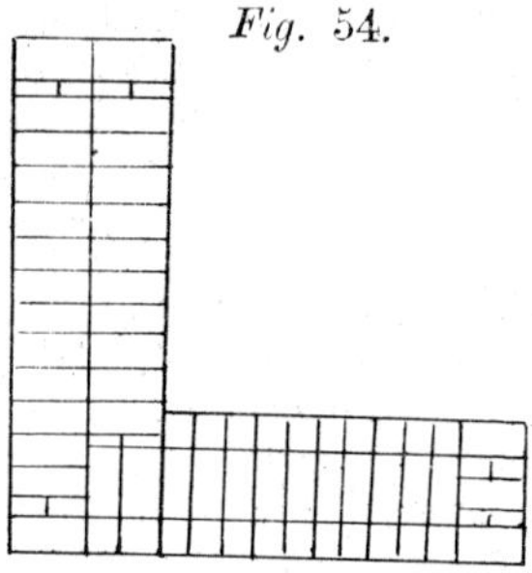

Fig. 54.

Fig. 55 is the elevation of part of a building thus executed. One corner of the building is shown with the walls extending on each side as far as two openings, which may be supposed to be doors or windows.

Fig. 55.

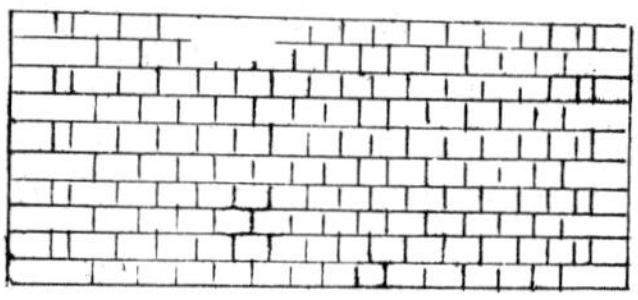

This style of building is preferred by Mr. Batty Langley to all others, and it certainly looks very neat, but I do not consider it so good as the common English bond; for, as compared with that, it merely consists in breaking every external stretcher into a couple of bats, which evidently cannot be an advantage.

Several very old buildings finished in this style may be seen at the Charter House School, in London, but I am informed that it is in contemplation to pull them down. There is a specimen of heading bond in this neighbourhood, in a very neat little house in Frindsbury.

This method is very rarely used at present, excepting when the outline of part of a building is laid out in a small circular projection. In this case, it is much easier to reconcile bricks laid as headers to the convex outline than it would be to accommodate stretchers to that form. At the same time it will be evident that, unless the sides of the whole bricks, as they appear in plan, be cut to a wedgelike form, radiating towards the centres of the supposed curves, the bond must be very imperfect.

Figs. 56 and 57 represent the plan of two successive courses of a circular projection of the description now under consideration.

Fig. 56.

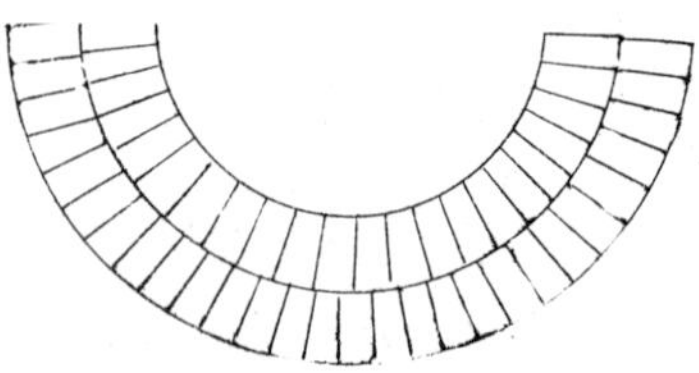

Fig. 57

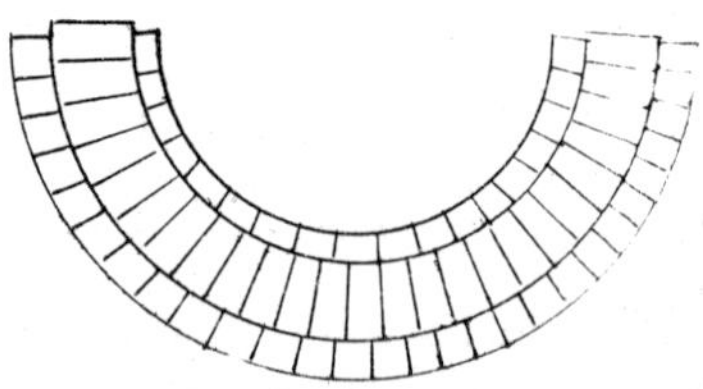

It may be observed that Mr. Batty Langley shows a different arrangement of heading bond from that which has just been explained by me; for in every course he uses a half brick and a whole brick alternately as headers on both sides of the wall, and from one end of it to the other; but his method is defective, for the reason before stated, because whilst he thus breaks his horizontal joints he causes his vertical joints partly to coincide, which is prejudicial to the strength of the wall.*

Of Flemish Bond.

Flemish bond implies brickwork so arranged that the bricks in every course present the appearance of a header and a stretcher alternately, from one end to the other and in successive courses; every header is placed over the centre of a stretcher in the course below it. Closers are placed next to the corner headers in alternate courses, in the manner that has already been explained.

Fig. 58.

Hence the elevation of part of a brick wall, constructed with Flemish bond, assumes the appearance represented in *Fig.* 58.

In the illustrations of Flemish bond which will be given in this article we will suppose a portion of brick wall equal to seven bricks in length and terminating with perpendicular ends.

Figs. 59 and 60 represent the plan of two successive courses of such a wall, supposing it to be only one brick thick. *Figs.* 61 and 62 are sections of different parts of the same wall.

Fig. 59.

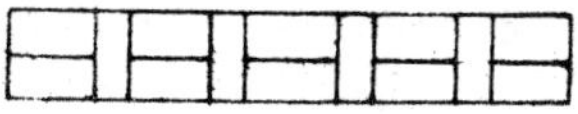

Fig. 61.

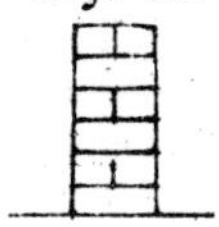

Fig. 60.

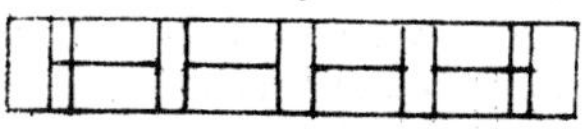

Fig. 62.

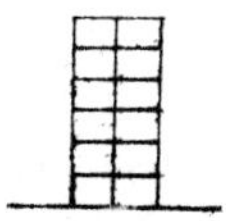

In walls of the above thickness both sides of the wall are necessarily alike, but in walls of greater thickness two different arrangements may be adopted, one of which is called single, the other double, Flemish bond.

---

* This arrangement produces the very same defect before explained, which applies to the imperfect kind of English bond as defined in certain architectural books. (See page 32.)

Of the Common Single Flemish Bond, now generally used in buildings, for all external walls exceeding one brick in thickness.

Single Flemish bond implies that a wall shows Flemish bond on one side only, whilst it shows English bond on the other, which arrangement is only applicable to walls exceeding one brick in thickness.

In this case, the Flemish bond, as being considered the most pleasing to the eye, appears, of course, on the face of the wall or external surface of the building, which is eventually the only side exposed to view.

The common single Flemish bond is merely a modification of English bond, which may perhaps be considered a little more ornamental, but which certainly is not so strong.

Like the other, it has heading and stretching courses alternately.

The stretching courses consist merely in using bats or half bricks instead of whole bricks, to represent headers, in every alternate course of the face of the wall.

In the heading courses real headers or whole bricks are used in the face of the wall, between which it is usual to back every stretcher with a second brick also laid as a stretcher. With this exception of using double stretchers in the heading courses, the same arrangement is followed as in English bond, of filling in all the remainder of a thick wall with headers.

The same rule is also followed of not breaking the horizontal joints.

Adverting to those particulars, the common Flemish bond, now in general use, as executed by the bricklayers in and about the metropolis, will be found sufficiently simple; and its similarity to the common English bond described in the foregoing article renders it probable that the latter was the only English bond ever used in this country, for the transition from the one to the other is such as might have been easily effected without doing any great violence to the former habits of the workmen.

Another reason also may have led to the use of false headers, which I shall not enlarge upon at present, until the common Flemish bond, now under consideration, shall have been more fully explained.

Fig. 63.

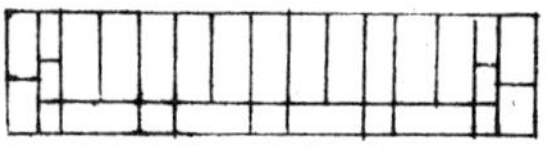

Fig. 65.

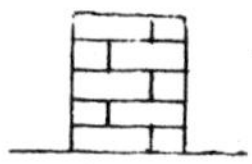

Fig. 64.

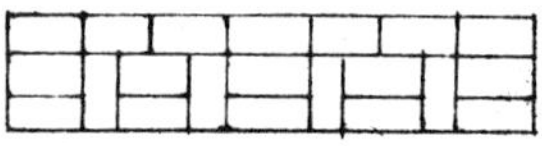

Fig. 66.

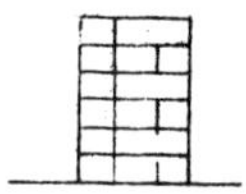

Fig. 67.

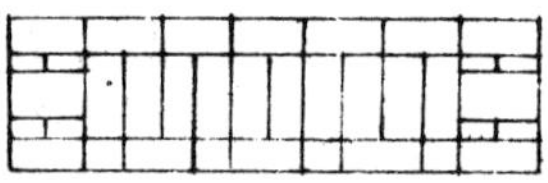

Fig. 69.

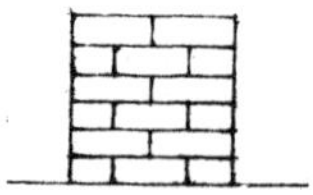

Fig. 68.

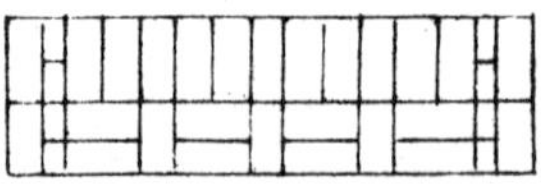

Fig. 70.

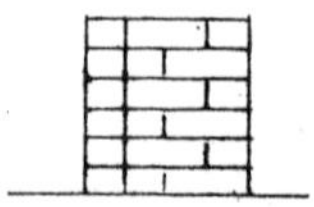

Figs. 63 and 64 represent the plan of two successive courses of a portion of a wall of the description before specified, built according to common Flemish bond, and one brick and a-half in thickness; and *Figs.* 65 and 66 represent the sections of the same wall.

Figs. 67 and 68 represent the plan of two successive courses of a wall of the same description, but two bricks thick; and *Figs.* 69 and 70 are sections of it.

Figs. 71 and 72 represent the plan of two successive courses of a wall of the same description, but two bricks and a-half thick; and *Figs.* 73 and 74 are sections of it.

Figs. 75 and 76 represent the plan of two successive courses of a wall of the same description, but three bricks thick; and *Figs.* 77 and 78 are sections of it.

Fig. 71

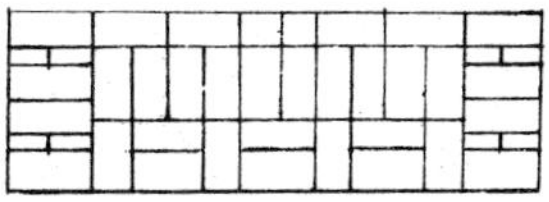

Fig. 73.

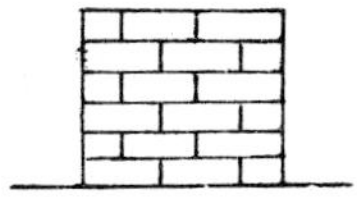

Fig. 72.

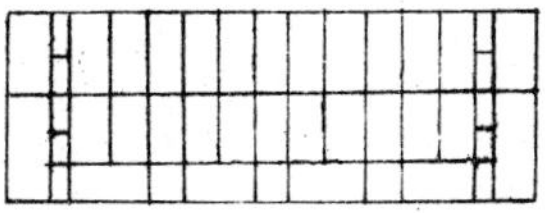

Fig. 74.

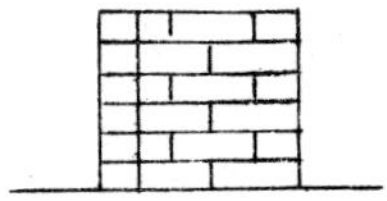

Fig. 75.

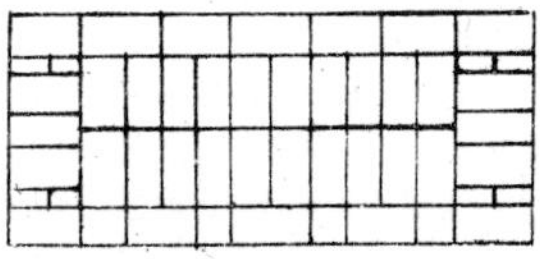

Fig. 77.

Fig. 76.

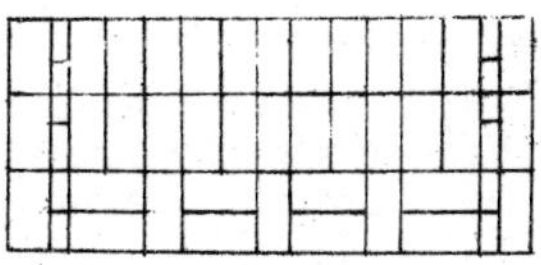

Fig. 78.

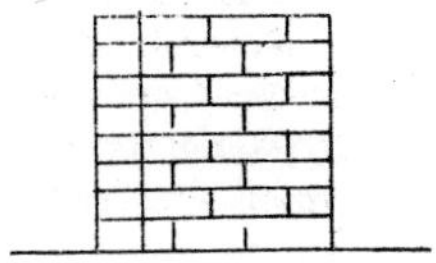

Figs. 79 and 80 represent the plan of two successive courses of the corner of a house, in which the walls are supposed to be three bricks thick, built according to the common single Flemish bond, and having the closers checkered in the usual manner.

Fig. 79.

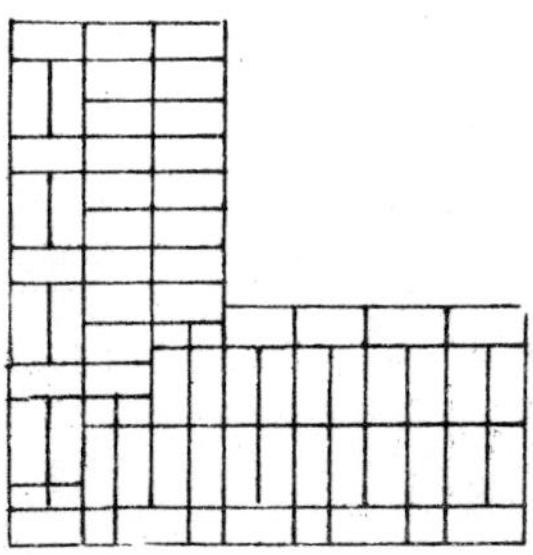

Fig. 80.

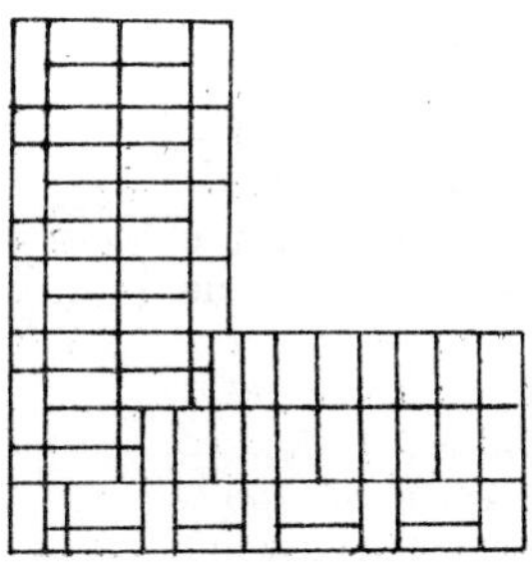

In all the examples of single Flemish bond which have been given, it will be observed that the section in one part of the wall is good, which is supposed to be taken over one of the true headers of the heading courses, whilst the section in other parts of the wall, supposed to be taken over or near one of the false headers or bats in the stretching courses, is defective, there being one unbroken vertical joint, or unfavourable fissure, from the bottom to the very top of the wall.

In the plan of a wall built according to single Flemish bond, these fissures extend longitudinally in one row, in a direction parallel to the face of the wall, and at the distance of half-a-brick from it. Each of these is equal to a whole brick, or nine inches in length, and they are separated from each other at intervals of about four inches or half-a-brick, being the width of the whole headers of the heading courses on that side of the wall.

Fig. 81, therefore, represents in plan the position of these fissures or unbroken joints in a three-brick wall, of which all the other joints are dotted.

Fig. 81.

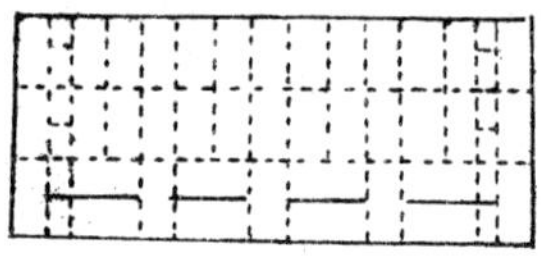

In reference to the system which now prevails, of always using false headers in Flemish bond, there seems to be a reason, before alluded to but not yet explained, which may partly have led to it—I mean the natural wish on the part of the bricklayer to save the superior and more expensive bricks used in the face of a wall, which object is, of course, in some measure, effected by false headers.

In former times still less attention appears to have been paid to Flemish bond than at present, for by Mr. Batty Langley's work, before quoted, it seems to have been the custom in London, about the middle of the last century, to use no headers at all in the face of a wall, excepting in every fifth course.

Fig. 82.

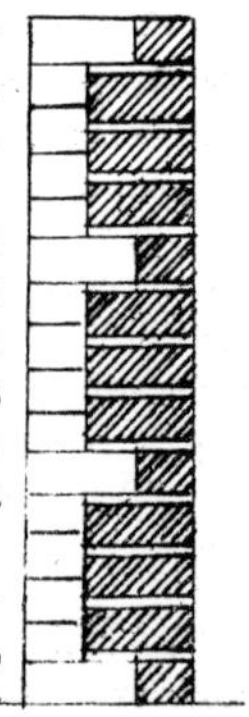

This arose from the circumstance of using fine mortar and very thin joints for the *stocks* or superior bricks which composed the face of the wall of a building, whilst the *place* bricks of an inferior quality, used for backing in, and which completed the remaining thickness of the same wall, were laid in rougher mortar, and with joints of double the thickness. Hence it followed that five courses of the stock bricks were only equal to four courses of the place bricks which were combined with them in the same wall, in the manner represented in *Fig.* 82, being a section of a one and a-half brick wall of this description.

It is evident that nothing could be worse than this arrangement, which, instead of a line of unconnected fissures, as before represented, produced one enormous fissure, extending the whole length of the building, and partially broken in every fifth course only; and which no doubt led to the absurd and pernicious result of the walls of dwelling-houses requiring to be new faced, which is mentioned as a common circumstance by the same author.

Of a variety of Single Flemish Bond, in which no false headers are used.

It will naturally occur to any person who considers the single Flemish bond now in common use, as described in the foregoing article, that Flemish bond of a much more perfect nature might be obtained by rejecting the bats or false headers of alternate courses altogether, and by using none but whole bricks as headers in every course.

This was my opinion at first, but after a full and mature consideration of the subject, I found that, upon the whole, the common practice of the bricklayers in and about the metropolis, such as have been described, yields a simpler and perhaps a stronger Flemish bond than the variety of it now under discussion.

It was before stated that the common single Flemish bond produces a row of vertical fissures or unbroken joints, each 9 inches long and 4 inches apart, at the distance of half-a-brick from the face of the wall. By excluding false headers, each of those vertical fissures is broken in the centre by the whole bricks of the course above, and thus a row of fissures is produced in the same position, each $2\frac{1}{2}$ inches long, and at intervals of about 4 inches apart.

It is evident that the face of the wall will thus be rendered rather stronger, but this advantage will be counterbalanced by a second row of similar fissures being produced at one brick's length from the former, and by a third or even a fourth row of such fissures being produced, if the wall be thick enough; all the rows being parallel, and at the distance of one brick's length from one row to another.

Fig. 83.

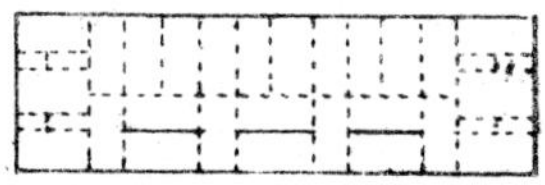

Fig. 84.

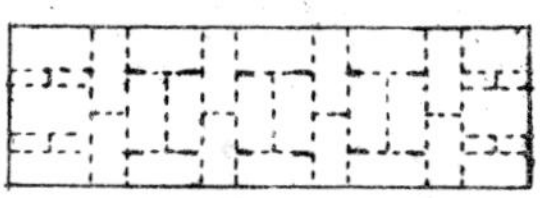

To illustrate the above remarks, *Fig.* 83 represents the plan of a two brick wall built according to common single Flemish bond, as described in the foregoing article, whilst *Fig.* 84 represents the plan of a wall of the same thickness, built also according to Flemish bond, but without any false headers. The vertical fissures or unbroken joints in each figure are marked by thick lines, the remaining parts of the joints, which are properly broken, being dotted.

On comparing these two figures together, it will appear that the common single Flemish bond has larger fissures towards the face or external surface of the wall, but is entirely exempt from any fissures or false joints near to the back or internal surface of the wall.

Now, as in a building the weight of the floors rests chiefly on the inside of the walls, the common single Flemish bond, which leaves that part of the brickwork perfectly sound and free from false joints, may, upon the whole, be considered better than that variety of Flemish bond which forms the subject of our present article, and in which no false headers are supposed to be used.

Of the Common Double Flemish Bond.

Double Flemish bond implies that this bond appears on both sides of a wall.

In double Flemish bond it is the custom of bricklayers in and about the metropolis to use alternate stretching courses on both sides of the wall, that is, courses in which false headers or bats are combined with stretchers.

The consequence of this arrangement is, that two rows of fissures or unbroken joints, each 9 inches in length, and at intervals of about 4 inches apart, are produced, one near the face and the other near the back of the wall, each being at the distance of half-a-brick from the adjacent surface of the wall.

Fig. 85 illustrates this circumstance, being the plan of one course of a portion of a two-brick wall built according to double Flemish bond. The unfavourable fissures or false joints are represented by thick lines: the other joints which are properly broken being dotted.

Fig. 85.

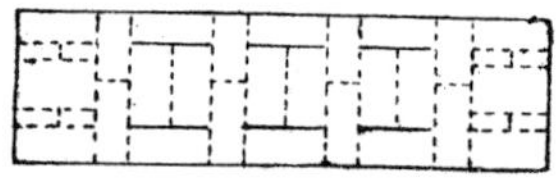

Hence, on comparing those two arrangements together, it will be evident that both sides of a wall are weak in the common double Flemish bond, whilst in the common single Flemish bond one side of the wall only is defective.

The common double Flemish bond is, therefore, a very imperfect arrangement, which never ought to be used at all, and to the best of my knowledge it never is used in the walls of a building by any skilful bricklayer. In the Ordnance barracks at this station, for example, English bond is used in all the inside walls, and the same appears also on the inner surface of all the external walls, which show Flemish bond on the outside only. I have observed that the same system is followed in the new British Museum, now in progress in London.

Of those particulars in which English and Flemish Bond agree.

Whether a wall be built according to English or Flemish bond, the chief resistance to a transverse fracture consists in the overlapping (or *lap*, as it is technically termed), of one brick over another, which is only equal to a quarter-of-a-brick's length, or to little more than two inches; and further, that as the horizontal joints are not broken, but continue in the same straight line from one side of the wall to the other, the resistance to a fracture of the nature above supposed is of course smaller than would be the case if those joints could be broken in that direction, without producing a greater inconvenience.

Fig. 86 represents in elevation the kind of fracture that may take place in a wall built according to common English bond, whilst *Fig.* 87 represents in elevation the similar fracture that may take place in a wall built according to Flemish bond.

Fig. 86.

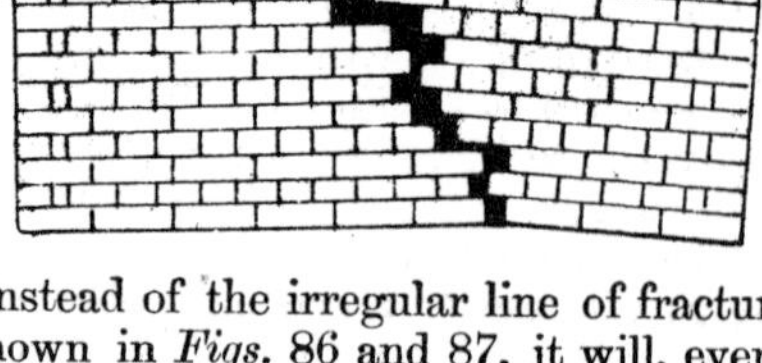

If the mortar should be stronger than the bricks, which sometimes happens, then instead of the irregular line of fracture always following the joints, as shown in *Figs.* 86 and 87, it will, every here and there, break one of the stretchers. If you are in the habit of

Fig. 87.

looking at brickwork with attention, there are few places in which some examples of such fractures may not be seen, although they will escape the eye of passengers in general.

It was before stated that the horizontal joints should never be broken either in common English bond or in Flemish bond.

The reason for this rule will appear evident, on a reference to your small models of bricks. If the horizontal joints be broken, the vertical joints will be sure to coincide in several parts of the thickness of the wall, and in a direction perpendicular to its outline, and this coincidence will greatly weaken the resistance of the wall to a transverse fracture.

Fig. 88 represents the plan of one of the stretching courses of a three-brick wall, built according to English bond, showing the unbroken joints or vertical fissures alluded to, such as may be produced by the injudicious arrangement of breaking the horizontal joints. These fissures are marked with a thick line, to distinguish them from the other parts of the same joints which are properly broken, and which are dotted according to custom.

Fig. 88.

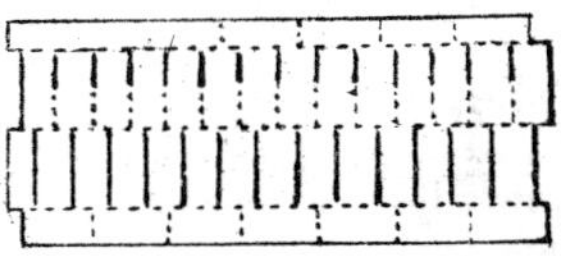

Fig. 89 shows the effect produced upon a wall of the same thickness, built according to common single Flemish bond. It must be very evident that so many new fissures, in addition to those which must necessarily occur in this kind of bond and at right angles to them, would render such a wall exceedingly weak.

Fig. 89.

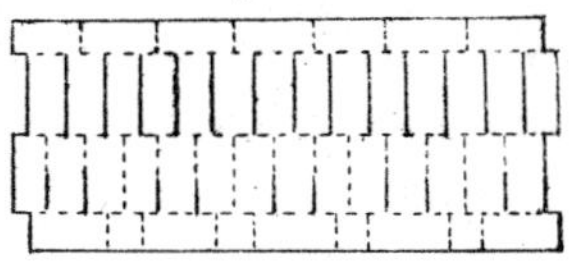

The best way to judge of any kind of bond by the models of bricks is to lay them, as it were, at open order, as in *Fig.* 90. On another course being laid over these, also at open order, all the unbroken joints, if there are any, will appear very conspicuous.

Fig. 90.

That English Bond ought in all cases to have the preference to Flemish Bond.

On comparing together the various kinds of brickbond that have been described, it appears to me that the common English bond is by far the simplest and strongest, and that it ought therefore to be always used in preference to Flemish bond, which is sure to leave one side of the wall weak, and, to a certain degree, liable to split by a longitudinal fracture.

I have remarked that the most skilful bricklayers of the present day use English bond everywhere, excepting for the outside surface of external walls which are afterwards to be exposed to public view, and where the fashion of the times, for there is a fashion in brickwork as well as in other things, compels them to use Flemish bond. But even in external walls which are afterwards to be plastered or coated with Roman cement, they use the English bond exclusively.

The only argument for using that imperfect arrangement of brickwork called Flemish bond at all is that it gives a handsome front to a building. This impression appears to me to arise from habit only; for I contend that if the eye were equally accustomed to English as to Flemish bond, the one would be considered as handsome as the other, for the same reason that the fashion and dress of one country, although considered to be the most becoming by those who wear it, often appears ridiculous to strangers.

I admit that in buildings whose fronts are composed of gauged bricks having uniform and equal joints all pointed and marked in a conspicuous manner from top to bottom, as is common in dwelling-houses and in shops, not only in London, but in Chatham and Rochester, and in all parts of the country where bricks are used,* the Flemish bond may perhaps have a more regular appearance to the eye than English bond would have, but where is the necessity or advantage of all this uniformity? It certainly tends to diminish the strength of the building, as will afterwards be explained, and whilst it greatly adds to the expense, in my opinion it adds nothing to the beauty; for it appears to me that the beauty of a brick front depends entirely upon the general proportions of the larger parts of an edifice, and upon the general color and effect of the brickwork in mass, not upon the arrangement of the particular joints of so small a material, which, when plain ones are used, absolutely escape the eye.

Compare, for example, the appearance of the Ordnance barracks at this station, in which plain and comparatively irregular joints are used, with the minute regularity of the nicely-pointed and sometimes painted brick fronts of Chatham and Rochester; and the former, in which the general effect only attracts attention, is certainly much more pleasing to the eye than the uniform and conspicuous joints of the latter.

Now, if gauged and nicely-pointed brick fronts be rejected, the difference between English and Flemish bond, in point of appearance, does not deserve a thought; for at a few paces distance it is scarcely possible to discover the difference between these two bonds in a wall built with plain joints. The truth of this observation may easily be proved by a visit to the British Museum. On inspecting the back wall of the first wing of the new building, now in progress, the difference between the appearance of the English bond below, and the Flemish bond above the offset, is so very trifling that I feel confident that none but a bricklayer would ever discover it.

In like manner, if you walk round the revetments of Chatham Lines, or the inclosing wall of Chatham Dockyard, which are both executed partly with Flemish and partly with English bond, you will never observe the difference, unless you go for the express purpose of endeavouring to find it out.

But English bond is not only much stronger in itself than Flemish bond; it has the great advantage, which the other does not possess, of being able to alternate occasionally, in a substantial manner, with one

* This is technically called *tuck* and *pat* pointing.

or more diagonal courses of bricks, in all walls of two bricks in thickness or upwards. This useful arrangement, which will be described in the next article, adds greatly to the strength of a brick wall.

Of Diagonal Bond.

In thick walls, but more especially in foundations, it has been the practice to lay occasionally a couple of courses of bricks, intersecting each other at right angles, and oblique to the line of the wall, with which they everywhere form an angle of about 45 degrees.

In adopting this arrangement, intermixed with English bond, the diagonal bricks in those courses do not extend so far as the face or back of the wall, where a line of common heading or stretching bricks is laid in the usual manner, and thus the wall, when finished, has all the appearance of common English bond.

Figs. 91, 92, 93, and 94 represent four successive courses of a three-and-a-half brick wall thus built, the first being common English bond, the two others being a couple of courses of diagonal bond, after which the common English bond again follows.

Fig. 91.

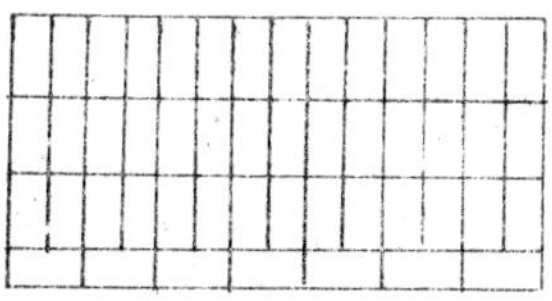

Fig. 92.

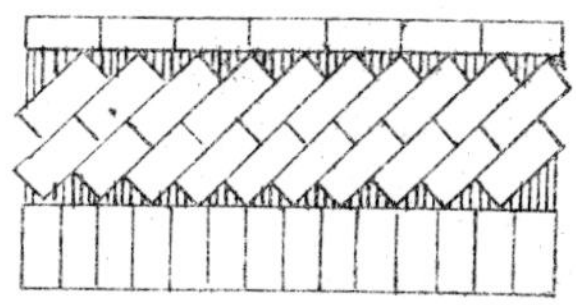

Fig. 93.

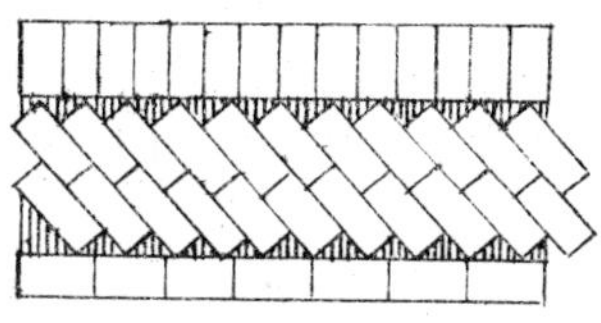

Fig. 94.

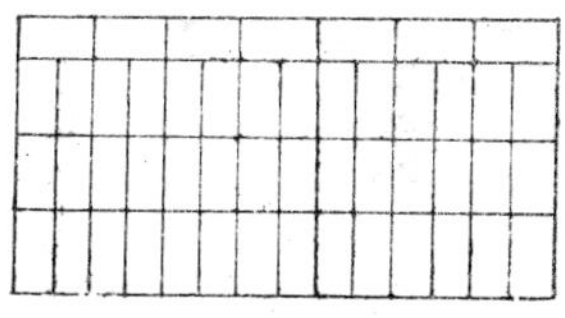

A two brick wall is the thinnest in which diagonal courses can be used combined with English bond, and in this case only in the stretching courses. *Figs.* 95, 96, 97, and 98 represent this arrangement, in which are introduced, first a heading course of common English bond; secondly, a diagonal course between the stretchers; thirdly, a second heading course; and fourthly, another diagonal course between the second set of stretchers, at right angles to the former.

Fig. 95.

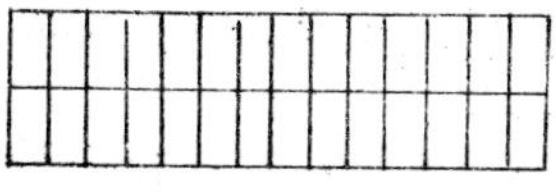

Fig. 96.

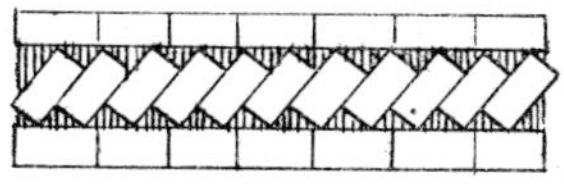

Fig. 97.

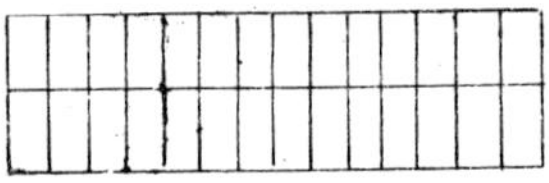

Fig. 98.

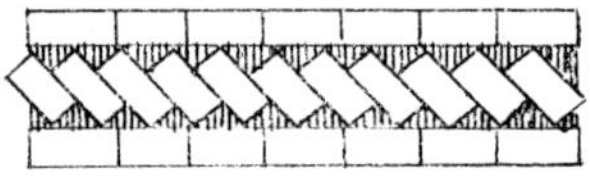

The triangular spaces at the extremities of the diagonal bricks, between them and the common headers or stretchers, at the external surface of the wall, must be filled in with bats or fragments cut to the same form.

Sometimes the bricklayers, instead of using all fragments, will cut the end of a whole brick obliquely, as in the annexed figure, so as to meet the outer row of headers or stretchers in which the diagonal bricks are inclosed. But this precaution seems scarcely necessary if the interstices arising from the former method be well filled up with solid pieces instead of stuffing them with mortar.

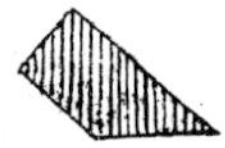

The use of diagonal courses consists in producing a more powerful bond than the lap of a quarter brick only, which is all that the common English or Flemish bonds are capable of yielding; and there is no possible way of laying the diagonal bricks wrong, for if they be intermixed with common English bond, according to the rule that has been stated, one, two, three, or more whole bricks, laid diagonally, will just fill up the core of a wall, in proportion to its thickness,* and all the successive courses of those bricks will break joint properly, whether in reference to each other or to the common courses of English bond above or below them. The brick models will prove the justice of this remark.

It will be observed that in thus combining diagonal courses with English bond, the same effect will be produced by throwing in the courses of diagonal bond singly, instead of using two in succession. For example, after one course of English bond might come one diagonal course, then another course of English bond, after which might follow the second course of diagonal bond. But whichever arrangement be adopted, the successive diagonal courses ought to be laid in contrary directions.

In a one brick wall of common English bond the number of headers and stretchers is edual, but in thicker walls, built according to the same method, the proportion of stretchers diminishes by degrees. Thus, for example, in a one and-a-half brick wall the stretchers are one-half; in a two brick wall, one-third; in a two and-a-half brick wall, one-fourth; and in a three-brick wall one-fifth of the number of headers, and so on. Hence in thick walls the common English bond becomes almost entirely heading bond. But against a transverse fracture stretchers would certainly be more effectual than headers, if they could be arranged so as to break joint properly in successive courses; and under this consideration the system of the entire stretching courses of the first-mentioned English bond, as defined in certain books of architecture, appears at first sight an advantageous arrangement, but on analyzing it in detail it proves defective for the reason before explained.

* One diagonal brick laid at an angle of 46 deg. 24 min. with the face of the wall, exactly occupies the same space in the thickness as one header. The bricklayers will easily and quickly find the proper angle to obtain this coincidence by trial.

The intermixture of diagonal courses with common English bond yields all the advantages that might be anticipated from this system without any of those glaring defects by which it is accompanied.

Upon the whole, it appeared to me that the common English bond, intermixed with occasional courses of diagonal bond, whenever the walls are thick enough to admit of it, is by far the strongest and best combinaton of bricks that can possibly be adopted.

The diagonal courses which have been described are sometimes called *herring-bone courses* by the workmen. They were used throughout the whole height of the thick brick walls which backed the stonework of the original front of the New Custom House, and which afterwards failed, not through any defect of the brickwork, which appeared to me to be as good as could be desired, but through the mismanagement of the foundation.

In one of the walls alluded to, now exposed to view, the common and the diagonal bond were intermixed in alternate single courses, one of each, and it was remarked that the lower part of the wall, which appears to have been grouted, seemed stronger than the upper part, built with mortar, the arrangement of the brick courses being alike in both.*

Of the arrangement most suitable for the Brick Revetments of Fortresses.

In the common sloping revetments of Vauban, and in leaning revetments, which may be considered preferable to them, a great difficulty occurs when brickwork is used. Either the outside bricks composing the face of the wall must be rubbed to suit the slope, which would occasion much extra trouble and expense, besides weakening the stretchers, or the joints of successive courses must be laid perpendicular to the external slope, in which latter case the rain will have a tendency to penetrate from the surface into the interior of the wall, to the prejudice of the brickwork.

Of these two arrangements the latter has, I believe, always been adopted, as being the simplest, and the unfavourable effect mentioned has generally been the result.

In many of the old fortresses of France the outer surface of the sloping brick revetments has completely peeled away to the depth sometimes of half-a-brick, but often of a whole brick, in consequence of the mortar having in process of time yielded to the effects of the rain lodging, and vegetable matter forming, in the external joints;† and a still more curious circumstance is recorded in respect to the revetments of Lisle, which were new faced with mortar of a superior quality in 1817. This mortar, although still proof against the weather, is said to have swelled in the joints in such a manner that the greater part of the new work has bulged out from the old in the extraordinary curve represented in *Fig.* 99.

The French Captain of Engineers, Bergère, from wnose rebort I have extracted those particulars, states as a fact, acknowledged by the corps to which he belongs, that such exfoliations of the external surface

---

* Diagonal courses are also thrown in occasionally in the new walls now building on the same spot, by direction of Mr. Smirke.

† Similar exfoliations have taken place in the brick revetments of Tilbury Fort.

of brick revetments occur only in those having a considerable slope, and more especially in those facing the rainy winds. He also states an opinion, which appears to be founded on just principles, that the great slope has a tendency to crack such revetments, as there is less weight at bottom, upon the front, than on the back of the brickwork.*

Fig. 99.

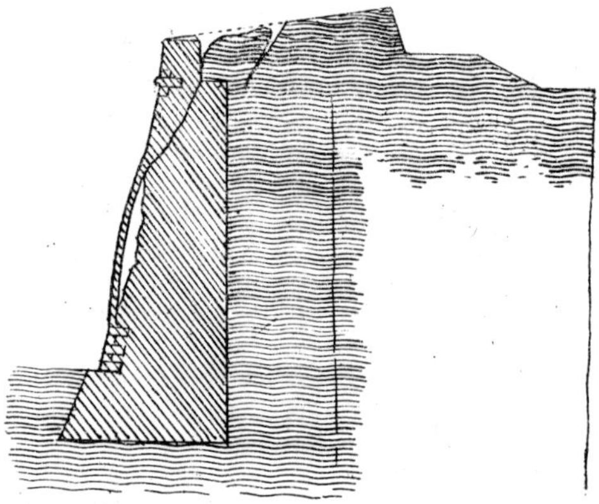

Thus the superior experience of our continental neighbours, who have perhaps 50 fortresses to our one, appears strongly to confirm the opinion I have given in my "Elementary Fortification" in favour of the countersloping profiles, which being perpendicular in front, or nearly so, the rain will have little or no tendency to penetrate into the brickwork.

The slope in rear, or diminution of the wall towards the top on that side, may be effected by offsets of half-a-brick each, set off upon the heading courses on the back of the wall, at nearly equal intervals of about five or seven courses, or of so many courses apart as will exactly or nearly agree with the given proportion.

Fig. 100.

* See an interesting article in "The Memorial de l'Officier du Genie," No. 7, entitled "Notice sur les causes de destruction des Revetmens de Maconnerie en brigues, et sur les moyens d'y remedier," par M. le Capitaine du Genie Bergere.

Fig. 100 represents the section of part of a countersloping revetment of this description, in which two successive heading courses are supposed to be used at every offset on the back of the wall, which breaks the bond better.

In fact, all offsets ought to be on heading courses, which rule I ought to have mentioned in treating of the footings of brick walls, and I here take the opportunity of correcting the omission.

After having written so far, a curious fact was communicated to me by Captain R. Thompson, of the Royal Engineers, which bears upon the present part of my subject. He informs me, that about thirteen years ago, the whole brick facing of the Circular Redoubt in Romney Marsh, separated from the body of the brickwork to the thickness of four inches, every header being broken in two, and fell into the ditch, although there was no exterior slope. This he ascribes to the circumstance of a more powerful cement having been used for the exterior surface of the wall than for the rest of the brickwork, which was built with common mortar a practice very usual in great works, but which Captain Thompson reprobates; and although in his communication to me he has not detailed his reasons, the opinion advanced by him seems founded on the justest grounds. It was before explained that all brickwork built with common mortar settles, whereas that built with water cements in general is said to be incompressible; and, from Mr. Smeaton's report, tarras will even swell in the joints,* and the same circumstance seems to have taken place at Lisle. Now, if one part of a wall sinks to a lower level, whilst the other stands fast or swells to a greater height, fracture is inevitable. I am not aware that this important rule, suggested by Capt. Thompson, of never using two different qualities of cement, a *face* mortar and a *backing* mortar for the same wall, has been noticed in any book on practical architecture. When fairly stated, however, its propriety is almost self-evident. The easy separation of the external coat of brickwork in the case alluded to may probably have been partly owing to common Flemish bond being used, in which there is generally only one header in four out of all the bricks employed in the facing of a wall.

In respect to the inequality of pressure upon the foundation of a sloping revetment, as noticed by Captain Bergère, that evil is of course completely done away with by throwing the slope to the back of the revetment, for the pressure of earth there will counteract the superior weight of brickwork towards the front of the revetment, which might have a tendency to fracture it if it were an insulated wall.

By this article it appears that the brick bond used in the old French revetments was exceedingly bad. Recently they have adopted all heading bond, with whole headers and half bricks appearing as such, alternately, in the face of the wall in every course, according to the method of Mr. Batty Langley, upon which I before commented.

In the same number of the above work there is an interesting paper on water cements, by Colonel Treussart, of the French Royal Engineers, He states that he has ascertained that certain qualities of lime are better slaked by exposure to air, others by water, and that some may be exposed to weather for a certain period without injury, whilst others ought to be worked up instantly: also, that finely pounded bricks will form an excellent substitute for tarras or puzzolana, but with this proviso, that if the earth of which such bricks are formed consist of silica and alumina only, in about equal parts, they can scarcely be burnt too much; but that if there be a mixture of carbonate of lime, say about one-tenth, combined with the former ingredients, they should be very moderately burned, for, if burned to hardness, they will lose their virtue as a water cement. I had not the advantage of perusing this paper when the articles on water cements were written, and I have not yet seen the experiments of M. Vicat, before alluded to, which are quoted, but with some difference of opinion, by Colonel Treussart, who further imputes less value to iron, as an ingredient in water cements, than has usually been allowed to it by former writers.

* This curious circumstance may now be seen in the lower courses of the Old Gunwharf wall, near the landing place, opposite to the Storekeeper's house.

But whether the countersloping revetment with horizontal joints, as recommended by me, or whether the common sloping or the leaning revetment with inclined joints, be adopted, it is essential in both cases that the common English bond, alternating with occasional courses of diagonal bond, shall be used, and no other, this being the only arrangement suited for a revetment or other substantial brick wall, which mixed bond I have not attempted to show in *Fig.* 100, as it is not easy to represent it clearly in a simple section not shaded.

As revetments must always be more or less liable to injury from wet penetrating through the earth of the parapet above, even when rain water cannot lodge on their external surface, it appears essential either that water lime should be used in building them, whenever it can be obtained at a moderate expense, or that some of the cheapest ingredients treated of in the articles on water cements should be used for improving the mortar when made of common lime. The same rule applies with still greater force to casemates and other bombproofs.

Of the Bond proper for Cross Walls or Party Walls of Buildings, and which also applies to External Re-entering Angles, when English Bond only is used.

In a street or continued row of buildings the walls which extend across from front to rear, so as to separate the several houses from one another, are called party walls, whilst the internal walls in any one house are termed cross walls, for they must necessarily be at right angles either to the front or back wall of the building, or to a party wall; and these cross walls are usually of inferior importance, for they only differ from a common partition in being more substantial.

The rule for bonding a principal wall and a party or cross wall together is sufficiently simple, the arrangement for alternate courses being as follows:—

In one course a line of closers is thrown into the principal wall, exactly opposite to the junction, and equal in length to the thickness of the cross wall. By this arrangement the bricks on that side of the principal wall are either caused to enter into the cross wall, or the bricks of the cross wall are caused to enter into the principal wall, and thus a toothing or lap of a quarter brick is obtained between these two. In the next course the bricks of the two walls merely meet, without entering each other, and these two arrangements are used alternately, from the bottom upwards.

This bond at the junction of two walls being equal to the usual bond of bricks in the same wall, it is evidently of no importance from which wall of the two the bricks enter the other, but the common bricklayers prefer the arrangement of causing the extreme headers of the heading course of the cross wall to enter one-half of their width into the principal wall, and in this case they meet the stretchers of the latter wall end to end, but not in the same alignment, as in *Fig.* 101, which represents the junction of a cross wall two bricks thick, with a principal wall of two and-a-half bricks in thickness.

Fig. 102 represents the next course in succession of a junction of this description, in which none of the bricks of the one wall enter into the there.

Fig. 101.

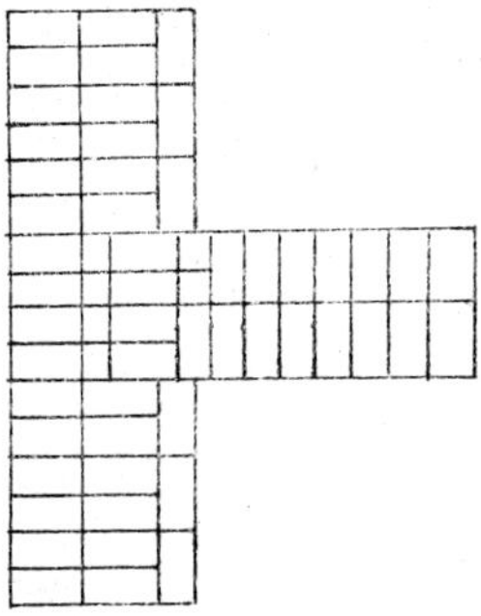

Fig. 102.

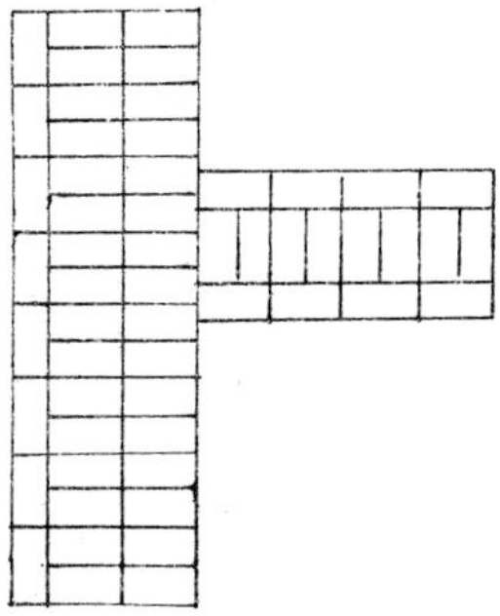

It is to be observed that in this second course the ends of the cross wall do not coincide with the transverse joints of the principal wall. In the first course represented they necessarily do, because there one wall is to be let into the other.

Of the External Re-entering Angles of Buildings when Flemish Bond is used.

The method described in the foregoing article is applicable to the junction of party or cross walls with the principal walls of a building, which junctions always take place internally, and, therefore, English bond is or may be used for all such walls, even when Flemish bond is adopted for the external surface.

Although I have recommended English bond to be used exclusively in brickwork under all circumstances, yet as Flemish bond is the prevailing fashion of the day for outside work, it appears proper to explain the method of using this species of brick bond at the junction of two walls which meet each other in such a manner as to form an external re-entering angle.

This circumstance, of very rare occurrence in the streets of cities, frequently takes place in country seats or in public buildings of importance, from the custom which architects have of making certain parts project more than others in an extensive front, in order to break the uniformity of one continued surface, as has been done in the Ordnance Barracks at this place, by introducing the towers at the ex-

Fig. 103.

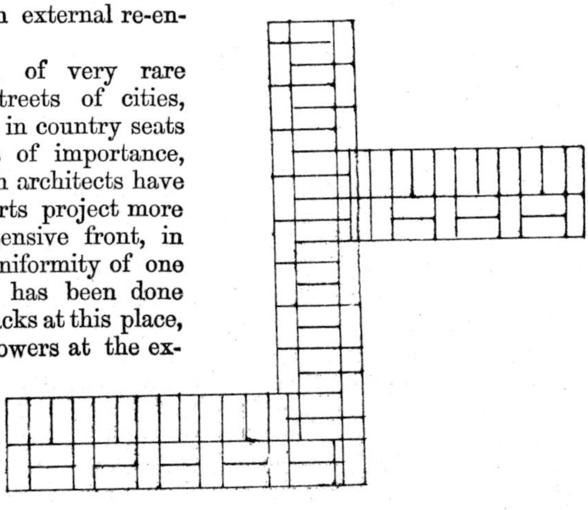

tremities of each side of the square, which produce a more pleasing effect to the eye than if each range had been entirely without any projecting or elevated parts.

Figs. 103 and 104 represent the plan of two successive courses of part of a building, in which a two-brick wall meets another of the same thickness in such a manner as to form an external re-entering angle, showing Flemish bond on each side of the angle, that is, of course, on the outside of the building only, which is supposed to be towards the bottom of the paper. Part of a cross wall appears internally, towards the top of the paper, opposite to the break on the outside of the building.

Fig. 104.

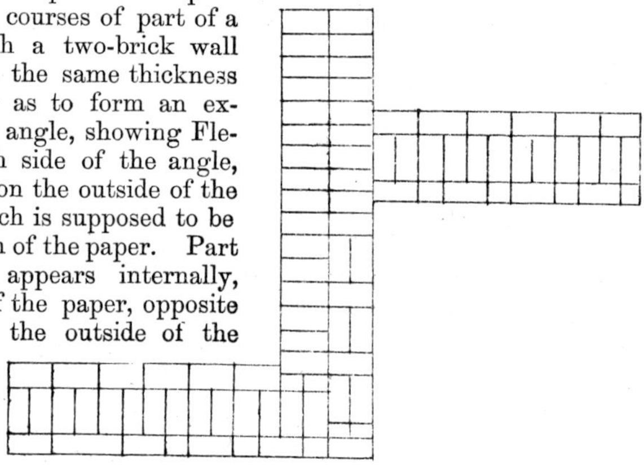

It may be observed that the bricks are always so arranged at an external re-entering angle, that a stretcher in one course and a closer in the next appear at the angle, with a header alongside of the closer, and this takes place in alternate courses in both of the two walls, which unite at the angle, and in such a manner that the corner stretcher of one wall meets the closer of the other in the same horizontal course. These closers are not actually such, but are headers half seen in the face of one wall, and partly buried in the other, so that at the junction of two walls in an external re-entering angle two headers always come together, even in Flemish bond.

Of the Openings of External Doors and of Windows in plain brick fronts.

In the internal doors of buildings the jambs or sides of the brickwork are always perpendicular to the wall, and are carried entirely across, in the same straight line, in the manner represented in the foregoing figures of portions of walls, given as examples of English and Flemish bond.

In the external doors of a brick building of any importance this simple form rarely occurs, unless the doorway is intended afterwards to be ornamented externally by columns and other architectural decorations of a different material.

The frame of an external door is therefore usually, and that of a window almost invariably, placed about half-a-brick or four inches from the face of the wall, within a recess, which is about four and-a-half inches wider on each side internally. At top the internal height of the recess also exceeds the external height of the brickwork by a space sufficient for the top of the said frames, which for a common sash window may be rather less, and for a common door need not exceed the last-named dimension of four and-a-half inches. These recesses are technically called reveals.

The frame for the door or window usually occupies a space of about six inches in rear of the reveal, and so far as this extends the brickwork must be rectangular, to suit the form of the woodwork, but in windows the remaining space behind the intended position of the frame is often splayed back on both sides, so as to diverge towards the interior of the building.

In doors, on the contrary, the whole of the brickwork behind the reveal is generally built in a rectangular form, without any splay or obliquity.

Figs. 105 and 106 represent these arrangements, one being the

Fig. 105.

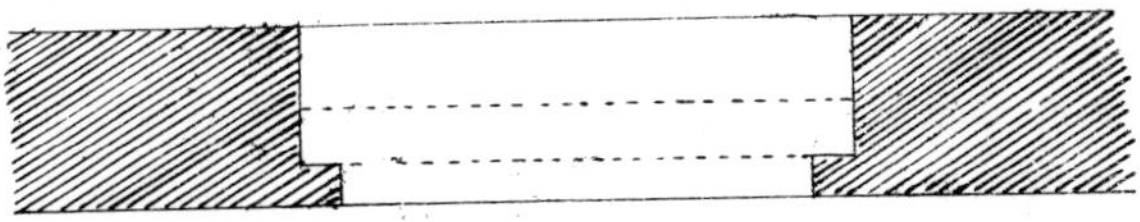

Fig. 106.

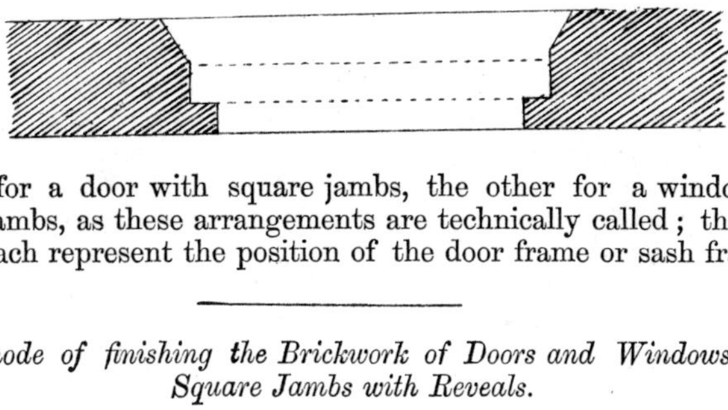

opening for a door with square jambs, the other for a window with splayed jambs, as these arrangements are technically called; the dotted lines in each represent the position of the door frame or sash frame.

Of the mode of finishing the Brickwork of Doors and Windows having Square Jambs with Reveals.

Figs. 107 and 108 represent the plan of two successive courses of a square jambed opening for a door or window, in which the brickwork is built according to common English bond, the wall being two bricks thick.

Fig. 107.

Fig. 108.

Figs. 109 and 110 represent the plan of two successive courses of the

Fig. 109.

Fig. 110.

same kind of opening in a wall of the same thickness built Flemish bond.

In the heading course of Flemish bond (*Fig.* 110) it will be observed that the half-closer in the front of the reveal, and the half-brick adjoining to it within the jamb, do not make good bond, and therefore it is usual to throw in a brick so cut as to form both pieces in one. This is called a *king-closer*, and if it were practicable, it ought to be formed as in *Fig.* 111, but as bricks cannot conveniently be cut in that manner, it

Fig. 111.

Fig. 112.

Fig. 113.

is cut obliquely, as in *Fig.* 112, and the stretcher adjoining to it in the face of the wall has its end also cut obliquely to the same level, as in *Fig.* 113.

Hence, if we suppose king-closers to be used instead of the half-closer and bat in *Fig.* 110, the plan of the opening will appear as in *Fig.* 114.

Fig. 114.

King-closers may also be used to advantage at the reveals of doors or windows in the heading courses of common English bond, as represented in *Fig.* 115, which arrangement may be substituted in place of that shown in *Fig.* 108.

Fig. 115.

Figs. 116 and 117 represent the plan of two successive courses of a

Fig. 116.

Fig. 117.

similar opening in a wall of two and-a-half bricks in thickness, built Flemish bond.

The projecting parts in front on each side of a door or window are certainly rather weak, being in themselves chiefly composed of bats, but they are afterwards well secured by the door or window frame.

It was stated, that the usual thickness of brickwork at the reveal or projecting part is half-a-brick, or four inches. This prevails in common buildings, such as dwelling-houses, from the ground floor upwards, but in the basement stories below that level the reveal is more commonly three-quarters of a brick, or sometimes one brick in thickness, because the face of the wall projects more below, there being generally an external offset above the basement story.

Figs. 118 and 119 represent the plan of two successive courses of an

Fig. 118.

Fig. 119

opening of the same description as before, but made in a wall of English bond three bricks in thickness, and having a reveal of one brick thick, to represent that of a basement floor.

It appears to me that the reveals of brickwork, such as have been described, should always be a whole brick, as in the last example, or a half-brick, as in the foregoing examples of this article. To make the reveal of the basement story three-quarters of a brick, in order that the sashes of that story may be precisely in the same vertical plane with the sashes of the upper stories, may be considered an unnecessary nicety, for it is such as could not be distinguished by the most accurate eye,* and to work with quarter bricks in the reveal would render that part of the work much more troublesome and less compact than when no such fragments are used.

It may be observed that reveals of a whole brick are generally used, even in the upper windows of chapels, or of other very large apartments having substantial walls.

In respect to bond, the above examples certainly exhibit the best method of finishing square jambed openings in brickwork, but I am informed that common bricklayers will seldom take the trouble of introducing the headers in alternate courses in the jambs which they ought to do, as shown in *Figs.* 107, 109, 116, and 118, and that king-closers, although useful, are frequently omitted altogether.

Of the mode of Finishing the Brickwork of Windows having Splayed Jambs with Reveals.

Having considered this subject with a good deal of attention, it appears difficult to lay down any rule that shall produce good bond at the jambs of windows of this description. The bond in the central part of each pier, between the two openings, may be perfectly regular, either according to common English or Flemish bond, but the brickwork of the splayed jambs must necessarily consist partly of bricks cut to the splay, and partly of fragments to fill in such irregular interstices as may occur, in consequence of this arrangement, at the junction of the oblique or splayed parts with the perpendicular or central parts of such piers.

For example, in the outline of the plan of part of the front of a building having splayed windows, shown in *Fig.* 120, those central portions of the piers between the supposed openings, which are represented by the rectangle *a b c d*, may have the common regular English or Flemish bond; but in the remaining irregular parts, at the extremities of the same piers, it will be difficult or impossible to obtain perfect bond, and therefore it must be left to the sagacity of the bricklayers employed to make as good an approximation to proper brick bond as is possible, in reference to the form and dimensions of the reveals and jambs.

Here it may be observed, that unless the windows of a building be placed much nearer to each other than usual, the central part (*a b c d*) of each pier, which, by supposition, is bonded together with as much

---

* In the Ordnance Barracks at this place all the windows of succesive stories are in the same plane, which is not the case in Gibraltar Place and Nile Terrace, in Chatham, and yet the difference would never be observed, unless by a person who went for the purpose of examining the point in question.

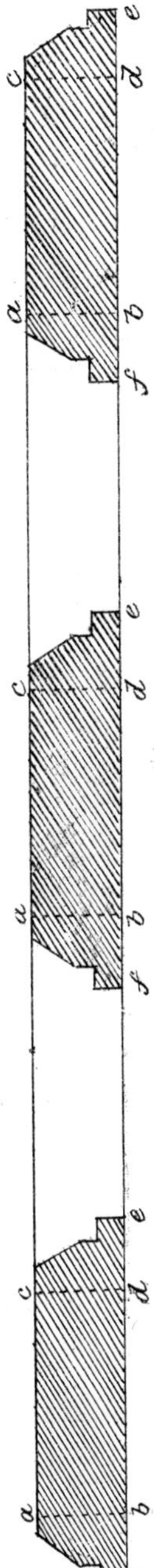

Fig. 120.

regularity as brickwork is capable of, will give sufficient strength to that wall of the building in which the windows are introduced, for in a two-brick wall each of the irregular portions of brickwork (*a b f* or *c d e*), which cannot be so well bonded, will not occupy more than the length of one and-a-half brick in the face of the wall (at *d e* or *f b*).

The only good practical rule appears to be to lay the oblique parts of each jamb alternately in heading and stretching courses, but let these heading courses of the jambs have no closers intermixed with them, such as were used in the square jambs treated of in the foregoing article, and let the ends of the headers be cut to the splay in such a manner that one side of the brick shall always retain its whole original length, as in the annexed figures, in which the portions cut off are dotted.

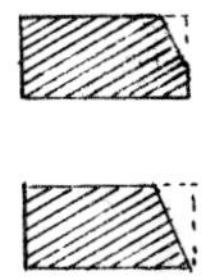

In the stretching courses of the jambs the bricks must also be brought to the form of the splay, either by cutting both ends, or one end and part of a side, as in the annexed figures, in which the portions cut off are also dotted; and by this means they will be reduced to the length of two-thirds of a brick nearly, which will enable them to break joint with heading courses destitute of closers, thus forming as good bond as can be expected; for, as was observed before, perfect bond must not be looked for in any openings having splayed jambs.

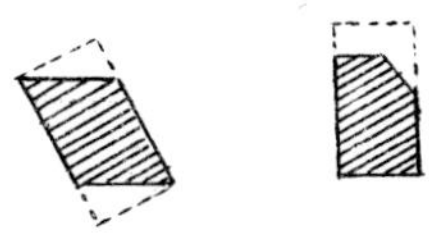

Fig. 121 represents the plan of the stretching course of a two-brick wall, English bond, having a window with splayed jambs, and, according to custom, that which is the stretching course of the face of the wall itself is the heading course in reference to the jambs.

Fig. 122 represents the plan of the next successive course of the same wall, in which the brickwork of the jambs is cut in an irregular manner, so as to obtain an approximation to good bond.

Figs. 123 and 124 represent the plan of two similar successive courses of a wall also built English bond, but two and-a-half bricks thick.

Figs. 125 and 126 represent the same arrangement in a wall also two bricks and-a-half in thickness, but built according to the common single Flemish bond.

In the above examples the splay is supposed to be in the proportion of one-half, and I do not think that this proportion ought to be exceeded.

Splayed jambs certainly weaken the brickwork, although in a very slight degree, if the obliquity be

Fig. 121.

Fig. 122.

Fig. 123.

Fig. 124.

Fig. 125.

Fig. 126.

moderate, but they throw more light into a room. It may be remarked, however, in two windows of the same height, one being splayed, the other not, that if the width of the square jambed window be equal to the mean width of the splayed one, the quantity of light obtained from both will be equal, or nearly so. In common dwelling-houses in cities, where the piers between windows are often very small, splayed jambs should not be used, and, under such circumstances, they seldom are; but in barracks or public buildings where the piers are more substantial, in proportion to the width of the windows, there is less objection to the latter being splayed.

Of Window Sills and Backs.

In buildings of any importance the external aperture of a window has a flat stone at bottom, called the sill, which extends about two inches into the wall at each end, and which serves as a coping to prevent the joints of the brickwork below the window from being injured by rain.

In well-sized windows this stone should be equal to two courses of bricks, or about $5\frac{7}{8}$ inches in extreme thickness, which is diminished towards the front within the aperture, by giving a gentle slope to the upper surface, in order to throw off the rain. In small windows the sill is often thinner than the above dimension. It projects about two inches in front, beyond the face of the building, and usually rests on a 9-inch brick wall, but does not extend quite so far as the back of that wall. Hence its extreme width is seldom more than eight inches. The bottom of the sash frame partly rests upon the stone sill, at the distance of about six inches from the front of it, in windows having the usual four inch reveal.

The thickness of brickwork immediately below the aperture of a window is always reduced to nine inches, even in the thickest walls, by an internal recess, which is made for the purpose of allowing persons within the room to look out of the window conveniently, in an oblique direction.

In the lower part of the square jambed windows the internal recesses formed by the back and jambs resemble the common external recesses of brick buildings, and the same method of obtaining bond, which is sufficiently simple, applies to both.

In splayed-jambed windows the same principle now alluded to must also be adhered to, as nearly as possible, for the arrangement of the brickwork within the recess, but it will not be possible to obtain quite such good bond there, on account of the necessity of cutting the bricks to the splay, as explained in the foregeing article.

Of the External Recesses of Brickwork.

The great projections or breaks which occasion the re-entering angles, that were treated of in a former article, are usual only in extensive public buildings, but small projections or breaks, or occasional recesses, are common even in dwelling-houses, where they are intended to produce the like effect, of relieving the uniformity of one continued surface of brickwork.

The recesses now alluded to are generally introduced about doors and windows, and are usually covered by semicircular or segment arches at some little distance above those apertures, which produce a handsome effect.

The space on each side, measuring from the aperture of the door or window to the extremity of the recess, seldom exceeds one brick or nine inches on the face of the wall.

Fig. 127 represents, in outline, a recess of the nature supposed, half-

Fig. 127.

a-brick in depth, and of one brick's length in the face, and the aperture is supposed to have a reveal for a door or window interiorly.

In working up the projecting part within the recess, which produces the reveal, it is usual for bricklayers to place alternately a stretcher in one course, and a bat and closer in the next. The stretcher is met at the re-entering angle of the recess by a corner header from the salient angle of the projecting part of the face of the building adjoining to it.

Figs. 128 and 129 represent the plan of two successive courses of a

Fig. 128.

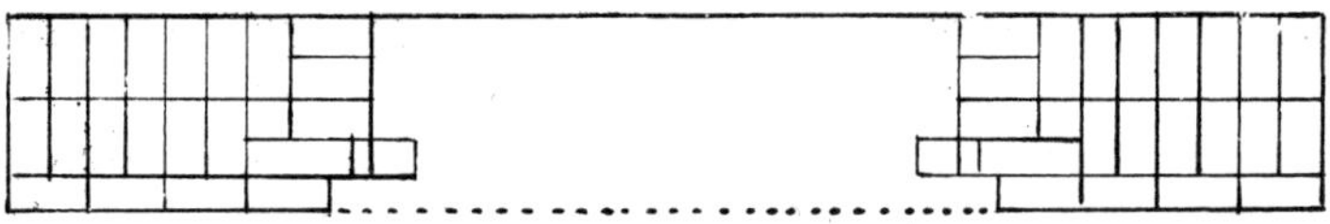

Fig. 129.

double reveal, as it may be termed, of the above description, built English bond, the wall being two and-a-half bricks thick at the pier or thickest part, but two bricks thick only within the recess.

In this arrangement no closers are used in the heading courses of the jambs, and accordingly three-quarter bats are supposed to be used in the stretching course of the same, behind the reveal in *Fig.* 129.

In large buildings, such as chapels, storehouses, &c., recesses are sometimes one brick deep, and the recessed part of the wall is sometimes also greater than was before supposed, such as one and-a-half or two bricks in the face.

This arrangement may be exemplified by supposing an external wall of three bricks thick diminished to two bricks in thickness by a recess of one brick in depth, and the recessed part of the wall to be three bricks in the face, and terminating in a square jambed opening without a reveal. Let us also suppose that a cross wall of two bricks in thickness is connected with the principal wall opposite to the coin formed by the projecting part or break in the face of the wall.

Figs. 130 and 131 represent the plan of two successive courses of brickwork thus arranged, in which English bond is used.

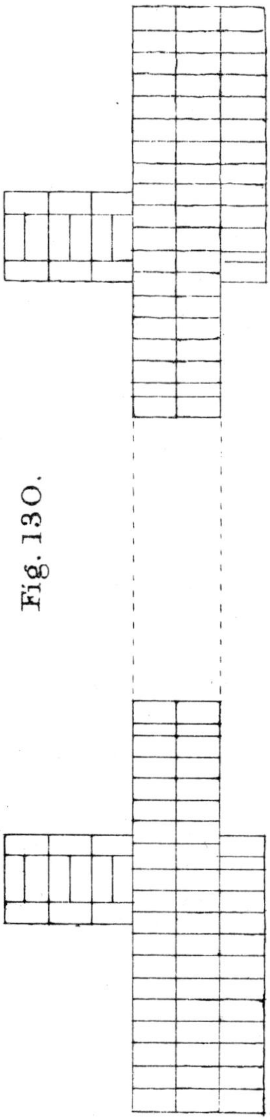

Fig. 130.

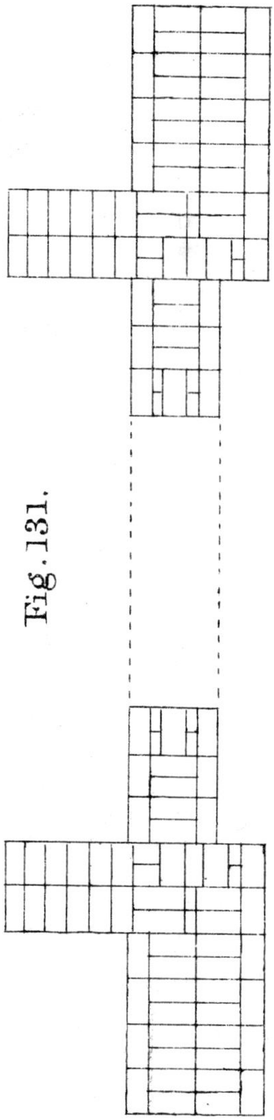

Fig. 131.

Remarks.—From the foregoing examples, it will have been observed that when a recess is one brick in depth good bond may be obtained, connecting the pier or projecting part with the recessed part of the wall. When the recess is only half-a-brick in depth the bond is not so good, but tolerable. When less than half a brick is used an inferior bond will be obtained, as will be explained in the next article.

But it may be remarked, that as those recesses are usually covered by arches resting on the piers, the small recessed parts of the wall might generally be removed altogether, with little injury to the strength of the building, so that, in fact, if there were no bond whatever between the projecting and recessed parts it would not be of much importance, excepting in respect to doorways, where the action of the door opening and shutting requires more attention to be paid to this point than is necessary in windows.

Of Blank Recesses in Brick Walls.

A blank recess in a brick wall may sometimes show on both sides of the wall, sometimes on the outside only. The depth of such recesses is usually half-a-brick, or about four inches.

As a recess in one part implies projection in another, alternate courses at the salient angle must have closers adjoining to the corner headers.

Figs. 132 and 133 represent the plan of two successive courses of a

Fig. 132.

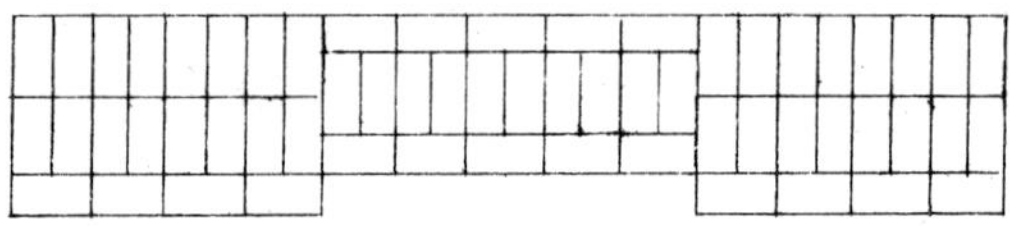

Fig. 133.

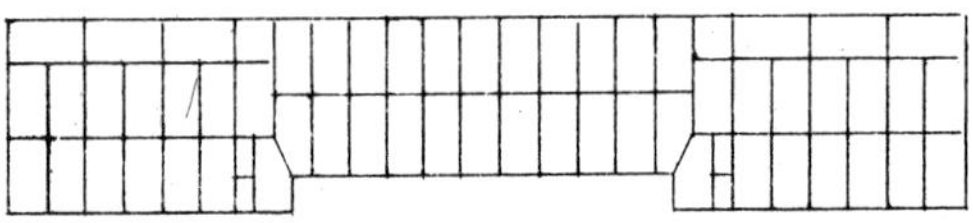

wall two and a-half bricks in thickness, diminished to two bricks by a blank recess, appearing on the outside only of the wall, which is built according to English bond.

In the last figure it will be observed that the corner header at the salient angle, and the brick which meets it at the re-entering angle of the recess, are both cut like king-closers.

Figs. 134 and 135 represent the plan of two successive courses of a

Fig. 134

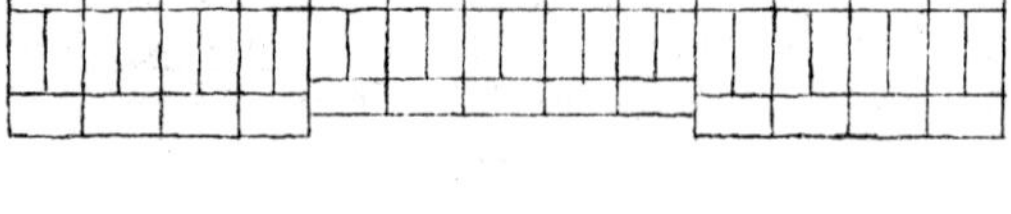

Fig. 135.

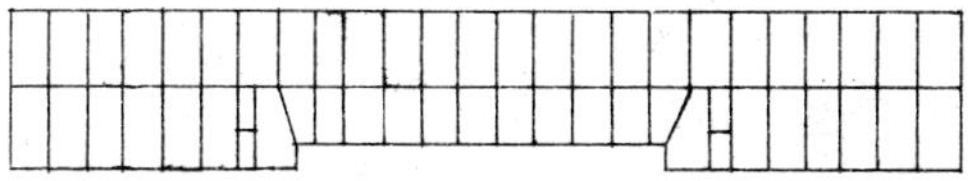

brick wall, built also according to English bond, which is two bricks in thickness, but diminished to one brick and three-quarters by a recess of a quarter brick.

In this case one set of headers in each course is supposed to be cut to three-quarter bats throughout the whole recess.

When recesses of a quarter brick are introduced on one side of a wall only there is no other mode of obtaining bond than by cutting the headers into three-quarter bats in both courses, or by introducing a continued line of closers within the recesses in the stretching courses.

Fig. 136 represents this arrangement, which may be substituted in

Fig. 136.

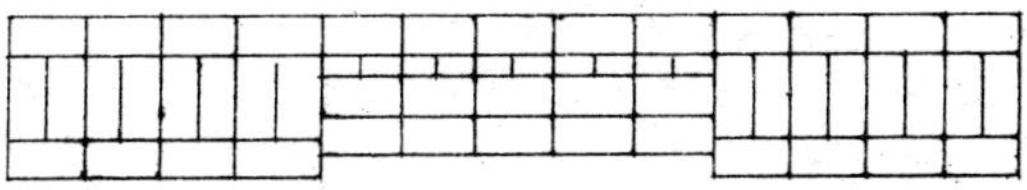

lieu of the stretching course of our present example, represented by *Fig.* 134, and you will find on trial, that it will alternate equally well with the arrangement shown in *Fig.* 135.

Of Brick Piers.

Piers are substantial pillars, for the purpose of supporting arches, or for hanging gates to, at the entrances of parks or grounds leading to a gentleman's residence. They are always rectangular, but may often be regular squares, so far as regards their plan. Those for gates have usually a wall attached to them on one side, and a projection or kind of pilaster on the other, forming a small recess both in front and rear, to the latter of which the gate is hung. Such recesses, called reveals in a building, are more commonly termed *rebates* in a brick pier.

Piers for arches are usually rectangular, excepting in vaulted store-rooms, in which case, groined arches being commonly used, every independent pier is a perfect square, whilst pilasters, connected with the external walls of the building, serve as abutments.

Of Square Piers, built according to English Bond.

Nothing can be simpler than this arrangement. Each course shows all headers and all stretchers alternately, the heading course of one side being the stretching course of the next side on the same horizontal level, and the closers of the heading courses are continued right across the whole pier, from one side to the other, by which means perfect bond is obtained.

In piers of one, two, or three bricks square, and in others whose sides are measured by whole numbers, all the bricks of the stretching courses are whole bricks; but in piers of two and a-half, three and a-half, or four and a-half bricks square, and in all others whose sides involve half-a-brick, it is necessary, either that a line of half bricks shall be introduced in or near the middle of each stretching course to make up the proper thickness, or in lieu of half bricks, one line of headers may be introduced into the middle of the stretching courses in the core of the wall only.

These particulars will be clearly illustrated by the following examples:

Fig. 137 is the plan of one course of a pier, two bricks square, in which all the courses are precisely alike, excepting that the side marked *a b* in this course is supposed to be surmounted in the next course by the arrangement marked *a c*.

Fig. 137.

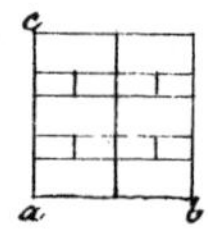

The same remark applies to all the other figures of square piers, and therefore will not be repeated.

Fig. 138 is the plan of one course of a pier two and a-half bricks square.

Fig. 139 is the plan of a pier three bricks square.

Fig. 138.

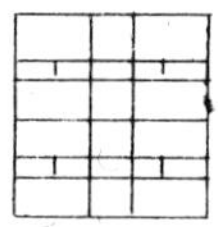

Fig. 139.

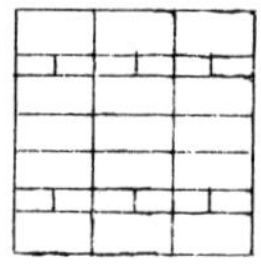

Figs. 140 and 141 represent the plan of a pier three and a-half bricks square.

In the former, half bricks are introduced into the middle of the stretchers. In the latter whole bricks are used for the same purpose, in conformity with the preliminary remark before made upon the subject of piers whose dimensions involve half a brick.

Fig. 140.

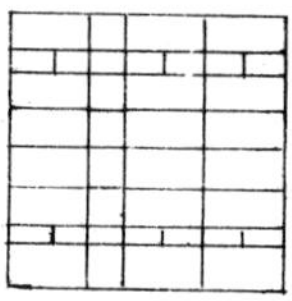

Fig. 141.

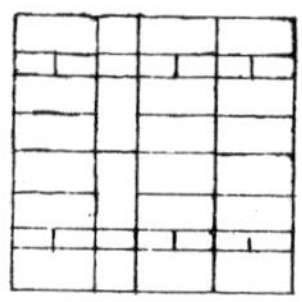

Fig. 142 is the plan of a pier four bricks square.

Fig. 142.

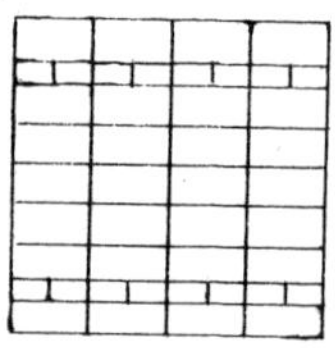

Remarks.—The kind of English bond represented in the above figures, as applicable to brick piers, which seems to me the strongest mode of building such piers, and in which the headers and stretchers are equal in number, is the only method by which thorough stretching courses can be used in a brick wall without falling into the inconvenience of the first mentioned kind of English bond, described in a former article (See Page 32).

But it may be remarked, that our present method, which suits the limited extent of a square or rectangular pier, would not apply so well to a common wall, for it will be obvious, that a line of closers, both in front and rear, continued in every stretching course all round the whole extent of a building, or along the whole outline of the scarp or counterscarp revetments of a fortress, would be exceedingly troublesome.

For the walls of buildings, therefore, and for revetments, I again repeat that nothing can be better than common English bond, alternating occasionally with diagonal courses, but in piers, such as have been described, the introduction of diagonal bricks is superfluous.

Of Piers for Gateways, built according to English Bond.

One example of this kind of pier will suffice, which we will suppose to be a pier four bricks square, attached to an inclosure wall of two bricks thick, and having a double rebate on the other side, formed by a kind of pilaster on that side of the pier, which we will suppose to project about six and three-quarter inches, or three-quarters of a brick from the pier, and to be one brick wide in the contrary direction. This projection is sufficient, because the gates attached to piers are usually double, and meet at bottom, in the centre of the gateway, against a stone or iron block, instead of being hung to one pier and shutting against the other.

Fig. 143 represents one of the heading courses, in reference to the face of the inclosing wall.

Fig. 144 represents the next successive course, being one of the stretching courses in reference to the inclosing wall. The bricks shown in the pilaster are supposed to be three-quarter bats.

Fig. 143.

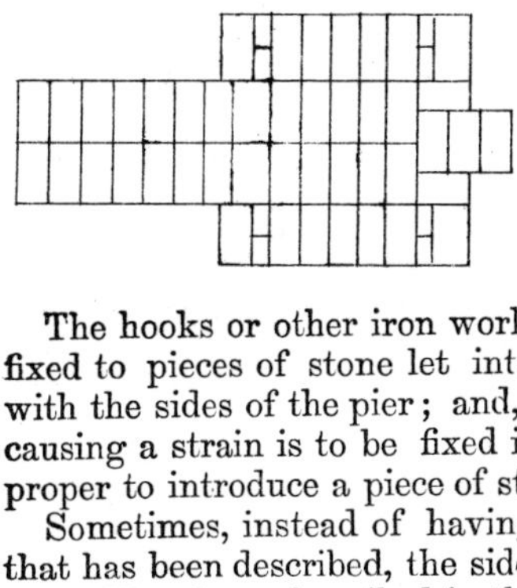

Fig. 144.

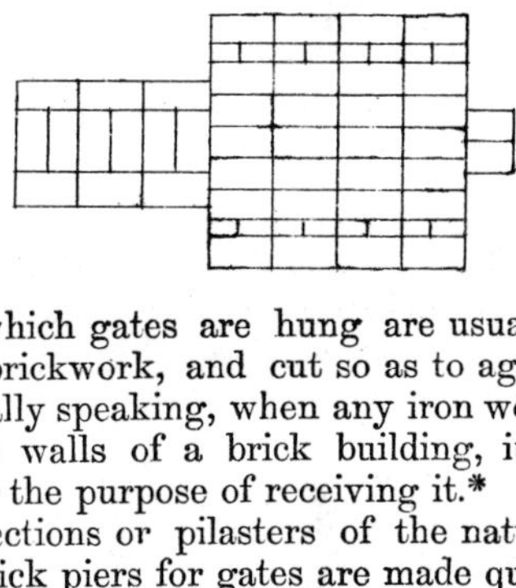

The hooks or other iron work by which gates are hung are usually fixed to pieces of stone let into the brickwork, and cut so as to agree with the sides of the pier; and, generally speaking, when any iron work causing a strain is to be fixed into the walls of a brick building, it is proper to introduce a piece of stone for the purpose of receiving it.*

Sometimes, instead of having projections or pilasters of the nature that has been described, the sides of brick piers for gates are made quite plain, like those described in the foregoing article, and in this case iron gates are usually fixed to the piers at top only, the bottom being let into stones below. In respect to wooden gates, they are not in this case attached to the piers themselves, but are hung to a couple of strong wooden posts, placed within the gateway, close to the brickwork on each side.

Of Piers built according to Flemish Bond.

There are very few dimensions of piers that yield the correct appearance of Flemish bond in every course. In fact, of square piers only, the two and a-half brick square, the four brick square, the five and a-half brick square, and the seven brick square afford that regularity.†

* Thus, for example, many of the lamps in the Ordnance Barracks at this station are suspended to iron brackets, which are nowhere inserted into the brickwork itself, but into rectangular blocks of stone, nearly square, and equal to four courses in height of the brickwork into which they are bedded, flush with the face of the wall.

† Larger piers than the last named dimension can seldom be required in practice.

This being the case it becomes necessary, if the appearance of Flemish bond be studied, to make the nearest approximation to it possible, of which a few examples will be given.

Fig. 145 shows the plan, and *Fig.* 146 the elevation, of a pier two bricks square, built so as to resemble Flemish bond, at least in alternate courses.

Fig. 145.

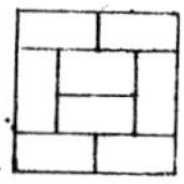

Fig. 146.

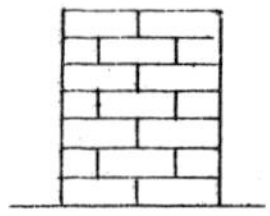

This is weak, compared with the same sized pier in English bond, for all the vertical joints of the core of the pier will evidently coincide from the bottom upwards.

Fig. 147 shows the plan, and *Fig.* 148 the elevation, of a two and a-half brick pier, affording, as was before mentioned, the perfect appearance of Flemish bond in every course.

Fig. 147.

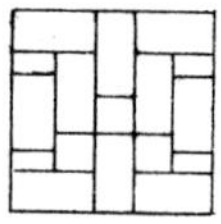

Fig. 148.

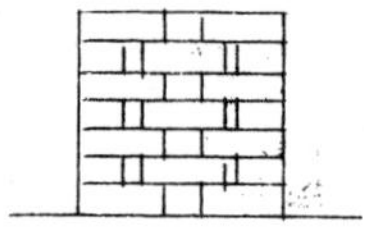

Fig. 149 shows the plan, and *Fig.* 150 the elevation, of a pier three bricks square, which, in a certain degree, resembles Flemish bond.

Fig. 149.

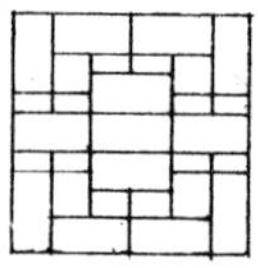

Fig. 150.

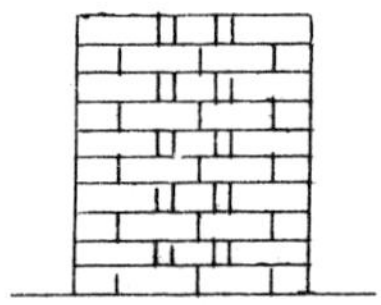

A pier of three and a-half bricks square cannot by any means be made to assume the appearance of Flemish bond, excepting in each alternate course, whilst the intermediate courses will appear quite irregular.

Fig. 151 shows the plan, and *Fig.* 152 the elevation, of a pier four bricks square, this being the second example of perfect Flemish bond in a square pier.

Fig. 151.

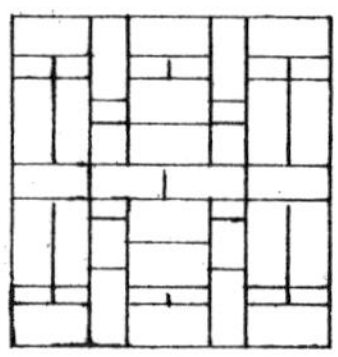

Fig. 152.

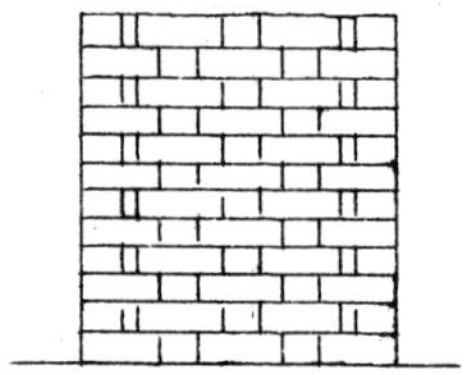

Even if false headers be used in alternate courses, it will be evident, after a little practice with your models, that Flemish bond, as applied to brick piers, is equally inferior to English bond in strength and simplicity, and in most cases it is even destitute of that kind of regular appearance which this species of bond is only capable of showing in the face of a common wall.

Of Brick Pilasters.

A pilaster in ornamental architecture is a rectangular pillar, projecting from the external surface of a wall, in which it is partly engaged.

A brick pilaster implies, therefore, any narrow rectangular projection from the face of a wall or building. In inclosing walls the buttresses, which are generally used, have, therefore, the appearance of pilasters, and may be considered as such

In the face of a brick building, pilasters are often used near the angles which answer the same purpose as recesses—namely, that of relieving the uniformity of one continued plane surface; and in those parts where recesses are used, the projecting parts between them have, of course, the effect of pilasters, at least so far as regards the appearance of the brick-work.

In private dwelling houses, where the piers between windows are often small, those piers, together with pilasters or other projecting parts used, are seldom so proportioned as to have the correct appearance of regular Flemish bond in each course. Indeed, it is impossible that it can be so, unless they are made either exactly two and a-half bricks, four bricks, five and a-half bricks, or seven bricks in width, or any greater number, increasing by one brick and a-half.

English bond, therefore, is generally capable of showing a more regular and uniform appearance in piers, pilasters &c., than Flemish bond, unless the width of the projection should happen exactly to coincide with one of the above named dimensions, which is seldom the case.

In English bond a two brick, three brick, four brick, or five brick pilaster will always exhibit perfect regularity; but if the dimension of the pilaster should involve half a brick, there will be a bat or false header in every stretching course, whilst all the heading courses will be perfectly regular.

Both in English and in Flemish bond, if the dimension of a pilaster involve a quarter brick, instead of having two closers in every alternate course, one at each end of it, there will be a closer in every course, at contrary ends.

In attempting to build pilasters according to Flemish bond, as the dimensions given seldom admit of perfect regularity, it often becomes necessary to place two stretchers, or two or more headers, next to each other in the same course, besides sometimes using one closer in every course, contrary to the common rule for that species of bond.

In building a portion of a wall, the bricklayers usually begin from coins or from the sides of doors and windows, in two directions, and work inwards until they meet. Hence the double stretchers, or double or treble headers, which sometimes occur in Flemish bond, are always near to the middle of the portion of wall in which they are introduced. This irregularity may be observed in all the piers, or spaces between doors and windows in the Ordnance Barracks at this station. The same deviation from strict Flemish bond may be observed in the smaller piers, or pilasters, or projecting parts of the brickwork, in some of the best houses of Chatham and Rochester, and in all other cities built with bricks.

The fact is, that the architect or person who plans a brick building will seldom give himself the trouble to calculate how many bricks and half bricks will go to a certain space, nor is it necessary, in deciding upon the general plan of a building, that such *minutiæ* should be attended to; but in the smaller parts of a brick building, such as pilasters, recesses, &c., it is desirable that the architect should keep those dimensions in view, instead of giving unnecessary trouble to the bricklayer by making them a fraction more or less than would agree with the materials used.

Figs. 153 and 154 represent the elevation of two brick pilasters, one of two and three-quarter bricks in width, the other four and a-half bricks in width, showing the impracticability of building pilasters of such dimensions according to regular Flemish bond.

Fig. 153.

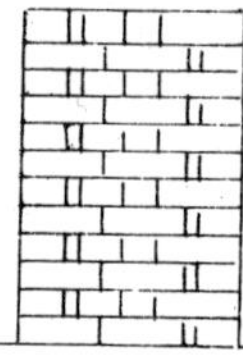

Fig. 154.

The latter evidently has a more irregular appearance than common English bond would have, and yet such irregularities occur in almost all dwelling houses where Flemish bond is attempted throughout; and wherever they do occur, it is of course in those parts which most attract the eye, and where, if ornament be studied at all, it is of the most importance—I mean in the narrow piers between windows, and pilasters, and recesses.

The matters discussed in this article again lead us to our former conclusion—that English bond is the best species of bond that can be used in brickwork.

Of Horizontal Projections, and of Offsets not connected with the Foundation in Brick Walls.

It is very common in brick buildings to have the walls of the basement or cellar story thicker than those of the ground floor, and to diminish partly by an external offset; and, as the former part of the building is usually concealed from view, being below the level of the ground or street, this offset being set off two or three feet above that level, assumes the form of a plinth to the face of the wall.

This offset is seldom more than a quarter of a brick in width, and in this case, an equal offset must be set off inside on the same level. In very large builings, it may be half a brick, and then it is not necessary that there should be any internal offset opposite.

It is also common in brick buildings, to have the walls of one or more of the upper stories thinner by half a brick, than those immediately below them. In this case, the offsets where such changes are effected are always set off internally, and are concealed from view by the floors, of which they support the wall-plates, which particulars will afterwards be explained.

There is another arrangement not so common in modern buildings, which consists in having two or three salient or projecting courses of brickwork in the face of a building, over the tops of the windows, between the several stories, and nearly opposite to the level of the floors.

These are sometimes called *string courses*, and they usually project from a quarter to half a brick beyond the face of the wall. If the latter dimension be used, both of the salient courses, if there be only two, should consist entirely of headers. If there be more than two, the lower and upper course only should be thus laid.

Fig. 155.

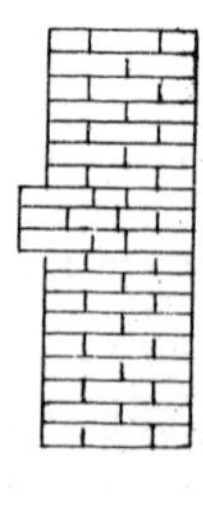

This arrangement may be seen above every story of the infantry barracks at this station.

Fig. 155 represents in section a portion of a two-brick wall, having three string courses of the above description.

Sometimes brickwork is made to project gradually, by gaining from one to one and a-half, and not exceeding two inches in each course, and the courses thus built are called *gathering courses*.

Fig. 156 represents this arrangement, which is supposed

to take place on two opposite sides of a party wall two bricks thick.

Fig. 156.

In common buildings this arrangement may be used with much propriety for supporting the wall plates of floors, as will afterwards be described.

External projections of the same kind may be seen in almost all cities built with bricks. There is a specimen of this arrangement in the gate of Chatham Dockyard, and also in the upper part of Gillingham Tower.

Of Brick Arches.

It is unnecessary to define an arch, but the student should know all the technical terms commonly used relating to arches, such as the *piers* and *abutments*, which support them; also the *span* and *rise* of an arch, the *length* of an arch, the *spring* of an arch; the *skewbacks*, when it does not spring from an horizantal plane; the *haunches*, or lower segments on each side, as distinguished from what common bricklayers call the *scheme* part, or upper segment of an arch; the *intrados*, or whole interior or lower surface; the *extrados* or whole exterior or upper surface; the *crown*, or summit of an arch; the *voussoirs*, or arch stones, and the *key*.

He should also know the common terms by which arches are distinguished, in reference to the curves by which they may be formed, such as *straight* arches, to which the term *soffit* applies, instead of intrados; *semicircular* arches, *segment* arches, or smaller portions of a circle, commonly called *scheme* arches by bricklayers; *elliptical* arches, *parabolic* arches, *gothic* arches, &c., &c.

The arches used in buildings of any importance are of two kinds, external and internal.

The former are always executed with great care, and in an ornamental manner; the latter are done roughly, being afterwards concealed from view by the inside finishing of rooms.

Of External or Camber Brick Arches, as applied to Doors and Windows of Buildings.

Outside arches are of two kinds, straight arches and semicircular, or segments of a circle.

Elliptic arches are also sometimes used, and other forms, of which specimens may occasionally be met with, but which I shall not notice, as they are not considered in good taste, according to the fashion of the present day.

Straight arches are perfectly horizontal above, and have the same appearance below, to an inexperienced eye. This, however, is not the case, for they have always a small rise of about half-an-inch in three or four feet, so that their lower outline is an arc of a circle, like that of a segment arch, but the joints always radiate to some point much nearer than the centre of the said arc.

The bricks used in arches always show their thickness in the face of the wall, which thickness being from two and a-half to two and three-quarter inches at the utmost, this dimension is used at the top of the straight arch, or at the extrados of the curved one, and from thence it diminishes, owing to the splay of the arch bricks, to the soffit of the straight arch, or to the intrados of the curved one.

This splay is by bricklayers termed the *sommering* of an arch.

The bricks used in external arches are always rubbed with great care to the proper splay or wedge-like form necessary, and according to gauges or regular measured dimensions. For this reason they are usually styled *gauged arches*, to distinguish them from inside or rough arches, in which common bricks are used, without being rubbed or shaped in any way.

Fig. 157.

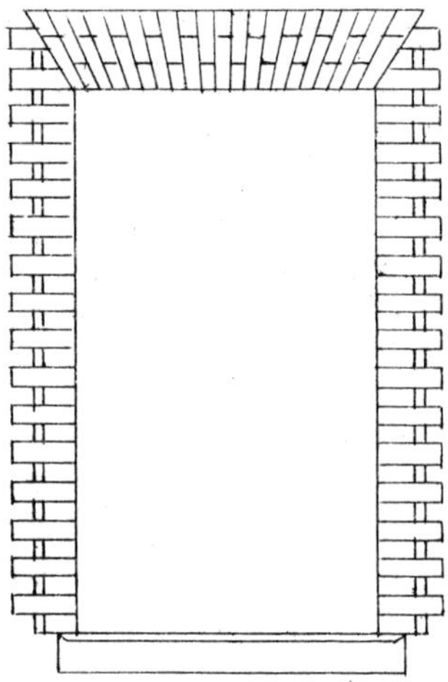

Fig. 157* represents in elevation a straight arch of the description now under consideration, such as is almost always used for the windows, and frequently for the doors of dwelling houses, and which at the width of four feet ought to have about 21 arch bricks. The vertical height of such arches is almost always made equal to that of four courses of common brickwork, and as the whole thickness necessary for such an arch, reckoning it obliquely, according to the lines of the joints of the arch-bricks, and which, therefore, varies according to the situation of those joints, cannot be obtained out of one brick, the depth is usually in two pieces. But the horizontal joints shown in the above drawing are false joints, marked for effect, alternately at one third and at two thirds of the height of the arch, measured from the soffit upwards. The real joints are not horizontal and oblique to the arch joints, but perpendicular to them; and this may easily be observed in brick buildings, unless the bricks be painted, for although the choicest bricks, both as to quality and colour, are always selected for such arches, after a number of years some of them will usually change, and then the diffierence of colour will distinguish the real joint, however well concealed at first.

It will be evident, that in a straight arch, no two voussoirs or arch bricks, as they may be termed, on the same side of the centre or key of the arch are alike, but those of contrary sides will all correspond with each other pair by pair.

In semicircular or segment arches, on the contrary, all the arch-bricks, between every two of the adjacent radiating joints, are exactly alike.

The number of bricks in a semicircular arch is determined by finding how many thicknesses of a common brick are required at the extrados or

* This figure is a correct elevation of one of the windows in the ground floor of the Ordnance Barracks at Chatham, which are four feet wide and seven feet high. The stone sill before treated of is shown below the aperture. The small rise of the flat arch is not however represented, this being nearly imperceptible on a small scale, and accordingly never included in an architectural drawing.

outside of the arch, and from thence drawing the lines of the joints towards the centre of the arch, which will determine the splay, and give the rule for rubbing the bricks to their proper form.

Semicircular or segment arches are commonly used over the recesses in which doors or windows are situated, and the aperture within the recess is also usually arched over by a concentric arch of a smaller span.

In this case, the upper or greater arch is usually made a brick and a-half in thickness, but the lower or smaller arch within the recess is usually made only one brick in thickness, being equal to the face of the thin wall on each side of the aperture within the recess.

Fig. 158.

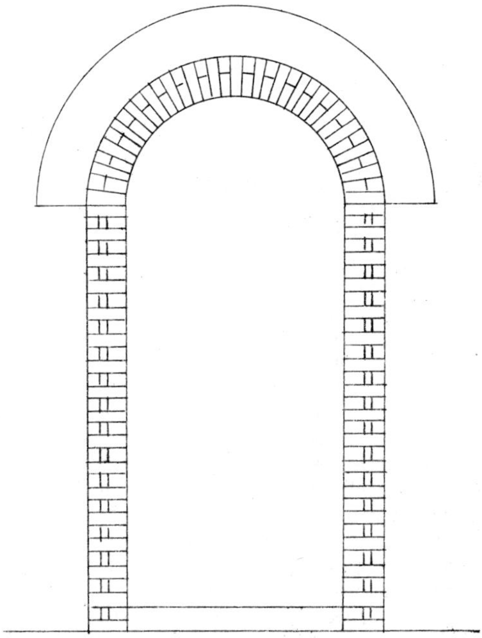

Fig. 158 represents a semicircular arch of four feet wide, and one brick thick, which requires about 39 bricks at top and which is a correct elevation of one of the arches over several of the doors of the Officers Quarters in the Ordnance Barracks at this station, which are of the above width, and seven feet six inches high, from the upper step to the spring of the arch.

They are each situated within a recess, the outline of which only appears in the figure. This upper or second arch has been omitted in the figure for the sake of clearness.

Fig. 159 is a correct elevation of one of the arches of the recesses about the same doorways, which are each one brick and a-half in thickness. The arch of the doorway itself, explained in detail in the former figure, appears only in outline in our present figure, for the sake of clearness.

Fig. 159.

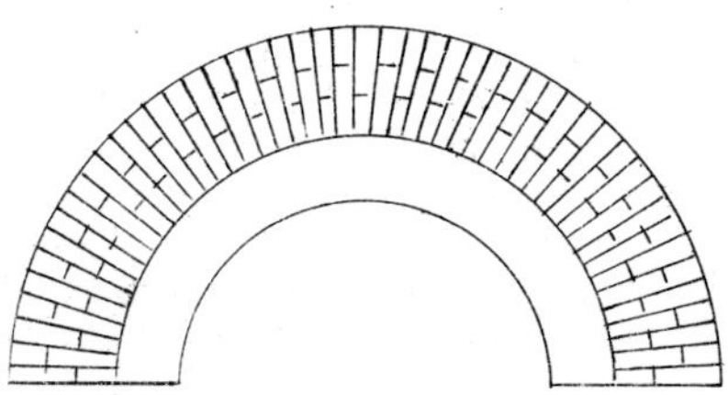

It is to be observed, that the length of ornamental arches of this description thrown over a recess, as measured from the face of the wall inwards, in a barrack or in dwelling houses, never exceeds half a brick, or about four inches, unless the depth of the recess itself should be greater, which is not usual.

In like manner the length of an ornamental arch above a door or window, whether situated within a recess or not, never measures more than half a brick, unless the reveal be thrown further inwards, which is not usual in buildings of the above description. Hence, such ornamental arches are mere shells, which support a portion of the wall towards the outside only. Accordingly, if, as in the recessed doors before alluded to in these barracks, there be a double semicircular arch, namely, one over the recess, and one over the door, these two between them support only eight or nine inches, or about one brick's length of the thickness of the wall, which being two bricks thick, all the remainder of it, towards the inside, must of course be provided for in some other way.

This is done by means of a rough brick arch towards the inside, concentric with the two others, but of rather greater span than the smaller of the two, in order to allow for a reveal of the dimension judged necessary.

Fig. 160 being a section of one of those doorways, represents this arrangement; *a* is the ornamertal arch over the recess, supposed to be a brick and a-half in thickness; *b* is the ornamental arch within the recess over the door,

Fig. 160.

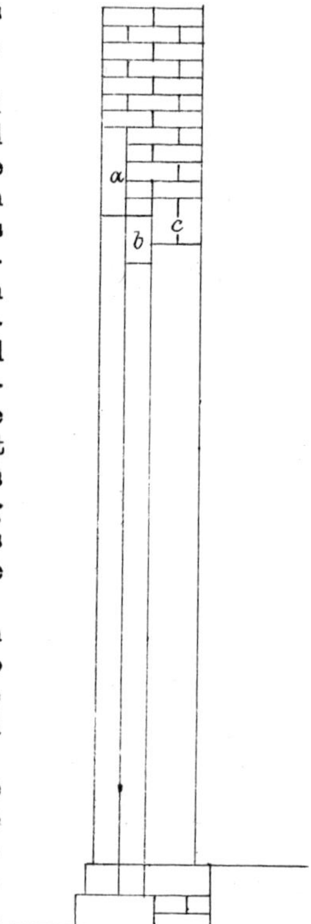

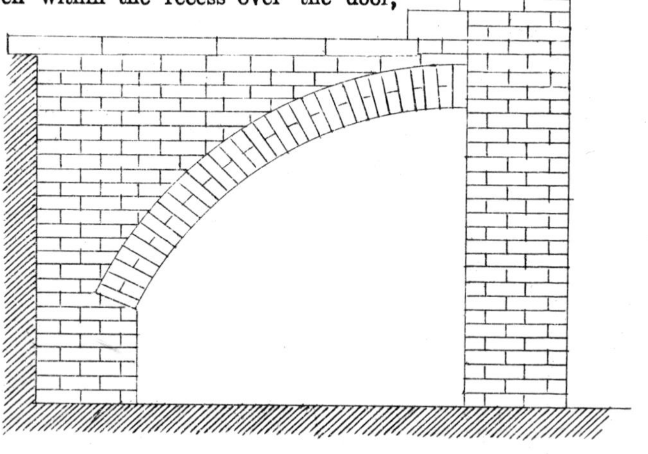

which is supposed to be one brick in thickness. Each of those two arches is half a brick in length, measuring from the outside of the wall inwards; *c* is the rough inside arch, afterwards covered with plastering, which is one brick in thickness, and one brick in length.

It will be evident, on considering the above figure attentively, that the two ornamental arches and the rough arch are by no means well connected. In fact, although each may support its own part of the superincumbent brick work, there is no common bond between them. If, however, they be respectively executed with proper care, and in a workmanlike manner, and if the bricks and mortar or grout used be good, there is no risk of failure. Indeed, the openings usual in dwelling houses are so small that after the mortar shall have set, such arches might be entirely removed, and yet the wall above them would not give way.

The execution of all the ornamental or thin arches in the face of a building, such as have been described should be most carefully attended to whilst in progress, because, although Bricklayers will always take sufficient pains to rub the outside of each arch brick properly, so that their work may have a handsome appearance to the eye, they will often slur over all the other parts of the work which become eventually concealed from view. Hence, in order to save trouble, they are very apt to cut away the inside of the bricks of those arches to such a degree, as may even deprive them of the proper wedge-like form, every where except at the external surface. This neglect produces cracks, and causes the arch to bulge forward, and may even cause one of the bricks of a straight arch to drop down lower than the soffit or bottom of the arch, which defects are frequently to be observed over the windows of common brick buildings.

In respect to ornamental arches of one brick in thickness, and only half a brick in length, of which a specimen is represented in *Fig.* 158 I should recommend using none but whole bricks in constructing them, because the use of the two bats, in every alternate voussoir, as shown in that figure, must necessarily weaken the arch in some degree. But, in this case, false joints may be represented in every alternate voussoir, should the builder prefer the appearance of bats, without actually using them, under the idea that a mixture of small joints will accord better with the general effect of the other parts of the brickwork.

Plain Brick Arches defined.

Plain or rough brick arches are those in which none of the bricks are cut or rubbed to suit the splay. Hence the joints, which are quite close to each other at the intrados, are necessarily wider towards the extrados of the curve. In illustrating this part of my subject, I shall first treat of semicircular or segment arches.

Of Plain Brick Arches having a Radius of Curvature not exceeding three or four feet.

When the *radius* of curvature of a brick arch is small, that is to say, when it is under or not exceeding *three or four feet*, the best arrangement is to build two or more concentric four-inch brick arches over each other, until the whole thickness necessary be obtained; each of these successive arches being built as an independent arch, without attempting to bond the upper or larger arch of any two into that which is immediately below it.

Fig. 161 represents a semicircular arch of this description, supposed to have a span of six feet, and consequently a radius of curvature of three feet, and consisting of four concentric four-inch arches, combined into one arch of two bricks in thickness.*

Fig. 161.

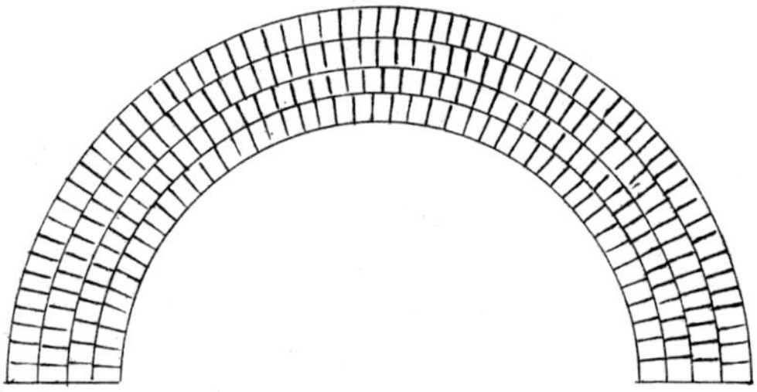

The bond of brickwork used in any arch is always judged of by the appearance which it exhibits to the eye of a spectator, looking up towards the intrados or lower surface of the arch. Hence, in our present figure, as all the bricks used, if so viewed, would have the appearance of stetchers, the whole arch is said to be built with stretchers, or all stretching bond, being composed entirely of whole bricks laid as stretchers, excepting only at the two extremities of the arch, where each alternate voussoir is a bat or half brick, without introducing which, the bricks composing a four inch arch would not be properly bonded together.

Fig 9 (in Page 29), which shows in elevation the appearance of the face of a wall, built all stretching bond, will, therefore, also afford a sufficiently correct notion of the appearance of the brickwork of part of the intrados or lower surface of an arch of the description now under consideration, which, as was just observed, is likewise built according to the same species of bond.†

The reason for building an arch of small radius with a series of four-inch arches, and without attempting to bond the successive parts into each other, is sufficiently obvious. The joints of such an arch diverge so very rapidly from the intrados outwards, that if it were built otherwise the brickwork would not be sufficiently compact; for, at the upper part of the arch, the bricks would be separated to an extraordinary and inconvenient degree.

For example, if we suppose the span of a semicircular brick arch to be only three feet, and that it is built in four concentric portions of four-inch arches, the radius at the intrados of the lowest of those small aches will, therefore, be 18 inches, whilst the radius of the extrados of the same small arch will be 22 inches, and the lengths of the two curves will, of course, be in the same proportion; and those numbers being to each other as one to one and a-quarter nearly, there will be open spaces at the upper part of the curve of one quarter of a brick in width between the several arch bricks, which will touch each other at bottom. Hence, no joint between two bricks will anywhere exceed about five-eighths of an inch, this being the one-fourth part of the width of a brick, as seen in the section of an arch.

In the upper portions of an arch thus constructed, the joints

---

* This is a correct representation of one of the arched passages leading from the back of the Ordnance Barracks at this station to the canteen.

† It gives a sufficiently just notion without being absolutely accurate, as the one is a plane surface, the other concave.

between the bricks will evidently diminish as each successive course is added. *Fig.* 162 is a correct representation of an arch of the nature and dimensions above mentioned.

Fig. 162.

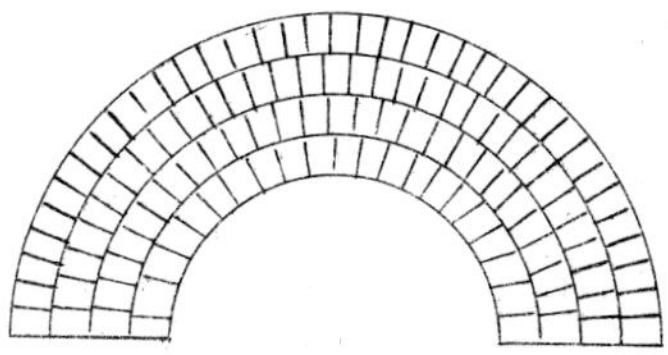

If, on the contrary, it were attempted to build a semicircular arch of three feet span, and of the same total thickness, with bricks laid as headers, which would form a couple of 9-inch arches, and if the radiating joints were continued from the intrados outwards, across the whole thickness, then, as the radius at the intrados is 18 inches, and at the extrados 36 inches, which numbers are in the proportion of one to two, there will, of course, be open spaces at the upper part of the curve equal in width to the bricks themselves.

Fig 163 represents this arrangement, showing the unfavourable effect that would be produced by building an arch of the small radius of curvature now under consideration, in the manner last described, which would occasion joints of no less than two and a-half inches in width at the extrados of such an arch. These joints, if filled entirely with mortar, would be prejudicial to the solidity of the work; and if filled with harder materials, such as pieces of stone, slate, or fragments of bricks, the arch would have a motley and unseemly appearance.

Fig. 163.

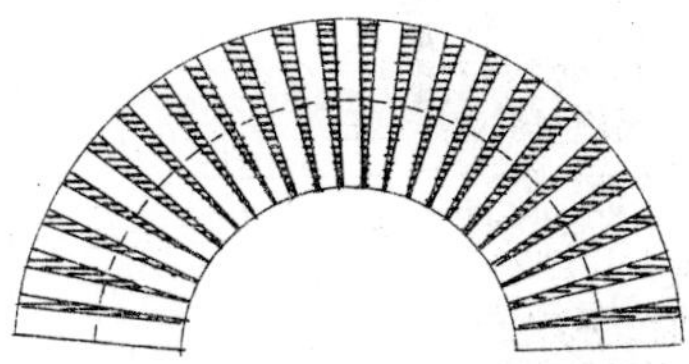

It would be superfluous to say more in explanation of the reasons which render four-inch arches the most suitable for a small radius of curvature. The term radius is used as being the most precise, for an arch of very small span may, of course, have a great radius, when in the form of a flat segment.

I shall conclude by observing that arches of the description treated of in this article, are seldom composed of more than three concentric four-inch arches, and this number will generally be sufficient in practice when the radius does not exceed three or four feet. I have shown a greater thickness in *Fig.* 162, not as being necessary for an arch of that span, but to illustrate the principle under discussion more clearly.

Of Plain nine-inch Brick Arches having a Radius of Curvature of from six to ten feet.

As the radius of curvature of an arch increases, the joints diverge less rapidly, so that the use of nine-inch arches, formed with heading bricks, ceases to be objectionable.

For example—in a semicircular nine-inch arch of 12 feet span (whose radius of curvature is of course six feet), the radius of the curve at the intrados is to its radius at the extrados, as six feet to six feet nine inches or as one to one and one-eighth. Hence, the joints at the upper part of the curve will not exceed one-eighth part of the width of an arch-brick or about five-sixteenths of an inch, which is neither inconvenient nor unseemly. If the radius of curvature be greater than six feet, the width of joint at the extrados of a nine-inch arch will be still further diminished, so that in a semicircular arch of 18 feet span, which is a very common dimension for casemates, it will scarcely amount to so much as one-fifth of an inch. A nine-inch brick arch may be built according to three several methods.

1st. The bricks composing the voussoirs may consist partly of headers and partly of stretchers, built according to Flemish bond.

Fig. 164 represents the end elevation or section of part of an arch of

Fig. 164.

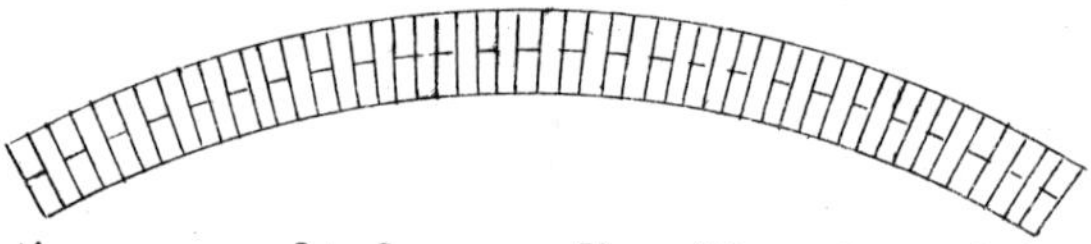

this description, supposed to have a radius of curvature of nine feet.

Fig. 58 (Page 41), which represents the appearance of Flemish bond in the face of a common wall, may also serve to give a sufficiently just notion of the appearance of part of the brickwork of the intrados of the kind of arch now under consideration, and, in the arch as well as in the wall, closers are necessarily introduced next to the corner headers at each extremity, in order to obtain the proper bond.

2dly. The bricks composing the voussoirs of a nine-inch arch may consist of headers and stretchers built according to English bond. In this case, the extremities of the arch have the same appearance as in the former arrangement, shown in *Fig.* 164, which likewise is a correct section of this method. But the appearance of the brickwork of the intrados of this arch is different, being similar to common English bond in the face of a wall, so that *Fig.* 12 (Page 30), gives a sufficiently just notion of it.

3rdly. The bricks composing the voussoirs of a nine-inch arch of the same span and curve, may be laid all heading bond, so that the transverse section of the body of the arch will every where be as in *Fig.* 165.

Fig. 165.

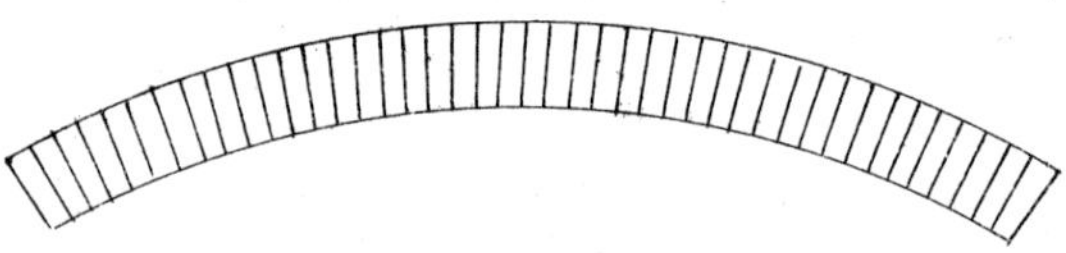

Fig. 55 (Page 40), which represents the appearance of the face of a wall built all heading bond, will give a sufficiently just notion of the appearance of the brickwork of the intrados of an arch, built according to the same arrangement. And, in an arch of this description, no false headers are of course used, nor are stretchers used anywhere, excepting at the extremities of the arch, where they are introduced in alternate voussoirs, for the sake of breaking the bond; the same being done in the face of a wall of this description, as shown in the figure last quoted.

Hence in a nine-inch arch, built all heading bond, the section is as *Fig.* 165, but the end elevation is the same as that of an arch of equal thickness, built according to Flemish or English bond (See *Fig.* 164).

Of the three methods of building a nine-inch arch, described in this article, the first method of Flemish bond, although the most commonly used in practice, appears to me to be the worst; the English bond seems better, but I am inclined to consider the *heading bond as the best of all*, for the strength of an arch depends most upon the length of the voussoirs, which constitute its thickness; and in great arches, such as those of bridges, where stones are used, the voussoirs are all in one piece, not in several lengths, or successive courses.

In an arch built with common bricks, the latter arrangement is unavoidable; but the fewer the number of parts to compose the entire thickness the better.

Of Plain Arches one Brick and a-half in thickness.

In this case, if the radius of curvature do not exceed three or four feet, such arches would, of course, be built in three concentric four-inch arches, according to the rule laid down in a former article. I shall now, therefore, confine myself to arches of a greater radius, such as from six to eight feet, this being the usual curvature of vaulted apartments in the basement stories of buildings, the arches of which are generally semicircular, as in the British Museum.

For such arches, the usual method, which seems as good as any other, is to build the voussoirs in alternate courses of headers surmounted by stretchers, and of stretchers surmounted by headers, from one side of the arch to the other.

Fig. 166 represents this arrangement, which is not only an end ele-

Fig. 166.

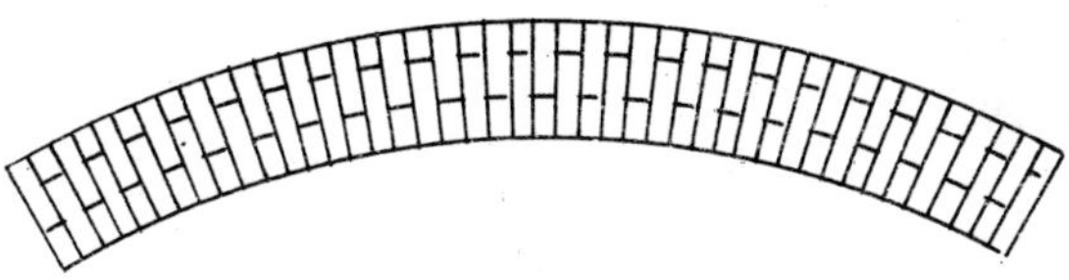

vation but a correct section of the arch throughout; and it may be observed that this thickness is scarcely ever exceeded for vaulted apartments, unless in the bombproof buildings or casemates of fortresses.

Of Plain Arches of two Bricks or more in thickness, such as are used for Casemates, &c.

Let us suppose as an example, an arch of three bricks in thickness and of *nine or ten feet radius* of curvature, such being the usual radius of semicircular arches when used in casemates, the former proportion being very common, and the latter never exceeded.

In this case, three or four methods suggest themselves—

1st. To form the whole arch out of two one and a-half brick arches of the nature described in the foregoing article.

Fig. 167 exemplifies this construction, as applied to a portion of an arch of ten feet radius.

Fig. 167.

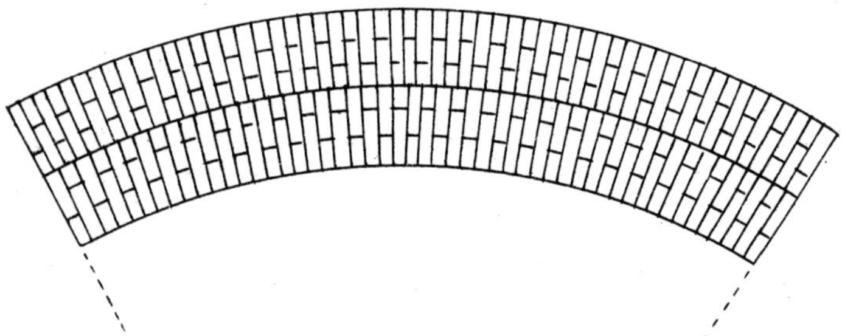

2dly. To form the whole arch out of three concentric nine-inch arches, in which both headers and stretchers are used.

Fig. 168 exemplifies this construction, which has been very commonly adopted in practice.

Fig. 168.

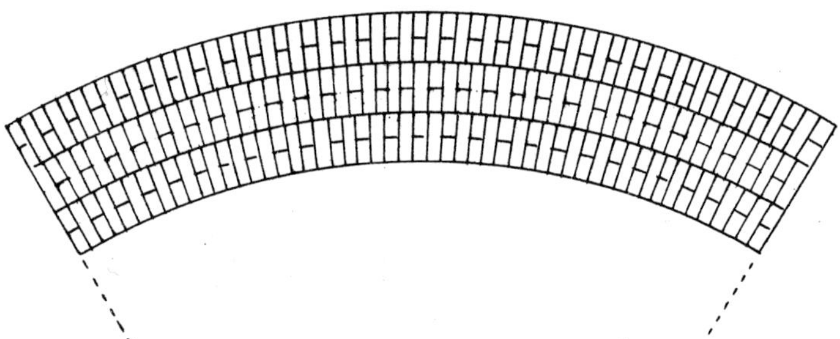

3rdly. To form the whole arch out of three concentric nine-inch arches, all heading bond.

Fig 169 exemplifies this construction, which, I believe, has been less frequently used in England than the former methods.

Fig. 169.

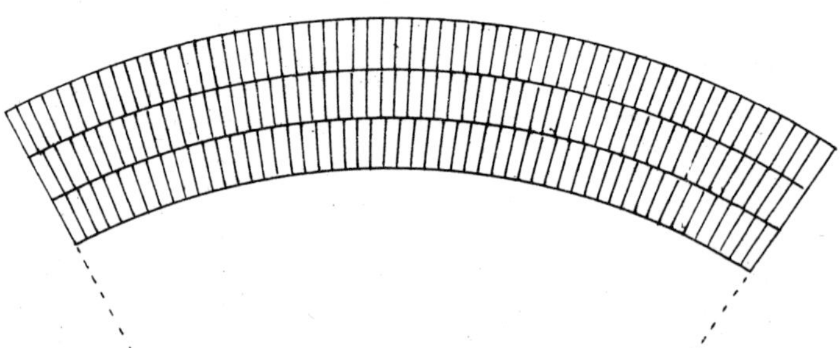

4th. To form the whole arch out of three concentric nine-inch arches, the lower one to be all heading bond, the others to be English bond, with headers and stretchers mixed.

Fig. 170 exemplifies this construction.

Fig. 170.

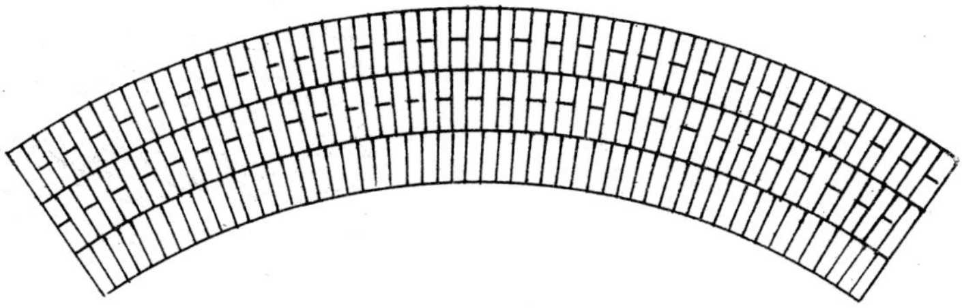

Of the four methods that have been described, it appears to me that the two last are the best, inasmuch as heading bond at the intrados of the lower nine-inch arch produces longer voussoirs at that part of the whole arch where the arch-like form is of most importance.

If the same rule applies as strongly to the two upper nine-inch arches which contribute, together with the former, to compose the whole arch, the third method, in which nothing but heading bond is used throughout, ought to be the strongest of all; but if length of voussoirs be less essential in those two portions, the last construction which has been exemplified may be considered that which bonds the brick work best, by an intermixture of stretchers and headers.

Of the Intermixture of Thorough Bonding Courses in Arches of two Bricks or more in thickness.

The thick arches which have hitherto been described are all composed of several independent concentric arches, built one over the other, and like the successive horizontal courses of a common brick wall, they are not bonded together; and it appears to me, that if bonding may be dispensed with in the courses of the wall, so may it be dispensed with in the courses of the arch; for if the piers and abutments of the arch be sufficient, every upper course must press upon the lower one, if it presses at all.

The celebrated Belidor is, however, of a contrary opinion. He recommends that the arches of casemates shall be built with thorough bond from the intrados to the extrados, as he observes that the independent concentric arches, used to form one whole arch of greater thickness, had sometimes been observed to seperate from each other.

It appears to me, that this defect recorded by Belidor could only have arisen in consequence of some glaring incongruity in the proportions of the parts, or in consequence of extremely bad workmanship; for our arches of casemates in the British service, where they have failed at all, have never failed in this singular manner.

Supposing an arch of three feet, or of four bricks in total thickness, to be the proper dimensions for a bombproof, which has been very generally allowed, and supposing the radius of curvature to be nine feet, which, as I before said, is an usual dimension for semicircular arches, the radius at the intrados will be to the radius at the extrados of such an arch as 9 to 12; or as 1 to $1\frac{1}{3}$. Hence, by adopting Belidor's principle of thorough bond throughout, the joints of the mortar at the extrados would be equal to one-third of a brick, or to five-sixths of an inch in thickness.

If, on the contrary, such an arch be built of four concentric nine-inch arches, the joints of mortar at the extrados of each will no where exceed one fourth of the above width, being only 5-24ths of an inch or 1-5th of

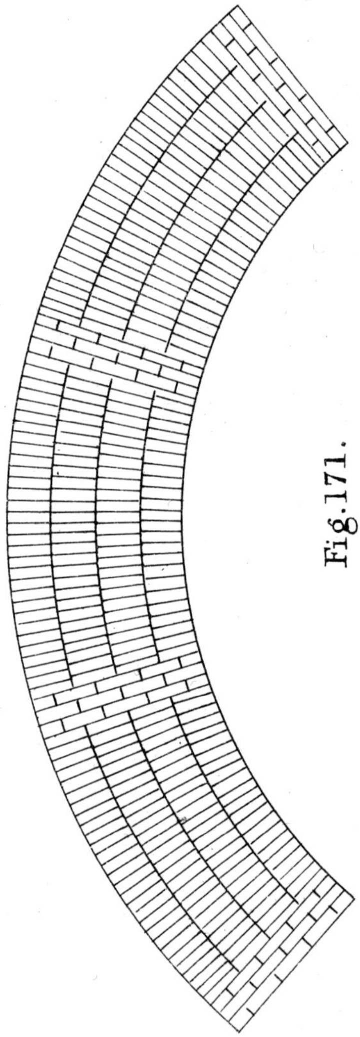

Fig. 171.

an inch nearly. It is scarcely necessary to observe how much more compact the brickwork of such an arch must be than that suggested by Belidor, if executed with plain bricks.

Instead of attempting thorough bond throughout, it has been common in this country to throw in here and there some three or four portions of thorough bond, at certain intervals throughout the width of the arch, whilst the great body of the arch has been built in successive portions, consisting of several concentric nine-inch arches, according to the system before described.

Fig. 171 (see preceeding page) exemplifies the principle of this arrangement, which has been attempted in several arches in the fortifications of Chatham; but it does not strike me that any very great advantage can result from it.

In this figure the thorough bond courses show, in the section of the arch, precisely the same arrangement that English bond does in the transverse section of a common wall of the same thickness.

The same Principles applied to the Construction of Parabolic Arches.

The width of casemates being usually 18 feet, if we suppose parobolic arches of that span and of nine feet radius to be used for casemates, as at Dover, the best mode of constructing such arches with common bricks shall form the subject of this article.

Fig. 172 represents the section of a parobolic arch four bricks thick, of the span and rise before stated, having all its joints correct tangents to the intrados,* and varying from concentric four-inch arches at the crown to thorough bond at the haunches, on the principle that has just been explained.

If it should be decided to use the parabolic arch in casemates I might perhaps venture to recommend the above construction as being the most suitable to the nature of that curve; but, for my part, I am most partial to the semicircle or segment, as being the simplest of all arches that can be abopted for this purpose.

It is true that some eminent mathematicians in this country, of whom Emerson was the first, have attempted to prove by theory that semicircular arches are infinitely weak, unless loaded at the haunches with infinite weight, which is a thing impossible; and those writers have accordingly rejected that simple form, and proposed what they call arches of equilibration &c., &c. Without entering further into a discussion of these doctrines, I shall merely observe that the experience of more than 2,000 years has proved, that if semicircular arches have piers and abutments of proper strength, they are as durable as any of the works of man; and it, therefore, has often astonished me that writers of such acknowledged talent as those alluded to, should so far have deviated from Lord Baccon's maxim as to set up any hypothesis in direct contradiction to the experience of ages.

Semicircular arches are certainly very inconvenient for the construction of bridges over great rivers, and they have, accordingly, been altogether set aside as inapplicable to that peculiar purpose; and ellipses or

---

* The rule for drawing the points of a parabolic arch is explained in my "Course of Practical Geometry" 2d Edition, Part 4, Prob. 45.

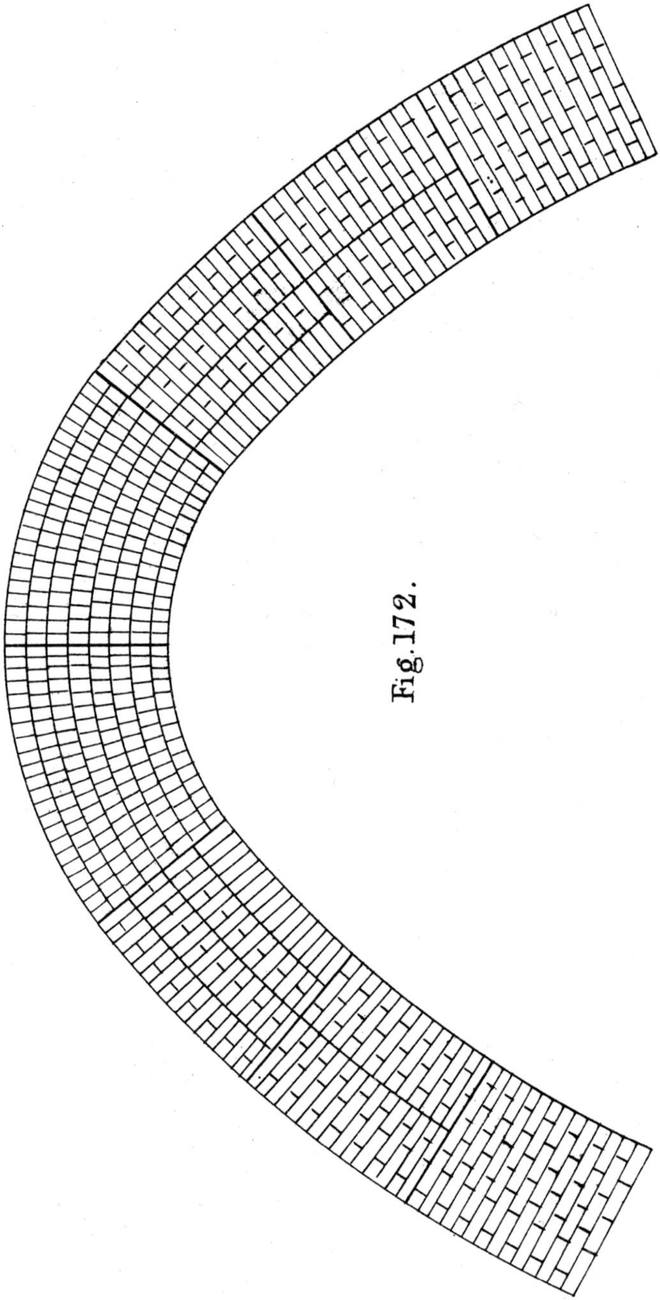

Fig. 172.

other flatter arches have very properly been adopted in preference. But for the common purposes of building, and for casemates and powder magazines, I am rather an advocate for the semicircle, or its segment, which, as I said before, are undoubtedly the simplest of all arches, and, at the same time, quite as strong as can be desired, provided that their piers and abutments be made of sufficient thickness.

Of Ornamental Arches of two Bricks or upwards, such as are sometimes used for Concealing the Extremities of Plain Arches.

Ornamental arches, which imply that every brick is gauged and rubbed to the splay or sommering, should never be used for vaults of any length, such as casemates, gateways, &c., as this process must greatly increase the expense of the work. They may, however, be sometimes employed to conceal the two extremities of plain arches, as has been done for example in the principal gateway leading from Chatham Lines into the town.

In this case the ornamental arch may be made concentric with the main arch, but of rather smaller span and somewhat lower; and thus it will conceal the plain arch better, on entering the gateway, and it may also afford a rebate for the gates to close against.

In such ornamental arches, instead of showing two or more nine-inch arches over each other, all the joints are not only made to radiate to the centre of curvature throughout the whole thickness of the arch, but the several vertical joints are broken transversely in such a manner that thorough bond prevails throughout.

Thus, for example, *Figs.* 173 and 174 represent in elevation two orna-

Fig. 173.

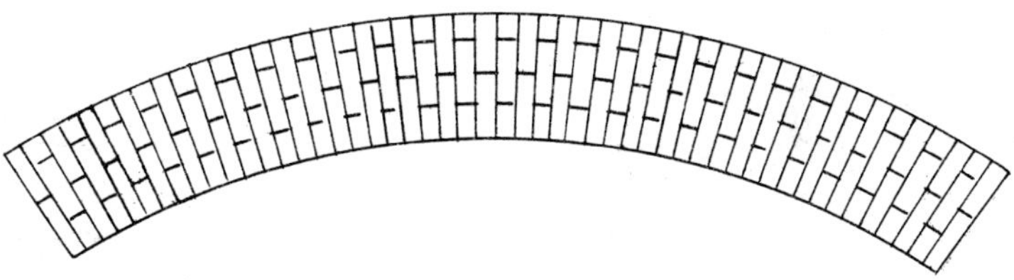

Fig. 174.

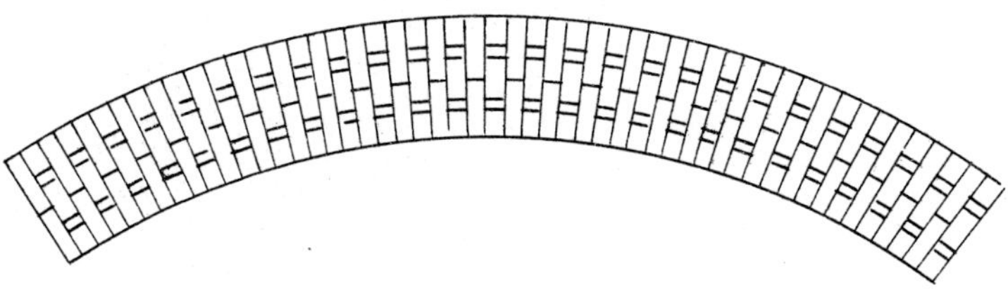

mental arches of this description, each equal to two bricks in thickness, such as may be used to conceal the extremities of a rough arch.

Of these two figures the first *(Fig.* 173*)* shows the best method of obtaining thorough bond in an arch of such thickness, for the bond shown in the section of this arch is the same as English bond in the section of a common wall.

The second figure *(Fig.* 174*)* shows a more complex method of effecting the same object, by introducing closers, without necessity. Specimens of this method may be seen in London, and it has also been frequently adopted in practice by the bricklayers in this neighbourhood, probably from an idea of its being the most ornamental, but I do not recommend it, as the first method is stronger and simpler, and appears to me to be as handsome as can be desired.

Gauged and rubbed bricks are seldom used throughout for arches of any magnitude. There is, however, an example of this arrangement in the archway leading from Chatham New Church towards Maidstone. This is not a system to be recommended. It adds greatly to the expense of an arch, as was before remarked, but nothing to its strength; and as for beauty, there is not one person in ten thousand, who, in passing through both, will ever observe the difference between this and the plainer and less expensive arch of the principle gateway of Chatham Lines.

Of Arch Bricks, or Bricks moulded to the Arch-like form. That it is scarcely worth while to make use of such Bricks for the Arches of Casemates, &c.

The example of the Romans, who are said to have used square bricks of three sizes, the largest two feet square and two inches thick, the medium size 18 inches square and one and three-quarters of an inch thick, whilst the smallest was seven and a-half inches square and one and a-half inch thick, sufficiently proves, that bricks of much larger size than the present English standard might conveniently be made;* and of all the processes of building in which a deviation from that small and simple form appears desirable, it might seem the most so in arches. Indeed, the idea of making bricks of the wedgelike form, on purpose for arches, appears so striking on a hasty consideration of the subject, that I believe it must have occurred to every person who ever gave any attention to the art of building.

The objections to such a system are, however, obvious, and are twofold; 1st, that the thing is unnecessary, because arches sufficiently strong for all practical purposes can be made out of common building bricks of the usual standard. 2nd, and here lies the chief objection, scarcely any two arches are alike, either in form or thickness; and, therefore, every arch would require bricks of a different mould, which would occasion great difficulties and delay, and extra expense, in the now simple art of brickmaking. Moreover, in arches of any considerable thickness, such as those of casemates, no less than four different moulds or patterns of arch bricks would be required.

---

* Without going so far back, the business of the tile-maker of the present day proves the same thing.

For example, if we suppose the arch of a casemate to be a semicircle of nine-feet radius and three feet thick, and built of large bricks made for the purpose, and bonded, as in *Fig.* 175, each brick being two feet long

Fig. 175.

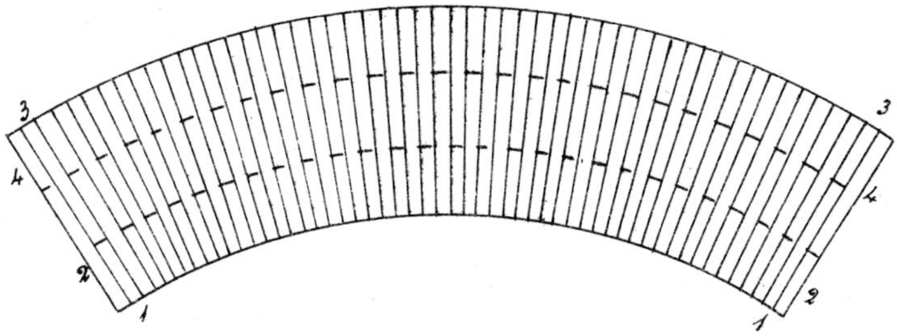

and one foot wide, to be used in two courses in the thickness of the arch, breaking joint with each other, and each consisting of headers and stretchers, their detailed dimensions would be as follows:—

First. The lower stretcher (1) would be two feet long and one foot wide, two inches thick on one side (the lower side), and 2 2-9ths inches thick on the other side.

Secondly. The lower headers (2) might each be two feet long and one foot wide as before; two inches thick at one end, and 2 4-9ths inches thick at the other.

Thirdly. The upper headers (3) to be two feet long and one foot wide, as before; 2 2-9ths inches thick at one end, and 2 6-9ths inches thick at the other.

Fourthly. The upper stretchers (4) to be two feet long and one foot wide, as before; 2 4-9ths inches thick on one side (the lower side), and 2 6-9ths inches thick on the other.

The use of arch bricks, moulded for that peculiar purpose, on the principle now under discussion, has nowhere come under my own personal observation; but I am informed that such bricks were made about 28 years ago, in the Royal Engineer Department at the Portsmouth Station, and used for thorough bonding courses, at certain intervals, in the construction of the arches of Tipner Powder Magazine, and I have also heard that the same expedient has been resorted to in the arches of tunnels in this country.*

The subject of Great Arches continued. Description of the Arches used in the Casemates at the Citadel of Quebec, and at Dover.

Having now given a general notion of the kinds of arches most suitable for casemates, I shall conclude this part of my subject by describing, in detail, the peculiar bond used in some of the casemates executed at Quebec and at Dover.

---

* Not in the famous Rotherithe Tunnel, now under execution by the ingenuous Mr. Brunel. There common bricks are used, all laid in concentric four-inch brick arches.

First. The arches of the casemates of the Citadel at Cape Diamond, Quebec, of which a section is shown in *Fig.* 176, are segments of 18 feet span and seven and a-half feet rise, having piers of four and a-half feet in thickness.

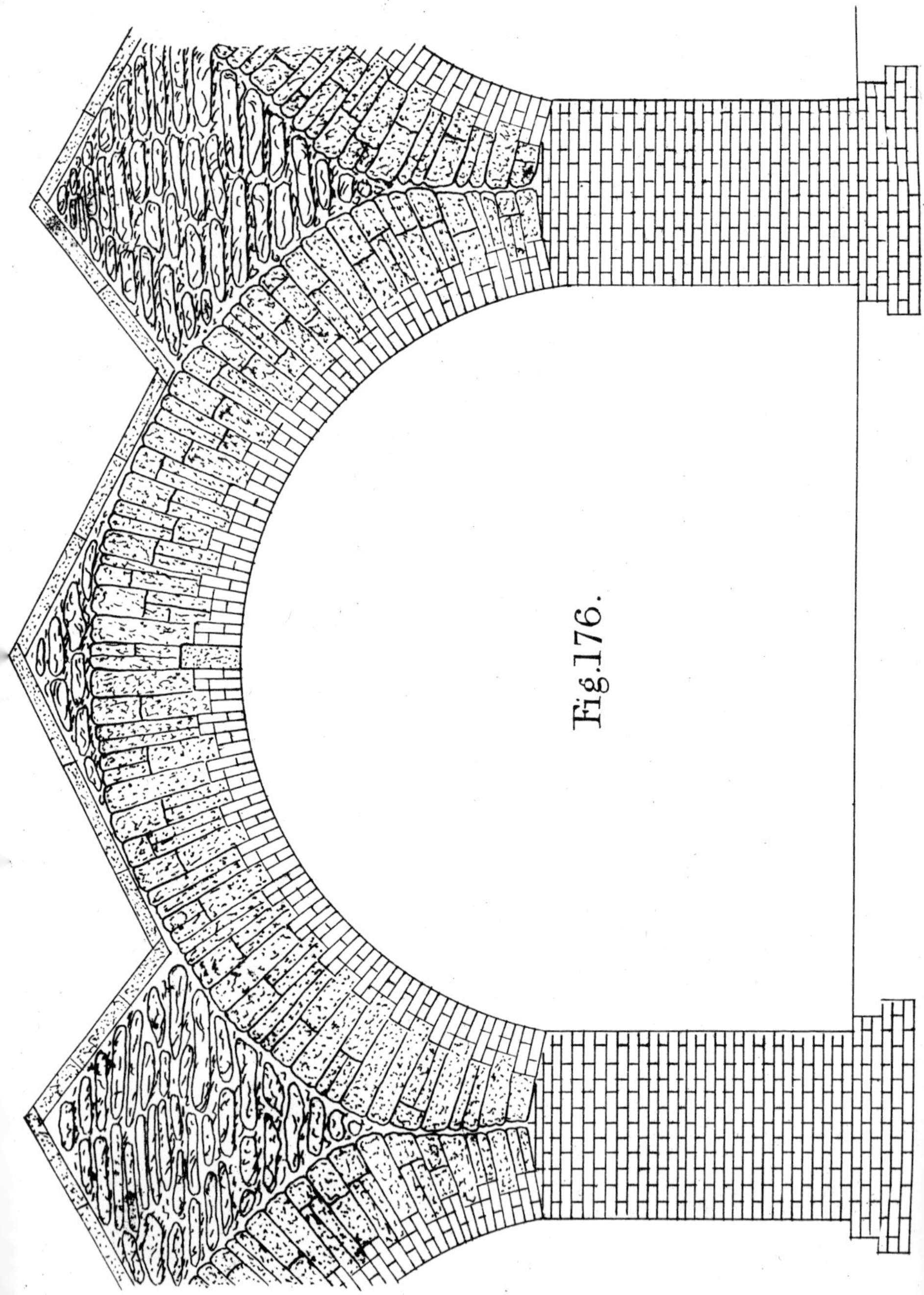

Fig. 176.

These arches were built with bricks surmounted by large stones from a quarry at Cape Rouge, in the neighbourhood, of the quality commonly used for paving stones. The bricks were laid partly in pairs of courses of one brick in thickness, and partly in pairs of courses of one and a-half brick in thickness, alternating regularly with each other from one side of the intrados of the arch to the other, and in both cases headers and stretchers were used.

Thus, a toothing was formed in the brickwork, into which the stones forming the upper part of the arch were bonded; and these were arranged so as to break bond with each other on the same principle as the bricks.

The total thickness of the arch thus formed was three feet six inches.

The ridge over each arch was formed of rubble work, and in the intervals between, smaller ridges were formed in the same manner.

The gutters in the valleys between the several ridges were of cut stone.

The whole of the masonry roof thus formed was covered with large flag stones, laid with cement of a singular description, composed of tar, whitening, and sand.* These particulars are explained in the figure.

Fig. 177 is a section of one of the arches used in the casemates of the New works at Dover. The curve is a parabola, having a span of 18 feet, and rise of nine feet. The whole of the arches are of brickwork as well as the piers. The latter are five and a-half bricks, or about four feet one and a-half in thickness.

At the vertex of the curve, or crown of the great arch, three successive nine-inch arches are used to make up the whole thickness. Lower down two successive arches make up the same, one an 18-inch, the other a nine-inch arch, whilst at the extreme haunches thorough bond is used.

At the crown of the great arch, and in some other parts on each side, thorough bonding courses, of three bricks in width only, are also used, as may be seen in the figure.

In these arches, much pains has evidently been taken in respect to the bond; but I cannot help thinking that they would have been better without the closers, which will be observed in many parts of this section. Near the crown of each arch, where the joints of the parabola are necessarily more open than in other parts, I am informed that pieces of slate, plain tiles, or other hard materials, were introduced instead of filling up the interstices with mortar alone.

Immediately above these arches, blocks of chalk were introduced, over which was a brick coping, forming the ridges and valleys. In the latter the gutters were covered with a quadrant arch one brick thick.

The roof thus formed was coated with Roman Cement, over which there were a great number of small brick channels or drains, leading down to the gutters. Above those little drains boulder stones and shingle were applied, and over all earth and rubbish were laid, to form the rampart and parapet.

Of Groined Arches.

Groined arches are those which, springing in two contrary directions, do, as it were, intersect each other, and meet in such a manner as to abut against, and afford each other mutual support.

---

* Three barrels of tar, two cwt. of whitening, and eight bushels of sand, well mixed, were boiled together and applied hot.

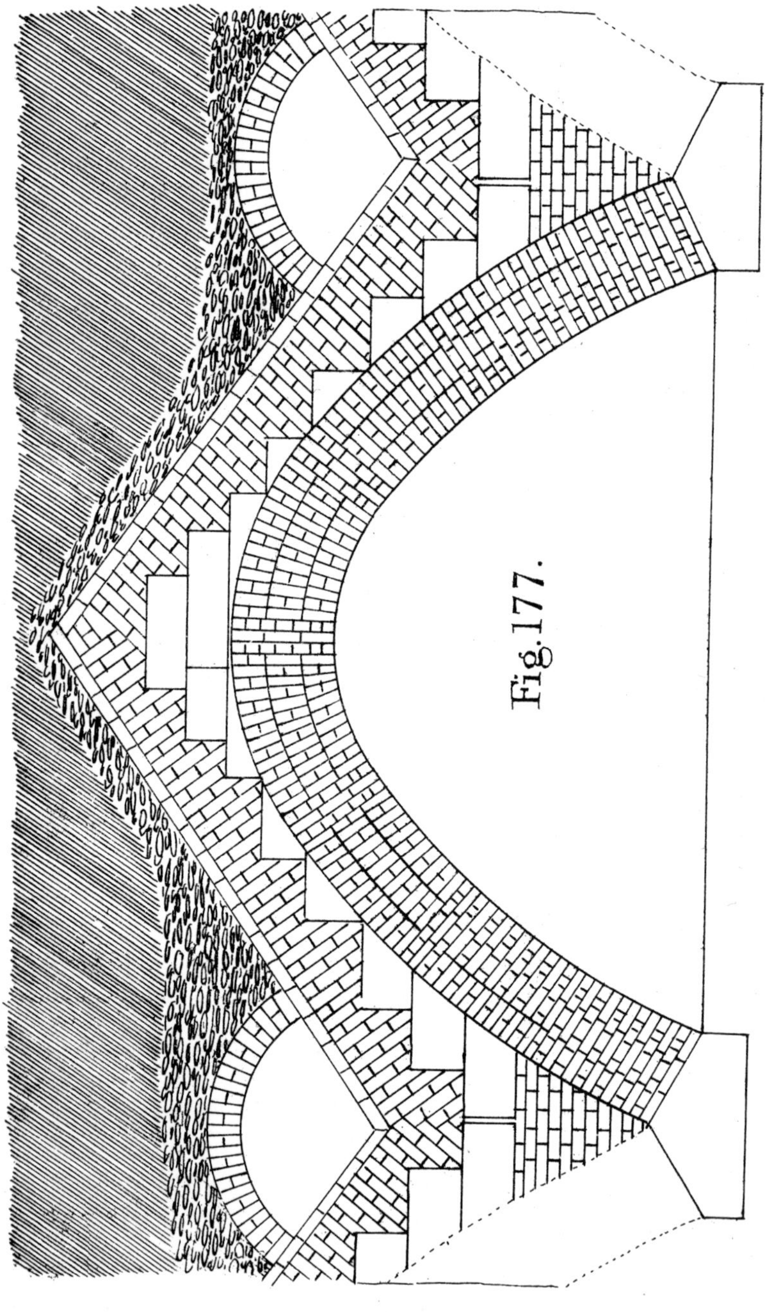

Fig. 177.

In this case square piers, of the nature before described, form the piers, whilst the walls, or pilasters engaged in them, form the abutments of the series of ground arches, and it is most convenient that all the arches should be of equal span and height.

Fig. 178.

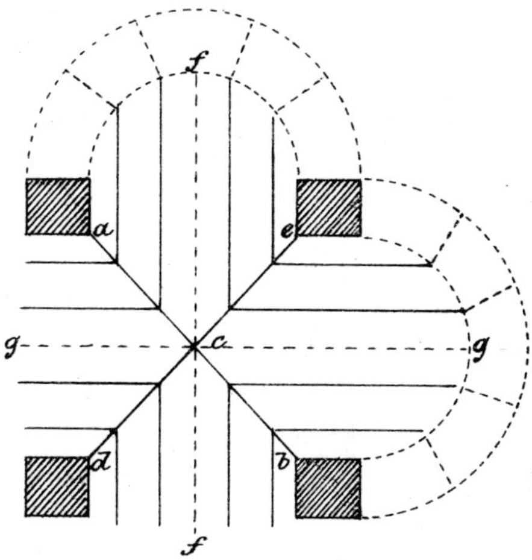

Fig. 178 represents, in plan, four square piers, supporting part of a series of groined arches, which we shall suppose to be all semicircular and equal to each other.

The four piers are shaded dark. Two semicircular arches, at right angles to each other, are shown in section by dotted lines, attached to the plan, of which they form no part; but they afford the means of determining the joints of the voussoirs in a plan of this kind. The dotted lines, *f f* and *g g*, show the highest parts, or the crowns of each of the two arches depending upon those piers, which intersect each other at right angles, in the point *c*, at the crown or summit of the groin.

The lines drawn parallel to *f f* and *g g*, which meet at right angles in the diagonals *a b*, *d e*, show the longitudinal joints of the intrados of the two intersecting arches, which are determined by means of the dotted sections in a very simple manner, as may be seen by inspecting the figure.

In brickwork the joints would be much more numerous, but too many lines would have created confusion.

The lines *a b*, *d e*, which intersect each other diagonally in the point *c*, represent the meeting of the two arches, and there a species of inverted ridges are formed, rising by degrees from the lowest points of each arch *a*, *b*, *d*, and *e*, to the level of the highest point, or crown at *c*.*

If the arches be supposed to be semicircular, these ridges will be regular ellipses, the right lines *a b* and *d e* being their several transverse axes, whilst the semiconjugate of each is equal to the radius of the supposed semicircular arches, by which the groin is generated.

A specimen of this arrangement may be seen at the North extremity of the range of Officers Quarters in the Ordnance Barracks at this Station, where the whole of the street in front of the mess room is supported by a series of groined semicircular arches, of which there are three ranges in width. Two rows of square piers, each three and a-half bricks, or two feet seven and a-half inches square, and placed at intervals of about nine feet apart,† serve as piers to those arches, whilst brick pilasters of the same width and projecting 18 inches from the area walls to which they

---

* By workmen, the ridges are technically termed the points of a groin.
† Some of them are two or three inches less than nine feet.

are attached from the abutments. These walls are four and a-half bricks, or three feet four and a-half inches in thickness, so that the total abutment is four feet 10 and a-half inches opposite to the spring of the transverse arches; but these abutments are strengthened by cross walls and lower arches, extending across the areas in certain parts, which it would be superfluous to describe.

The height from the floor of those vaulted apartments to the crown of each groin is 15 feet, and below the spring of each arch there are two projecting courses of brick at the top of the pier, forming what is technically called an impost.

These groined arches are one and a-half brick in thickness, above which there are about two feet of rubbish and pavement.

Fig. 179 is the plan, and *Fig* 180 is the transverse section, of a part of the vaulted apartments which have been described.

In those figures feet and inches are marked like minutes and seconds of a degree, which is a method commonly adopted by architects, as it admits of those dimensions being written more clearly in drawings on a small scale.

Groined arches afford much greater space between the exterior walls of a building than common arches, and, therefore, the former are used in preference, whenever the saving of room is an object.

Fig. 179.

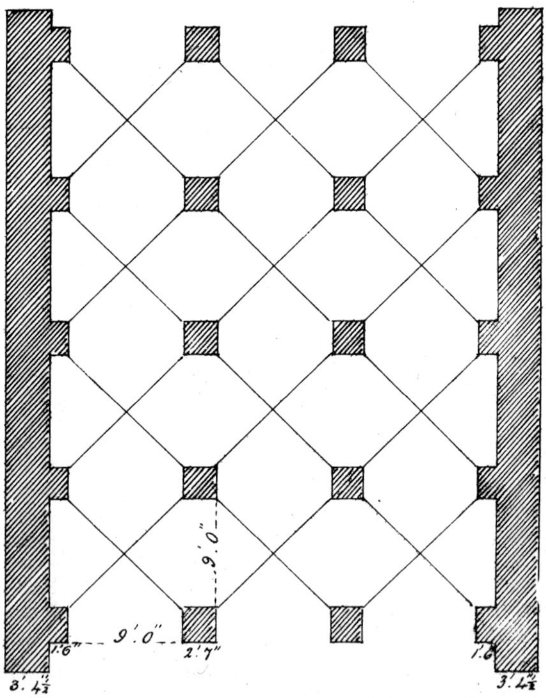

Fig. 180.

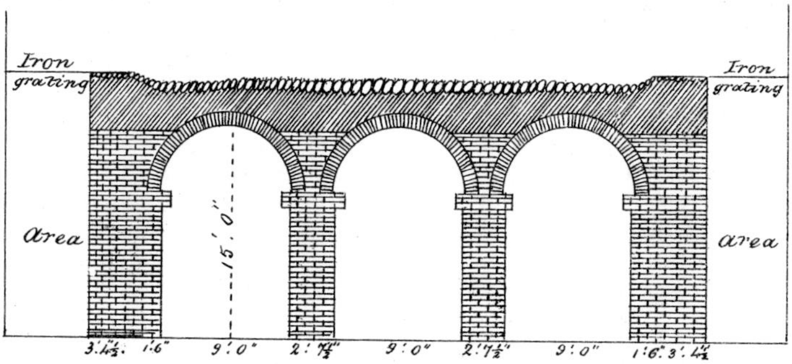

In the plan of the series of groined arches, given in *Fig.* 179, the intersecting diagonal lines represent the ridges of the groins. If the groins were ribbed—that is to say, if the spaces from pier to pier were covered by concentric arches of rather smaller radius, making narrow breaks resembling ribs, which sometimes occur in ornamental groins, as in ancient cathedrals, &c., then those ribs also would be shown in plan, by right lines, as represented in *Fig.* 181.

Fig. 181.

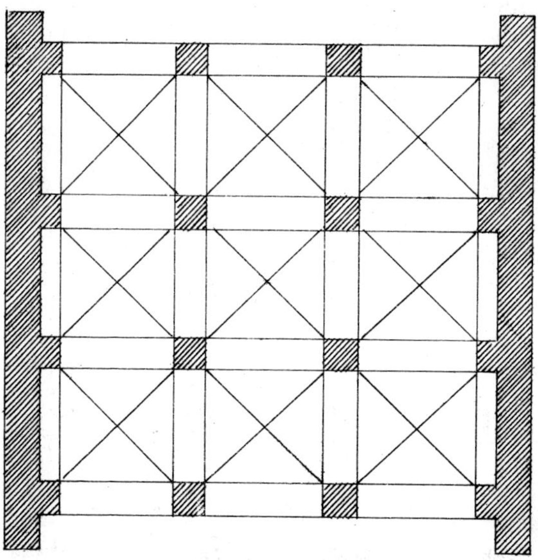

Another example of groined arches, very neatly executed, may be seen in the basement story of the North Gunshed.

They are segment arches, having a span af seven feet six inches, and a rise of two feet three inches, and spring from oblong rectangular brick piers, only 18 inches thick but seven feet six inches long in plan, and seven feet high from the floor to the spring of the arches.*

These vaults were originally intended as a store for forage, and are only accessible at the back of the Gunshed, where the arches are shut by batten doors, represented by dotted lines in the plan. On the other side they terminate on a retaining wall, filled up with earth and rubbish to the level of the floor above, and on that side there are pilasters of smaller dimensions than the piers. (See *Fig.* 182).

Fig. 182.

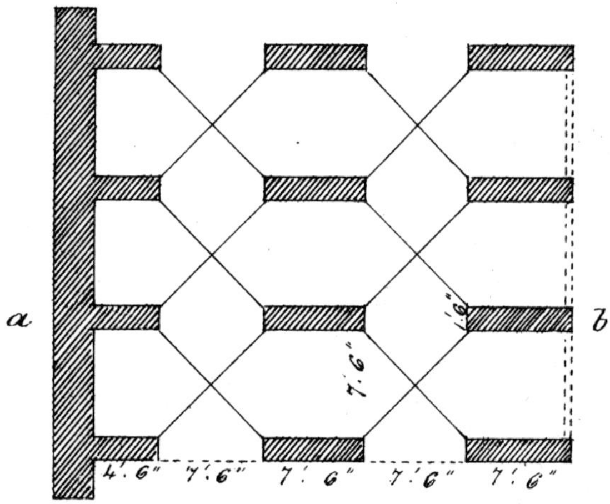

It will be understood that the oblong piers increase the strength of resistance to the thrust of the groined arches in one direction, namely, in the transverse direction *a b*. In the contrary direction, or in the longitudinal direction of the building, the thrust of the series of groined arches is provided for by strong pilasters, abutting against the end walls of the building, from which they derive aid.

The above are the simplest kinds of groins in use, upon which it has been remarked that the ridges being very acute, as they meet at right angles, the bricks are necessarily cut away in a manner that weakens the work, and, accordingly, octagonal piers, instead of square or rectangular, have been recommended, which, by multiplying the number of intersecting arches, would take off the sharpness of the ridges, as represented in *Fig.* 183.

---

* The basement story of the Officers Mess room in these barracks is also composed of a series of groined arches, each having a rise of six and a-half feet, but of unequal spans. The main arch is 18 feet wide, and springs from pilasters two and a-half bricks or 22 inches thick, and projecting four feet on each side from the walls of the building. As this arch runs longitudinally the total width of the mess kitchen between the side walls is 26 feet. Those walls are two and a-half bricks thick. A tiebeam, six inches in depth and nine inches wide, extends at the spring of the arch between those pilasters.

The transverse arches, springing in a contrary direction from the same pilasters and groined into the main arch, have each a span of 10 and a-half feet only, with a rise of six and a-half feet, as before mentioned. The height from the floor to the crown of those arches is 18 feet. The main arch in these groins is a segment of a circle, the others are parabolas.

Fig. 183.

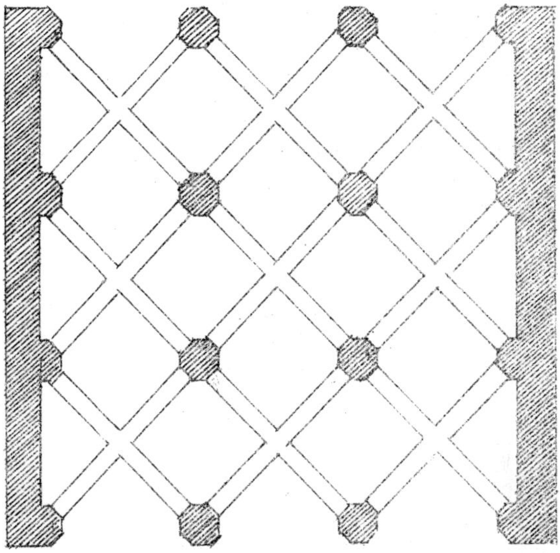

But, considering all circumstances, I should not recommend this arrangement, as it would be infinitely more troublesome and expensive than the common method of groining, which is strong enough.

The subject of groins might be greatly enlarged upon, for they admit of almost infinite variety, as it is evident that two arches of the same height but of unequal span, or of the same span but of unequal height, or two arches varying in both those particulars, or, in fine, two arches altogether dissimilar and unequal, may intersect each other, as also, that more arches than two may intersect each other, all of which will form groins of a more intricate nature than those which have been described, but none of which can be of much use in the Royal Engineer Department.

I shall therefore conclude this article by noticing only one more particular respecting groins, which frequently occurs in practice. I allude to those groins which are formed in a series of common arches, parallel and similar to each other by arched doors of communication left in the pier walls.

This circumstance usually occurs in a range of casemates, where the several arched doorways of communication are always groined into the great arches, unless the pier walls of those casemates be of very extraordinay height.

For example, if we suppose the arched doorway in the pier of a casemate to be four feet wide, and six feet high at the sides, with a semicircular arch, increasing the height of the doorway at the crown to eight feet, and if the great arch of the casemate itself have a span of 18 feet, and a rise of nine feet, and spring from piers of six feet in height, it will be evident that the whole of the arch of the doorway will cut into the lower part of the great arch, and thus groins will be formed, the ridges of which will not extend across the whole casemate, but will advance partially into it, assuming the appearance of a curve.

N

Fig. 184 represents this circumstance in plan, showing in what man-

Fig. 184.

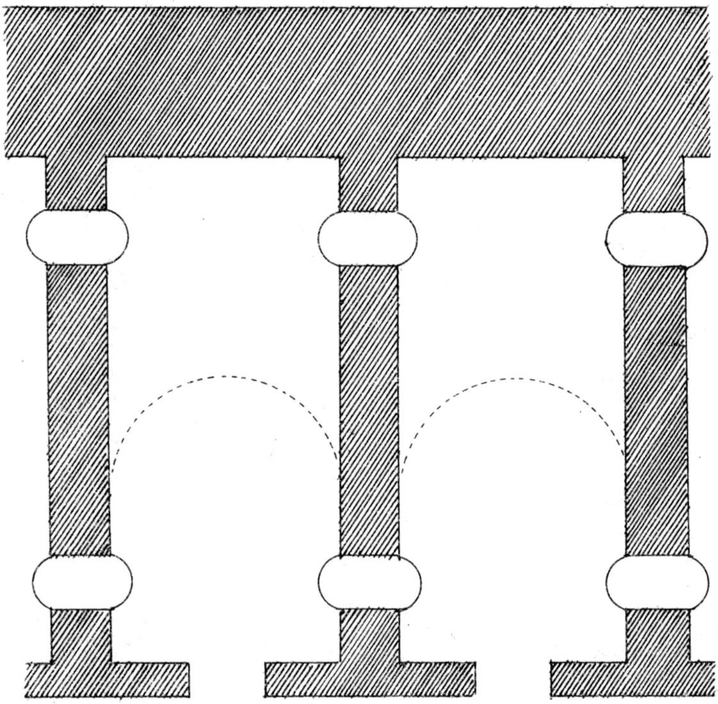

ner the small doorways of communication would become groined into the great arches of adjacent casemates. The small curves are the ridges formed by the intersection of the smaller arches with the greater. The arch of the casemates itself is represented by a dotted curve between the pier walls.

This method of representing a common arch in a plan* is preferable to that which is commonly, but improperly, used in this country, of representing all arches alike, by intersecting diagonals, whether groined or not.

Groins formed by the intersection of small arches with larger ones, in the manner that has been described, are technically termed Welsh groins, or under pitch groins.

Of the Bond of Groined Brick Arches, with Preliminary Remarks on the Centering of Arches.

In the execution of all arches, a wooden mould called a center or centering is used, which enables the stones or bricks to be laid according to the proper curve, at the same time that it supports the arch until it is

* The same is used by Mr. Rondelet in his "Art de Batir."

finished and consolidated by the setting of the mortar, after which this woodwork is withdrawn.

The upper part of the centering, composed of planks in small arches, or of timbers in large ones, must be precisely of the same form as the intrados of the proposed arch. Those planks or timbers rest on frames placed at certain intervals, and supported by upright stanchions or posts, or in large arches by an assemblage of posts or piles braced together.

In the common small arches of buildings, light centers, made altogether of pieces of plank nailed together, are used, which are supported by small light stanchions.

In the arches of casemates, and still more in those of greater span, the construction of the centering is an interesting problem of carpentry, often involving the use of greater masses of timber than in the larger roofs, and, therefore, I shall abstain from treating of it at present.

In groined arches, it is usual to make a centering complete for the principal arches, called the body arches, as if they were common arches, not intersected by any others.

Then centerings are made for the cross arches, which are cut off at those points where they meet the continued centering of the body arch, and thus hollows are produced of the exact form required for the inverted ridges or points of the groin.

Fig. 185.

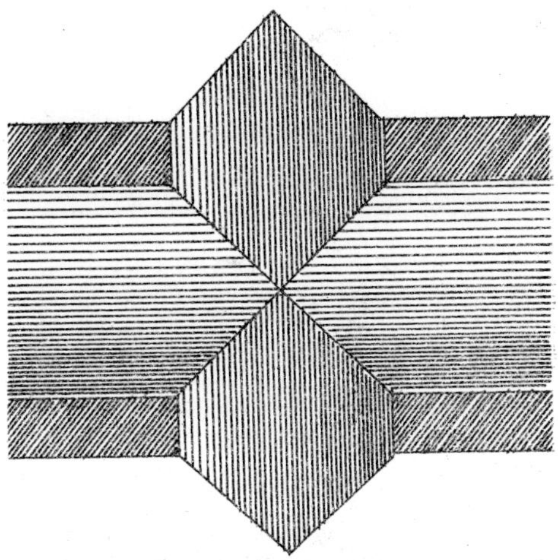

Fig. 185 represents the top of the centering for groined arches, proposed to be formed by the intersection of two equal and similar curves, springing from oblong rectangular piers.

In building such groins, the body or central part of each of the intersecting arches is commenced in the usual manner and the courses all round are made of the same height. At the groin or intersection the bricks are cut so as to suit the form of the hollow marked by the centering, but instead of cutting them diagonally, and thereby forming the joint, called by carpenters a mitre, which would be exceedingly weak, they are laid as stretchers and headers, alternately overlapping each other at all the angles, and the ends of those bricks which show as headers in alternate courses are usually cut in three different ways.

1st. They are cut to the proper angle of the groin, so that whilst one side of them rests on the woodwork of one centre, without cutting, the end of them is made to fit the other centre, and enter closely into the hollow between.

2ndly. In the lower part of the groin, until within about 45° of the crown of the arch, the back or under side of the heading brick is cut away for about half of its original length, in order that the next brick, which meets it at right angles, may be notched into it. This notch is usually in the form of trapezoid or triangle, and the brick fitting into it is usually a closer, which necessarily assumes the form of a trapezoid.

3rdly. As all the arch bricks on both sides are necessarily inclined at top, so that the natural tendency of the courses is to meet each other, like two slopes of a glacis, the upper part of the end of each brick, which appears as a header on one side of a groin, is cut off so as to agree with the slope or level of that particular side until it meets the other slope in the direction of the ridge.

Towards the summit or crown of a groin less cutting is necessary, and there the intersecting arches nearly agree, so that headers are commonly used in preference to the alternation of headers and stretchers, which applies to the lower parts of the same arches.

Fig. 186 represents in elevation one pier of a groin, formed by the

Fig. 186.

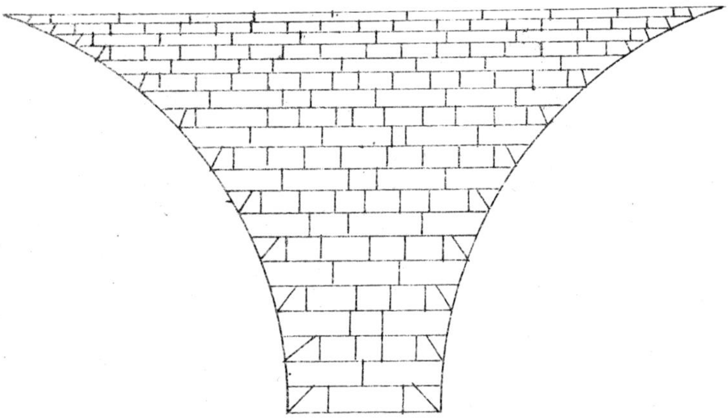

intersection of two equal semicircular arches. The side of one arch is also shown in elevation, at the extremities of which are the ridges formed by its intersection with the two collateral arches.

At the lower parts of the groin it will be seen that the ends of the headers of alternate courses, after having been notched, as before stated, are all in the form of a triangle, and they have trapizoidal closers next to them.

In the lower part of the arch, the header (1) has its lower side cut away in a trapezoidal form, to fit the closer (2), so that the whole brick (1) would appear as in *Fig.* 186, if the remaining part of it could be seen; but the notched part, which is shaded in the annexed figure, is concealed in the former figure by the closer (2).

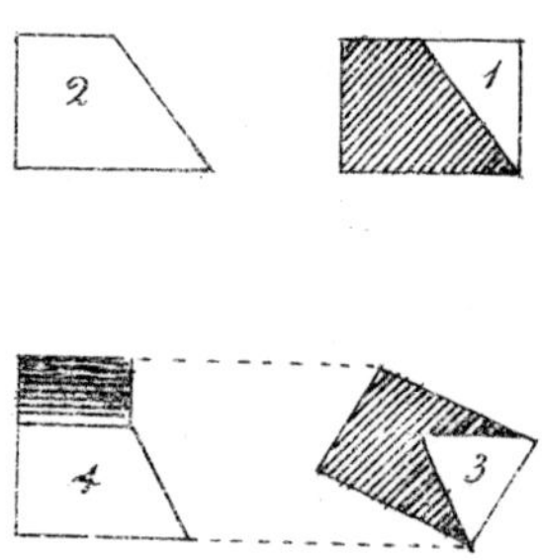

At a higher level, the header

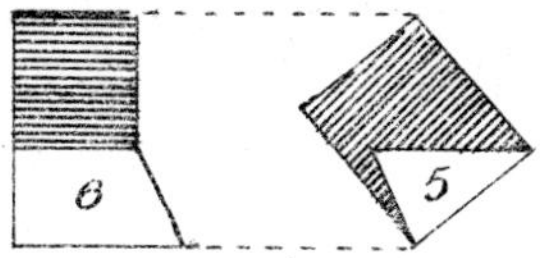

(3) has less of its lower part cut off, as in *Fig.* 186, whilst the top of the closer (4) is cut according to the slope of the same course in which that header appears.

At a still higher level, the under part of the header (5) is very little cut, because it then begins nearly to agree with the angle, at which the voussoirs of the collateral arch are laid. In this case also the upper part of the closer (6) is sloped off.

The system of cutting the ends of headers into a triangular form might be dispensed with at the bottom or spring of the groin, and seems to have been adopted by bricklayers merely to give a more uniform appearance to the whole of this part of the work.

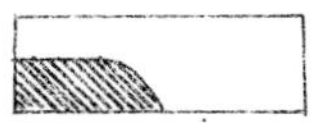

The closers add little to the strength, for it is impossible to cut the notches out of the heading bricks in a rectangular form, as may be done in woodwork; and, therefore, the sides, instead of forming a right angle, are roughly rounded at the re-entering angle, as in the annexed figure. Hence, the largest end of the trapezoidal closer in groined arches is downwards, whilst its point is towards the extrados, contrary to the principle of the arch. These closers, being small fragments of bricks, are therefore kept in their places by the mortar and by the pressure, as they are too light to drop out, in consequence of their imperfect form.

In Welsh groins the courses of intersecting arches do not come on the same level, excepting near to the spring of the arch; and, therefore, instead of the triangular headers in alternate courses all the upper bricks of the smaller arch, which coincide with the groin, appear as headers in the elevation of one side of the greater one, and they show two or more oblique sides, as in *Fig.* 187.

Fig. 187.

Groins are much more simple in stone than in brickwork, for in stone the lines may easily be found by geometry, and the stones cut by moulds; and at the groins of arches, stones large enongh to embrace both sides are commonly used without having any joint at or near to the ridge, and the ends of those stones are cut at right angles, the sides being sloped and the bottom curved. *Fig.* 188 shows the primary form of a stone for a groin about to be cut. The dotted diagonal is to agree on the lower side with the ridge of the groin.

Fig. 188.

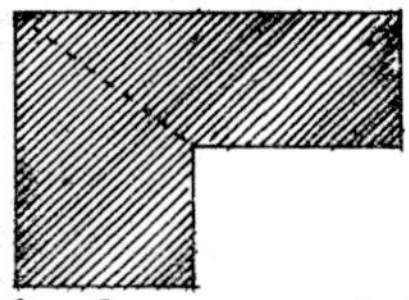

To explain brick groins by means of models, the best way is to have a model of a groin, showing all the joints, and not to take to pieces, but having the bricks at the ridge numbered. Then let a wooden centering, proper for the same groin, be prepared with separate bricks, also numbered, and let them be all notched and cut in the proper manner, so as to form a similar groin to the former. By putting together and taking to pieces this second groin, and comparing it with the other, it will elucidate this subject as fully as models are capable of doing.

I shall conclude this article by observing that it will now be evident that the execution of groined brick arches, such as have been described, is sufficiently complex, without adding to the difficulty and expense of it by adopting hexagonal groins. Indeed, groined arches so seldom occur that many good bricklayers have never been employed at all at this kind of work, but the student of practical architecture will be able to find specimens of them in every great town.

Of Coved Arches.

Fig. 189.

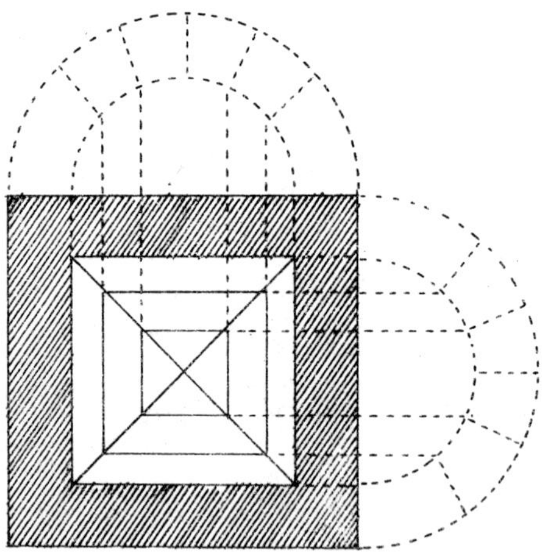

Coved arches, like groined arches, are formed by the intersection of two or more arches in contrary directions, but with this difference, that coved arches spring from opposite walls, whereas groined arches spring from opposite pillars or piers. Hence, the joints of the voussoirs which always meet each other diagonally, are arranged in a contrary direction in those two kinds of intersecting arches, in consequence of which, the coved arches form hollows, projecting upwards at their intersection; whilst groined arches form ridges, projecting downwards, as before explained.

Fig. 189 represents the plan of a coved arch, or compound arch, formed by the intersection of two semicircular arches springing from two pairs of opposite walls, at right angles to each other. Dotted sections, which form no part of the plan, are introduced, in order to explain the method of finding the longitudinal joints of the voussoirs.

Coved arches are not commonly used in this country, but ceilings are frequently formed in imitation of them. In this case, however, it is more usual to have the center part flat, and to confine the curves, whereby the covings are formed to the sides of the room only. Of this arrangement there is an example in the Officers Mess-room in the Ordnance Barracks at this station.

Of Plain Arches, as applied to Common Brick Buildings.

The plain or rough arches used in common buildings are generally concealed from view.

They are applied under the following circumstances :—

1st. To cover the aperture of an external door, or of a window, in addition to the ornamental arch, which is generally a mere shell, as was before explained.

When the arch of an external door, or of a window, is a semicircle or segment, the rough internal arch follows the same form, and no woodwork is used in aid of such an arch.

But when the ornamental arch of an external door, or of a window, is a straight or flat arch, pieces of timber, called lintels, are laid over the remainder of the aperture in rear of the ornamental arch; and in the internal doors of a building these lintels cover the whole space with a bearing of about one foot or 18 inches at each end.

In this case, a rough arch is also usually thrown over the aperture, above the said lintels, in order to relieve them of their weight of the superincumbent brickwork.

This arrangement, which is concealed from view by the papering [in the officers apartments, may be seen in the soldiers quarters in the Ordnance Barracks at this station. *Fig.* 190 represents the mode in

Fig. 190.

which such rough arches ought to be formed in aid of a lintel.* The intermediate space between the bottom of the arch and the woodwork, is filled with bricks and bats cut to the curve, and laid in mortar.

2dly. Rough arches are used for fireplaces.

In this case, a bar of iron from $\frac{1}{2}$ to $\frac{3}{4}$ inch thick, and usually not less than four inches wide, is prepared before hand, suited to the form of the arch, and extending across the chimney breast from one side to the other, so that it comprehends the width not only of the proposed arch, but also of the piers on each side, or of part of them.

The ends of this bar are split, and turned back in contrary directions, like the capital letter T. Upon this bar, technically called a *turning bar*, the arch is built, which bar, however, is not removed like a common center, but remains permanently in the same place, in aid of the arch.

Fig. 191

Fig. 191 represents a chimney with its rough arch, turning bar, &c., on the above principle, being the usual mode of finishing the brickwork of fireplaces in this country; but these details are afterwards concealed from view, not merely by the plastering and papering of rooms, but usually by an ornamental chimney piece of marble or stone, added in front of the brickwork. Wood, also, is sometimes used for chimney pieces, but with less propriety.

3rdly. Rough arches are used beneath the hearths of fire-places.

It will afterwards be seen, in treating of floors, that no woodwork is allowed below a fireplace; and that joists, extending in that direction, are framed into a transverse piece of timber, called a *trimmer*, which is usually two feet distant from the chimney breast in buildings of importance. Between this trimmer and the bottom of the fireplace a four-inch segment brick-arch, rather less than a quadrant, is constructed, which rests on one side against the upper part of the trimmer, and this is the vertex of the curve. On the other side, which is on the level of the bottom of the trimmer, it springs with a skewback from the wall. The spandrels are usually filled in with brickwork.

Fig. 192 represents this arrangement. The trimmer joist and wall, to which the fire place is attached, and the small arch abutting on these two are shown in section. The chimney jamb, and part of a common joist, supposed to be framed into the trimmer, appear in elevation.

Above the small arch, a hearth stone is supposed to be laid on a level with the flooring boards which are to cover the joists. This kind of arch is called a *trimmer arch*.

4thly. Rough arches are used beneath the entrances of all buildings having an area round the basement story. In this case a leaning arch

* This mode is not usually followed in practice. The relieving arch is often made too narrow, so that it rests on the ends of the lintel. Hence, if the wood decayed the arch would lose its abutments.

is used which on one side abuts against the area wall; on the other it leans against the wall of the building. Above this arch and extending partly across the area, and on the same level with the pavement, or sometimes one step above it in a city, large flag-stones are laid beyond which a flight of two or three steps at least usually leads from that level to the entrance door of the building. *Fig.* 160 in Page 76, exemplifies this arrangement according to the system adopted in the Ordnance Barracks at this station, in which the rough arch, which is one brick in thickness, is a circular segment rather less than a quadrant.

Fig. 192.

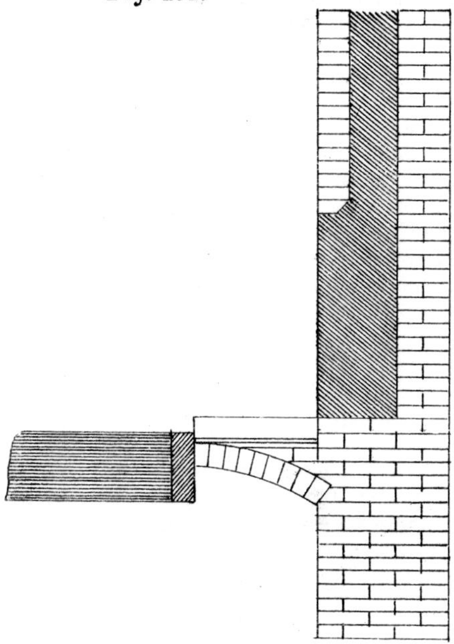

These arches are technically termed *bridgeways*.

The above are the purposes, to which rough arches are applied in almost every brick building without exception, that is, in all buildings of importance.

The following circumstances only occur occasionally.

Fig. 193.

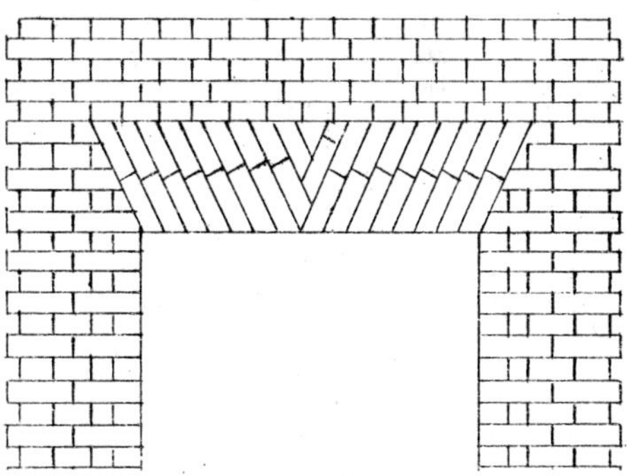

5thly. When the whole or part of a front is to be covered with cement in that case straight arches are frequently built rough in a peculiar manner represented in *Fig.* 193. At the key or crown of the arch, fragments of bricks are laid cut to the form of the letter V. All the others are whole bricks parallel sided, but diverging inwards in contrary directions and cut at top and bottom only to suit the splay.

6thly. Inverted arches are generally used in bad or doubtful foundations as was before mentioned, and sometimes even in good foundations immediately under the apertures of the doors and windows.

Figs. 194 and 195 represent in elevation arches of this description, which may either be segments or semicircles. The parabola has been recommended by some writers, but I prefer the former figures, as being simpler.

Fig. 194.

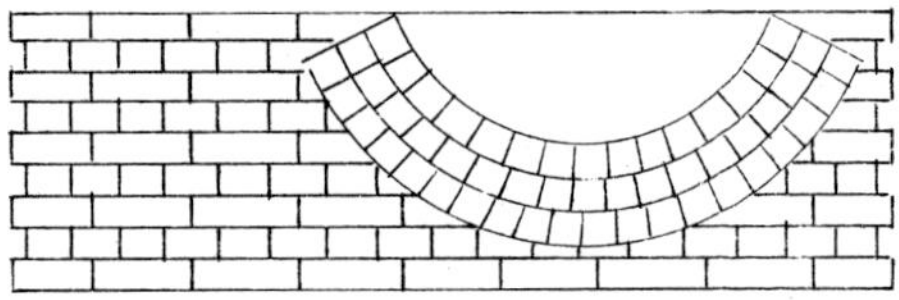

7thly. In storehouses, stables, &c., when much light is not required, semicircular openings are sometimes made, arched at top, with a sill at bottom, as shown in *Fig.* 196.

Fig. 195.

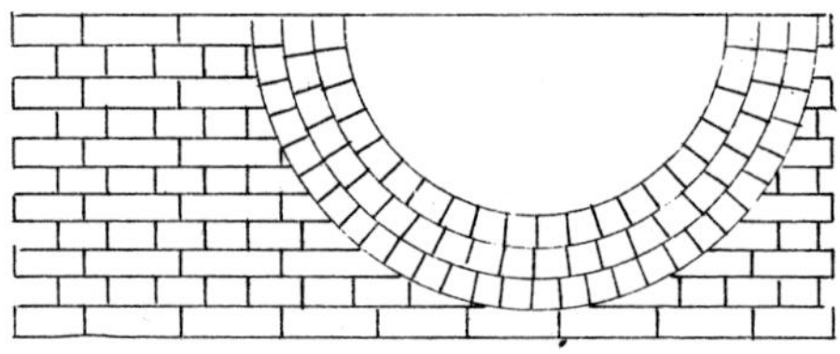

8thly. In the basement story of buildings, small circular apertures technically called bull's eyes are sometimes made, which are arched all round so as to possess the properties both of the common, and of the inverted arch; and which seldom exceed 3 feet in diameter, as shown in *Fig.* 197.

Fig. 196.

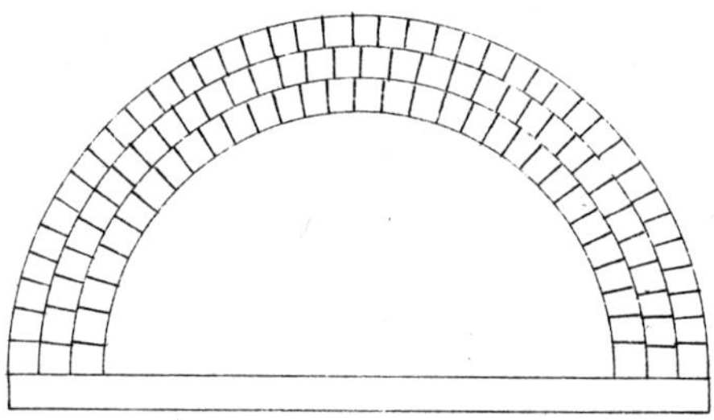

Fig. 197.

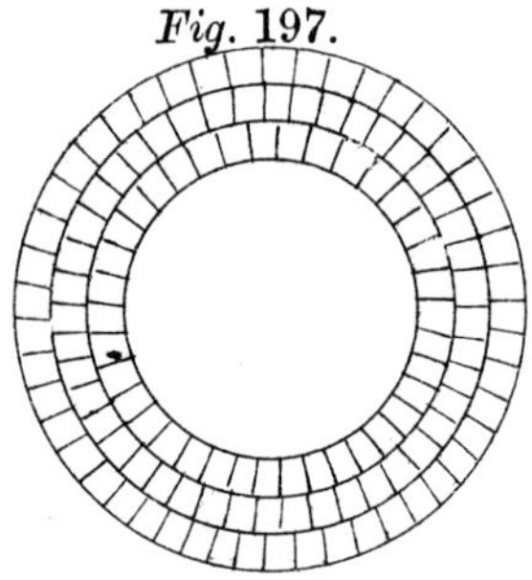

Specimens of this arrangement may be seen in the New British Museum.

9thly. Arches are often used to support the flues of chimneys whenever it is required to change the direction of those flues, in order that those belonging to distant fireplaces may be brought up in the same shaft at the top of a house. Such arches will afterwards be treated of in the article chimney.

10thly. Arches for coal holes, &c. are often made opposite to the basement stories of houses beyond the area, so as to extend below the pavement or part of the street in cities. These arches are commonly semicircular, one brick thick, not exceeding nine feet in span, and having pier walls of one and a-half or two bricks in thickness, that is, when there is a continued range of them, they being supported at the extreme ends by abutments of greater thickness, when necessary.

In plan the pier walls of such arches usually serve to support horizontal arches forming the extreme ends of the same vaults, which are thus enabled to resist the pressure of earth of the street, or external ground.

Fig. 197 and 198 represent the plan and elevation of such arches, the latter as they would appear before the backwalls are added.

Fig. 197.

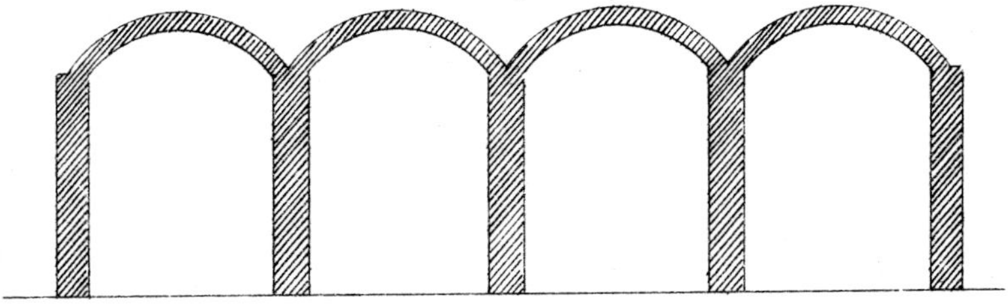

Plan.

Fig. 198.

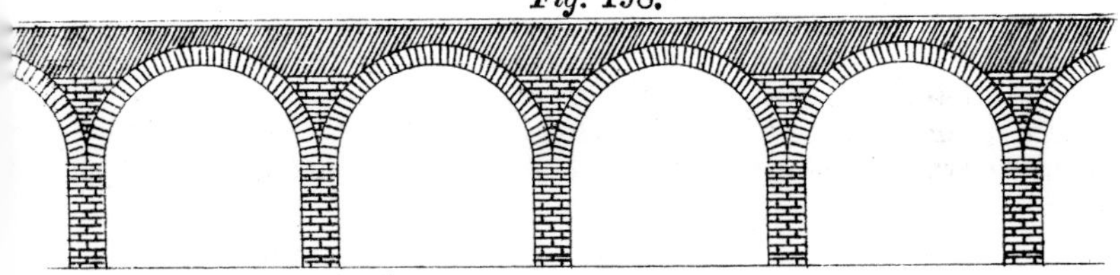

11thly. Arches are used for drains and for all subterraneous excavations of every description such as common sewers, culverts, galleries of countermines, tunnels, &c., &c., but I shall not enlarge upon these at present, my object being to illustrate the arches that are peculiarly applicable to buildings.

General Remarks on the Proportions of Arches.

In considering the proportions of arches, in this and in some of the succeeding articles, it will be my object to state some few principles, which are generally acknowledged, and to lay down some practical rules, on perusing which, I am persuaded, that those whose opinions may differ from mine, will acknowledge, that I have chosen the safe side of the question; for if I err, it will be on the side of strength. I shall commence by treating of the semicircle.

Supposing an arch to be built of this form, with its voussoirs, all of equal depth, so that the intrados and extrados are parallel to each other, it will depend upon the thickness of the arch, or in other words upon the depth of those voussoirs, whether such an arch will stand or not, admitting that in all cases its piers or abutments are sufficient.

For example, if a semicircular arch be built in the manner represented in *Fig.* 199, having voussoirs equal in depth to $\frac{1}{8}$ of the span, or upwards, such an arch will stand, after the centering upon which it was built is removed, provided that the abutments be sufficient.

If on the contrary, a semicircular arch be built, on the same principle, but of less thickness, as for example, with voussoirs only equal in depth to 1-16th part of the span or thereabouts, as shown in *Fig.* 200, this

Fig. 199.

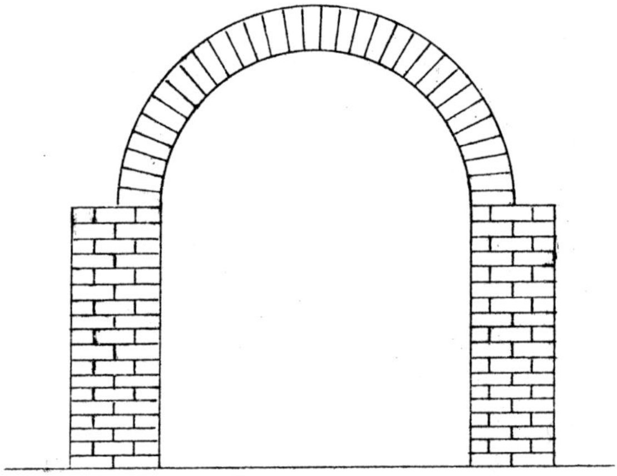

Fig. 200.

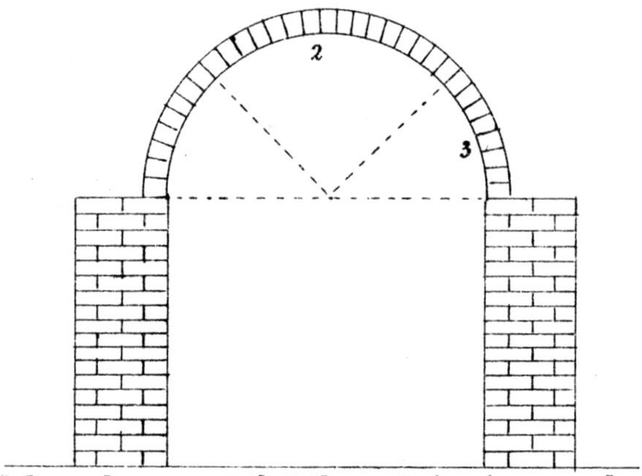

arch will fall down the moment that the centering is removed; for the upper segment (2) which may be an arc of 90°, or more or less according to circumstances will force the two lower segments (1 and 3) outwards, and overset them. In fact, in this case the two lower segments on each each of the haunches of the arch may be considered as abutments, in relation to the upper segment, about the crown of the arch.

Whilst this operation is in progress, the voussoirs in the upper segment, about the crown of the arch open in the first instance at the intrados, and are held together only by their upper edges, whilst the voussoirs about the haunches of the arch, near the meeting of the upper and lower segments, on each side open at the extrados, and are held together only by their inward edges, as shown in *Fig.* 201, after which the whole falls down.

The truth of these observations may be demonstrated by wooden models, for which purpose I would first recommend having a model of

Fig. 201.

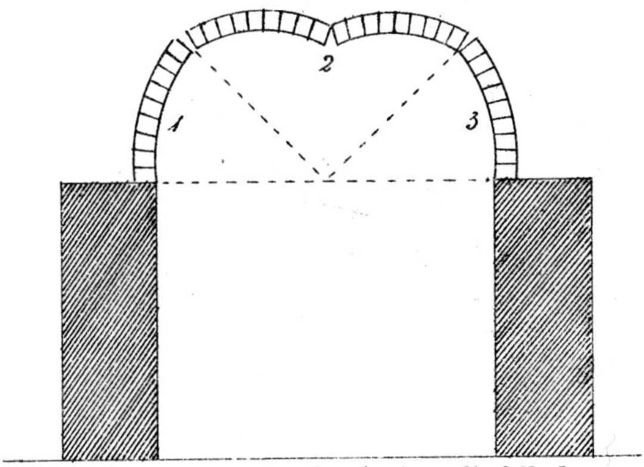

the centering of a semicircular arch made to a supposed span of six feet. If to this centering, you apply wooden models of voussoirs cut to the arch like form, and having a thickness of nine inches or upwards, by the same scale the arch will stand, after removing the centering without loading it at the haunches, but if to the same centering, you apply voussoirs only equal to four inches in depth, it will fall down the moment that the centering is removed.

These models of the centering and voussoirs may be made on the same scale as the models before recommended for studying the bond of brickwork, but the models of bricks themselves, will not answer for experiments upon the voussoirs of arches. Being parallel sided, their imperfect form renders them inapplicable to this purpose because the cement which binds real bricks together cannot be allowed in such experiments.

Models of voussoirs made of pieces of Bath or Portland stone or of chalk, and also fitted to a wooden centre, may be used in like manner, to prove the truth of the above principles and of others that will be laid down in this article.

These results differ widely from the doctrine of certain mathematical writers before alluded to, who have commenced with supposing the voussoirs of arches to be perfectly smooth bodies without friction, and consequently without lateral resistance. This hypothesis, upon which they have condemned the semicircle, and all other arches springing from an horizontal plane, is so contrary to the truth, that in the year 1805, when I tried experiments in Malta, to investigate this subject, I found that square blocks of different sizes of the common freestone of that Island, which nearly resembles bath stone, when laid upon a horizontal plane of the same substance, without mortar, both surfaces being rubbed quite smooth, required each a power of from *two thirds* to *three fourths of its own weight*, to overcome the lateral resistance arising from the mere friction of those two surfaces, that were in contact with each other. But in a book professedly practical, I shall take no further notice of such speculations.

To return to our former subject, it is to be remarked, when a thin semicircular arch, falls down, under the circumstances before supposed as illustrated by *Fig.* 201, that the upper segment about the crown of such an arch, would be in itself strong enough, as an independent arch, and would have no tendency whatever to separate, provided that it either abutted against adequate piers as in *Fig.* 202, or what is the same thing that the lower segments (1 and 3) of the semicircle before supposed, were backed in at the haunches or spandrels, with a sufficient quantity of brickwork or masonry, to act the part of adequate piers, as in *Fig.* 203.

Fig. 202.

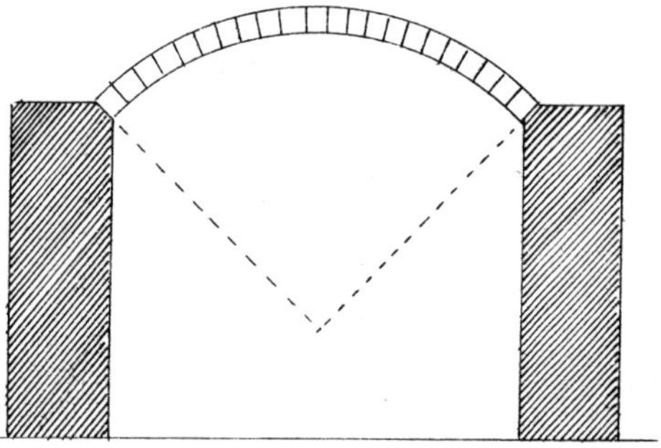

By those authors who have investigated the theory of arches, the extreme points of the upper segment where the tendency to separate takes place are called the points of rupture.

It is evident from what was before said that the position of these points must not only vary in different curves having voussoirs of the same thickness, but in equal and similar curves having voussoirs of different thicknesses.

Fig. 203.

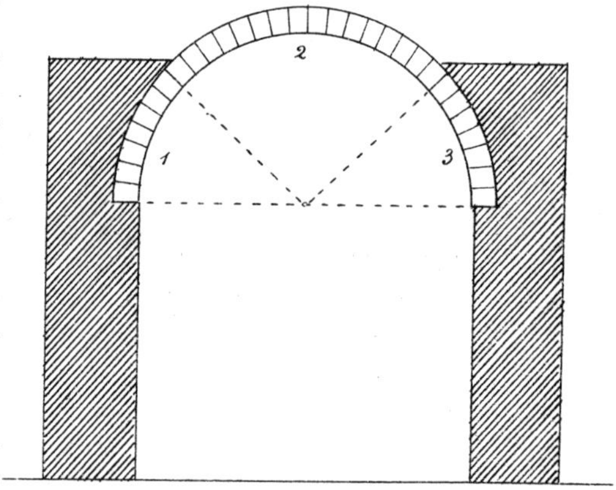

If a stone or brick be laid upon an inclined plane not exceeding a certain angle with the horizon, instead of sliding downwards by the force of gravity as the perfectly smooth voussoirs of the mathematicians ought to do, it will remain at rest, in consequence of the great friction of such materials. This angle is usnally termed *the angle* of *repose*, by writers on the theory of arches. It varies according to the nature and polish of the material used. Mr. Rondelet states that polished marble will slide at an angle of about 15°, but that common building stone, or marble not polished will not slide at a less inclination than about 30°.

Taking into consideration the circumstances in respect to semicircular arches which were before stated and which as I said are capable of demonstration by models, as the haunches of a semicircular arch of uniform thickness are generally in themselves insufficient to resist the outward pressure, or thrust of the upper segment; and as the said thrust as well as the resistance opposed to it, must in some measure depend upon the weight of the several parts acting, or acted upon; it has been judged useful in great arches such as those of bridges, to make the length or depth of the voussoirs gradually decrease on each side from the spring towards the crown of the arch as shown in *Fig.* 204 which arrangement may be observed in the most celebrated modern stone bridges, both in this country and in France.

Fig. 204

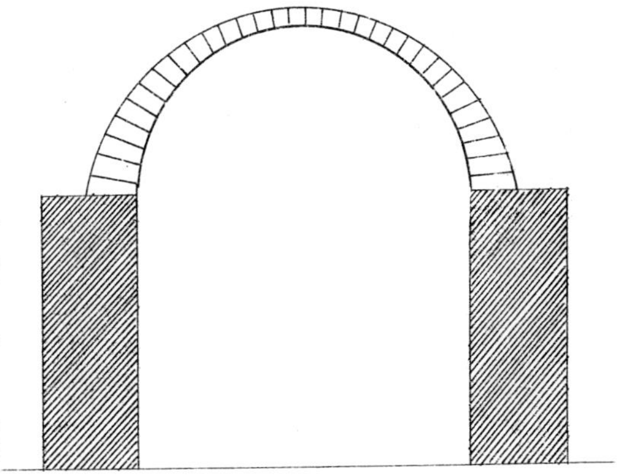

In acting upon this principle it has not however been usual in such bridges to make the extrados of the voussoirs a regular curve, as in the above figure. They have more generally been laid out at top so as to combine in a handsome and strong manner with the horizontal courses of stone above them, which are not cut to the arch like form.

The following properties of arches may also be demonstrated by models.

Fig. 205.

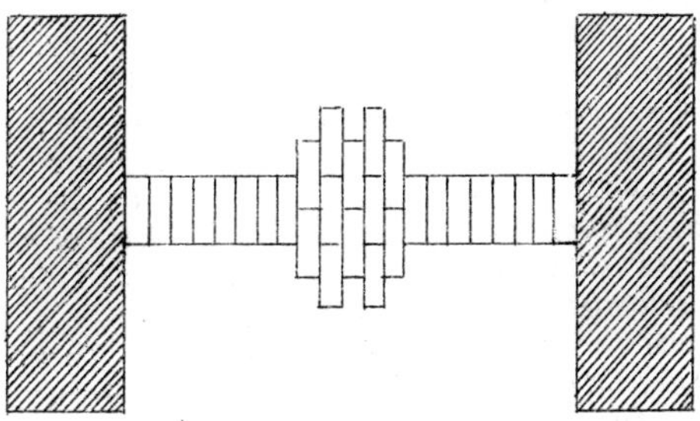

You may construct a segment or semicircle with a set of models of voussoirs, consisting of single ones at the springing of the arch on each side, but of two or three voussoirs bounded into each other about the crown, in the manner represented in plan, in *Fig.* 205.

On removing the centering, the crown of the supposed arch will stand, although the whole of it does not press in a direct line upon its abutments. This experiment shows the utility of good bond in the intrados of an arch, and at the same time, it illustrates in some degree the properties of groined arches, since this arrangement resembles one arch of a groin, without its intersecting arch.

Of the proper proportions of the Piers and Abutments of Semicircular Arches.

A single arch has of course only two piers, which act also as its abutments; and upon the strength, or in other words upon the proportional thickness of those, in reference to the span and nature of the curve, the stability of the arch must necessarily depend.

The intermediate piers of a series of arches, may be made of much less thickness, than is proper for the abutments. This is a principle now generally acknowledged, and indeed there can be no doubt of the truth of it. For it is almost self evident, that if the two abutments or extreme piers (A and D) of a series of arches be capable of support- the extreme half arches, all the other pairs of half arches, by pressing upon the intermediate piers (B and C) in contrary directions, counterbalance each other in such a way, as to destroy the lateral thrust upon those piers. See the following figure.

Fig. 206.

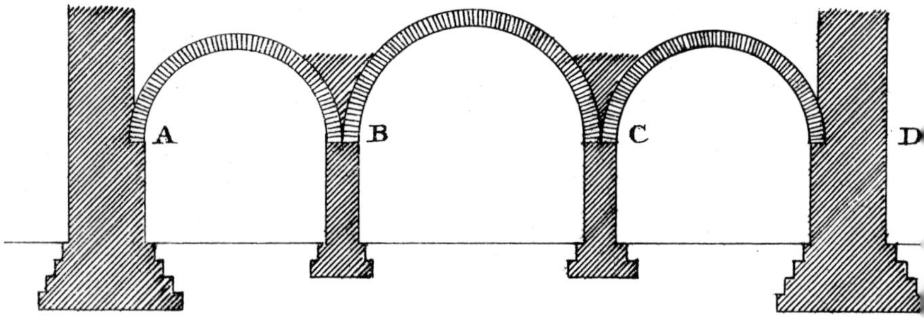

As an example of this arrangement, *Fig.* 206 represents a transverse section through part of the basement story of the east wing of the New British Museum. Three semicircular arches one brick thick, and varying in span from about 11 to 14 feet, are supported by the front and back walls (A and D) of the building as abutments, each of which is six bricks or four feet six inches in thickness, being about one third of the average span of those arches, whilst the intermediate piers (B and C) are only two bricks or 18 inches thick, being only from $\frac{1}{8}$th to $\frac{1}{9}$th of the said average span. Arched doorways of communication about five feet wide are left in the piers, and form with the principal arches, the peculiar species of groins called Welsh groins, explained in a former article. The height of the piers of all those arches is about six feet.

The weight of the upper part of the massy walls of that building must of course have a considerable effect in strengthening the lower parts of the same, which act as the abutments of those arches, but I should not advise any great diminution of the thickness of abutments on this account.

The thrust or outward pressure upon the abutments of arches, is more or less in proportion to the nature of the curve. High arches, for example, whose rise is greater than half the span, press in a more vertical direction upon their abutments, and consequently have less tendency to overset them, than low arches whose rise is less than half their span. The semicircle is the medium between those two. In order to insure its stability, it therefore requires thicker abutments than the higher arch, but will stand with thinner ones than the lower arch of equal span.

For example, in a semicircular arch as before represented in *Fig.* 200 Page 108, the upper segment of it (2) presses entirely downwards and would fall in that direction if not counteracted. Hence the thrust of this part of a semicircular arch is all outwards tending to overset the abutments.

The two lower segments (1 and 3) of the semicircular arch, on the contrary, have little or no outward pressure. From the spring of the arch up to an angle of 30° with the horizon, they will nearly stand without support. Above that height, their tendency is to fall inwards.

Hence, if a segment arch of 90°, a semicircle, and a Gothic arch, composed of two equal segments of 45°, meeting each other at the vertex of the curve, be all constructed of the same span, the semicircle, which is the medium curve between those two, will tend to overset its abutments with greater force than the Gothic arch, and, consequently, will require thicker piers than the high pointed arch; whereas it will act with less thrust upon its piers, than the comparatively low segment of 90°, which segment will therefore require the thickest piers of all.

Fig. 207 represents the high Gothic arch above supposed whose rise exceeds the span by about one-sixth part. *Fig.* 208 is the semicircle, whose rise is half the span. *Fig.*209 is the segment of 120°, whose rise is rather less than one-third of its span; and *Fig.* 210 is the low segment of 90°, whose rise is rather more than one-fifth of its span.

Fig. 207.

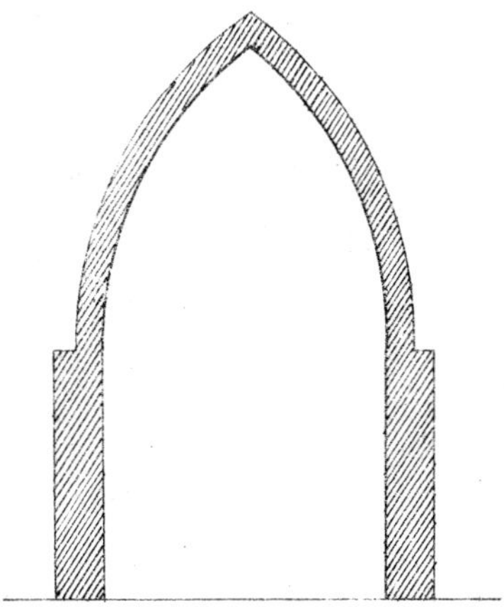

Fig. 208.

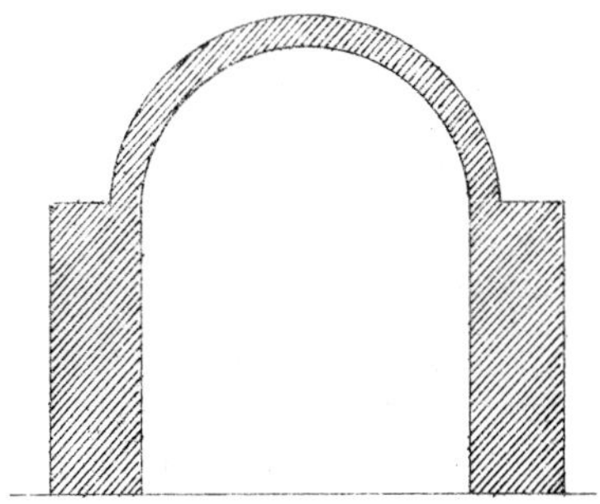

Fig. 209.

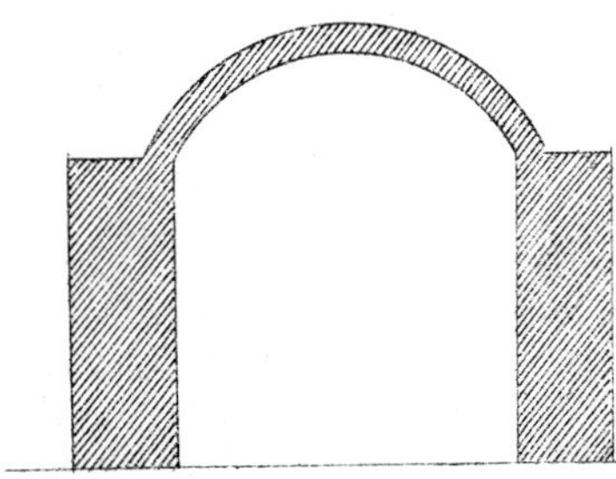

Fig. 210.

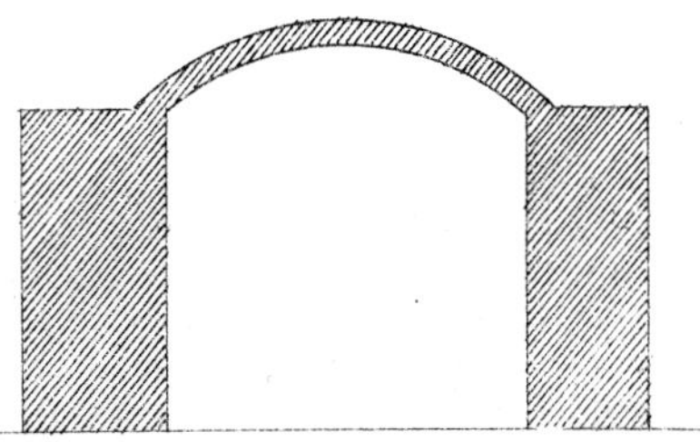

In all those figures the arches are by supposition of equal span, and drawn to the same scale, and the piers increase in thickness according to the diminution of the rise or height of the arch.

In like manner the high semiellipse, whose smaller diameter is horizontal, may have thinner abutments; but the low semiellipse, whose greater diameter is horizontal, requires thicker abutments than a semicircle of equal span.

The high parabola, whose rise is greater than half its span, may have thinner abutments than the parabola commonly used, whose rise is equal to half its span, and both of these may have thinner abutments than a semicircle of equal span. I am not aware of the low parabola, whose rise is less than half its span, ever having been used as an arch, but if such were used it might, in proportion to the diminution of its height, require thicker piers than the semicircle of equal span.

Fig. 211 represents a high parabola, whose rise is equal to three fourths of its span. *Fig.* 212 is the high semiellipse of the same proportions. *Fig.* 213 is the common parabola, whose rise is equal to half its span. *Fig.* 214 is the semicircle of the same proportions. *Fig.* 215 is a low ellipse whose rise is equal to one third of the span. *Fig.* 216 is a still

Fig. 211.

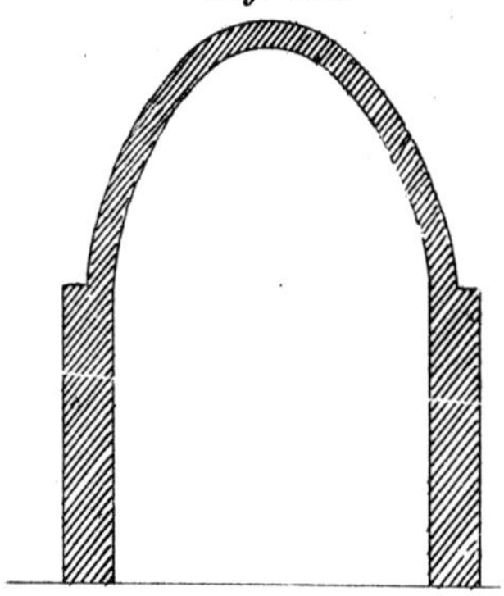

Fig. 212.

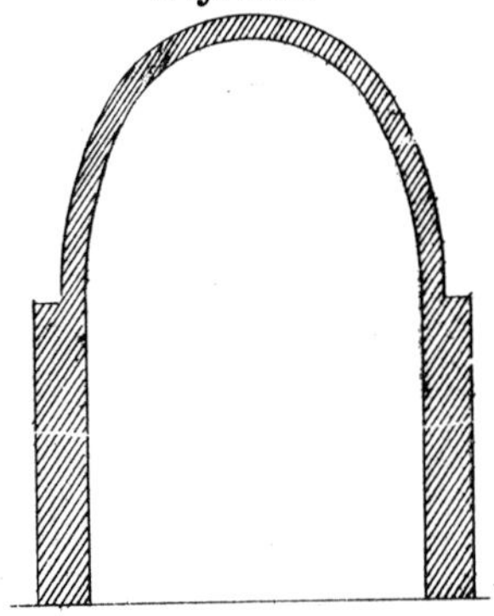

Fig. 213.

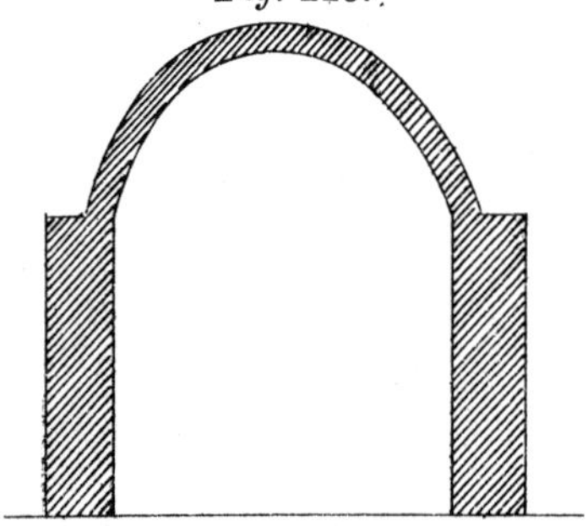

Fig. 214.

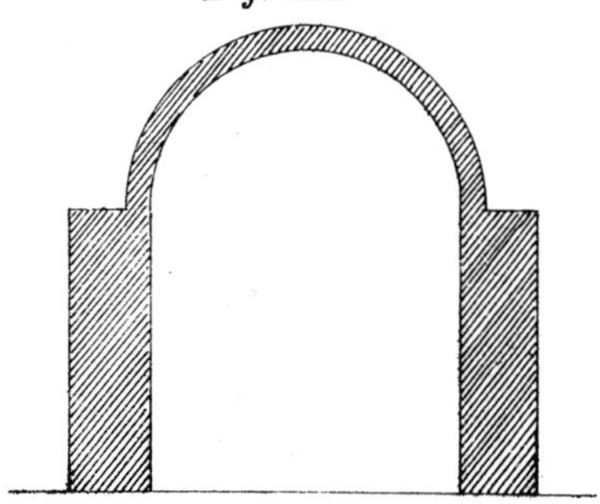

lower ellipse, whose rise is two-ninths of the span. All these arches are drawn on the same scale and of equal span, but the abutments are represented varying in thickness according to the principle that has been explained. For the sake of clearness, the haunches are not filled in, as they ought to be, to a certain height in practice.

As the semicircle may thus be considered an intermediate curve, between the high pointed arches of ancient cathedrals, and the low segments, or semiellipses, used in modern bridges, our first object

Fig. 215.

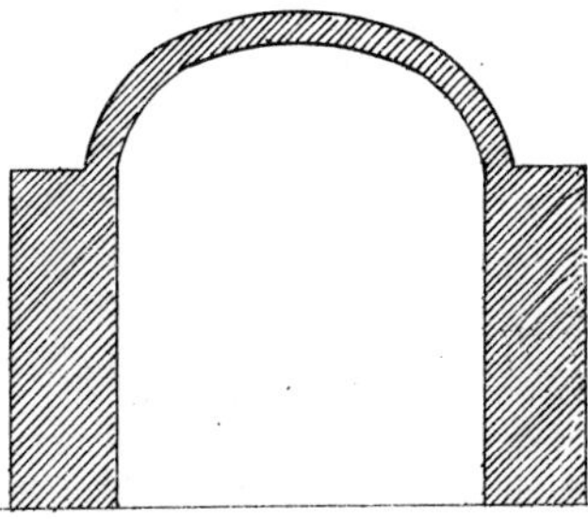

Fig. 216.

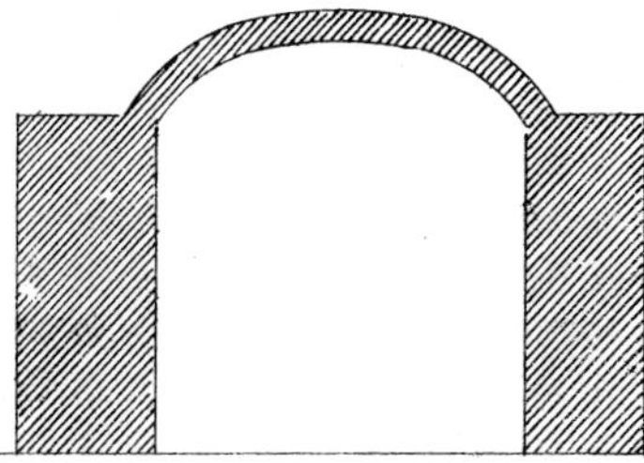

will be to investigate the proportional thickness, proper for the abutments of the semicircular arch, which will be a sort of medium thickness for the abutments of all arches of the same span, but of different heights, being greater than the thickness proper for some, but less than the proper thickness for others.

Not being satisfied with any of the numerous theories of those writers who have treated this subject mathematically, which are not only very discordant from each other, but some of which have been founded on erroneous hypotheses, I consider that in this, as in all other practical arts, Lord Bacon's rule of questioning nature by experiment, is the only possible mode of determining the subject in a satisfactory manner.

The most useful experiments that could be tried as such, would be to build arches of various curves and proportions, all of dry materials, such as soft stone or chalk, or of bricks rubbed and gauged, the span of which ought not to be less than four, and need scarcely exceed six feet, and which ought to have abutments of various thickness, also built dry of the same materials.

Not being possessed of such experiments, accurate drawings and descriptions of arches that have been executed would be still more satisfactory, if they could be collected of sufficient variety, and in sufficient number. But above all, records of failure and accidents that have occurred, from time to time, would be by far the most useful, and would afford the most instructive lesson upon the subject that could be desired; for so long as arches stand, it may be doubted whether their abutments may not possess an excess of strength like those of Rochester Bridge, or of Old London Bridge, the latter of which is now about to be pulled down; but when arches fall the experiment is conclusive. I shall relate a case of this kind, which was reported to me by persons on whose accuracy I can rely.

Two semicircular arches, of four bricks or three feet in thickness, and of 19 feet span, springing from three equal piers of four feet in thickness and ten feet high, two of which served as abutments, had just been finished. The haunches of the extreme half arches were raised a little above the top of the spring of these curves by courses of bricks in steps, rising to the extreme height of about four feet on the inward side of each, as represented in section in *Fig.* 217.

One day only after these arches were finished, the weather having been rainy for some time so that the mortar must still have remained in a very soft state, one of the centres was struck, upon which the whole

Fig. 217.

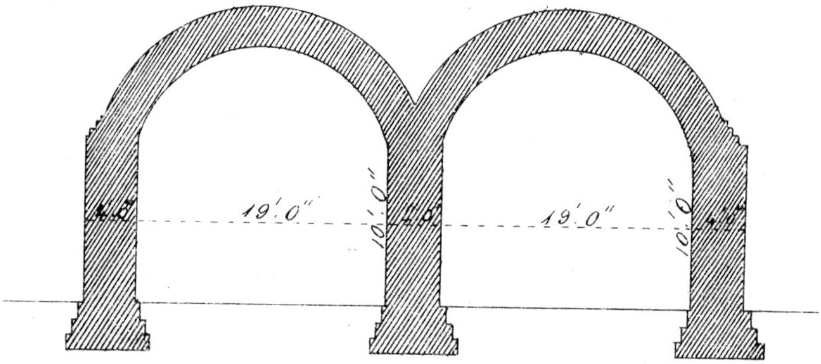

arch depending upon it gave way, by falling down at the crown. What is remarkable, the abutment pier below the spring of the arch was scarcely injured by the shock; for, on being examined one day after the accident by an experienced mechanic, it was found not even to have lost its verticality.

This proves that the haunches of semicircular arches should always be loaded before the centering is struck; and that in brick arches in particular, in which the parts do not possess the wedgelike form, the centering ought to be gradually eased, but not struck or removed all at once whilst the mortar is green. For it is my opinion, that if this arch had been built with stone voussoirs, cut to the proper sommering, it would not have fallen. The same circumstance also seems to prove, that abutments of less than one-fourth of the span in thickness may suffice for the semicircle. In fact, in the above instance, they were scarcely more than one-fifth of that dimension.*

To return to the narrative of this failure, the second arch, of which the centering had not been removed, was afterwards pulled down; and two new arches, of the same span and rise, but of the parabolical form, for the above accident had rendered the semicircle obnoxious, were afterwards built upon the same piers; but the two extreme piers serving as abutments were further reinforced by attaching smaller transverse arches to them, which, with their own piers, act as external counterforts to the principal arches.

After Westminster Bridge was completely finished, the foundation of one of the piers was injured and sank to a lower level than the others, in consequence of which the two arches depending upon it, as well as the greater part of the pier itself were pulled down, and rebuilt in a more secure manner. In the mean time, the other arches remained firm, which, of course, proved that every pier in that bridge was capable of acting as an abutment. The arches were semicircular, and the thickness

* In this case, the courses at the spring of the arch may possibly have slided outwards according to the theory of the mathematicians, for the fluidity of the mortar may have facilitated this operation, like the greasing of the ways in launching a ship. But when the mortar is set, the abutments and arch become as it were one body, and the fall of the latter must be by fracturing and oversetting the former, not by sliding along the upper part of them.

of each of the piers was as nearly as possible two-ninths, or rather less than one-fourth of the span of their collateral arches.

In Blackfriars Bridge, the arches of which are low semiellipses, the centre one having a span of 100 feet, and a rise of 43 feet, and which, consequently, have a greater lateral thrust than the semicircle, the thickness of the piers is one-fifth of the span only, and it is stated that all of those piers possess the strength necessary for abutments, because the arches are said to have been built in succession, not simultaneously.

In Llanrwst Bridge, in Denbighshire, which consisted of three segment arches, built about 150 years ago, the middle arch had a span of 58 feet, with a rise of 17 feet, which corresponds with a segment of about 120°. The piers of this arch were only 10 feet thick, being little more than $\frac{1}{6}$th of the span. One of the side arches gave way, owing to some accident in the year 1703; and thus the pier of the centre arch on that side became an abutment, which was found perfectly adequate to resist the lateral thrust of the arch, although a lower segment and, consequently, more oppressive to its abutments than the semicircle would have been.

From these circumstances combined, it seems probable that the writers of the excellent article "Bridge," in the Edinburgh Enclycopædia, may be correct in their opinion, that one-sixth of the span, or even less may be a sufficient abutment for a semicircular arch; and two very useful and elaborate French writers on the same subject, Messrs. Gauthey and Rondelet are still bolder in their opinions; but, for my part, I should be sorry to apply the principles derived from the proportions of the arches of great bridges to the smaller arches used in common buildings, and for three reasons.

1st. All writers on arches agree that they act upon their abutments with a certain leverage; and that, therefore, in equal and similar arches, exerting precisely the same thrust or outward pressure, that which has the highest ought also to have the thickest abutments, in order that both may meet with equivalent resistance. Now, the height of the piers of Blackfriars and Westminster Bridges is extremely insignificant, in comparison with that of the piers of common arches, which is usually equal to half the span. Indeed, the height of the piers of the above bridges is almost null, for they may be considered as foundations, being nearly buried below the level of the bed of the river, about which level the curves of those arches begin to spring.

2ndly. The arches of great bridges are usually so much more massy than those of common arches, that no weights which pass over such bridges, are likely to bear any any considerabe proportion the weight of the superstructure of the bridge itself. Hence, they are never liable to any very great or unusual strains. Small arches, on the contrary, such as are used for basement stories in England, and still more generally for floors in other countries, are of such moderate thickness, that they may often be liable to the action of weights, greatly exceeding their own, and which, of course, must produce an unfavourable effect upon their abutments.

3rdly. The abutments of bridges invariably terminate on the natural ground, whether rock or earth; and the former substance, by its superior tenacity, the latter by acting on the abutment as on a revetment, and both in direct opposition to the thrust of the arches, render a much smaller thickness of masonry effectual for the abutments of bridges than ought to be hazarded under other circumstances. Moreover, the abutments of bridges usually have walls diverging from them on each flank, called wing walls, each of which acts as a powerful counterfort, in opposition also to the thrust of the arches.

On these grounds, I do not consider it prudent to assume the light proportions of those piers of great bridges, which have been known to act as abutments, nor even those of the actual abutments of bridges, as a proper standard for the abutments of common arches: and I beg leave to lay down the following rule in preference, which has, I believe, been adopted by many practical builders about the metropolis.

The thickness of the abutments of a semicircular arch should never be less than one-fourth, and need not exceed one-third, of the span, unless under peculiar circumstances.

The above rule, however, applies to common arches not to bombproofs. In the latter, as for example, in the arches of casemates, powder magazines, &c., in which an excess of strength is always desirable, I would recommend, and, in so doing, I merely follow the custom of the department, that the thickness of the abutments should be made greater than one-third of the span, instead of reducing it to a lower proportion; and even the piers should not be less than one-fourth of the span.

In those arches which appear in the face of a building, such as arched windows, doors, or gateways, the abutments are almost always made much greater than one-third of the span, for the sake of beauty, as well as out of regard to strength, for it would be preposterous, in a building of any importance, to place an arched door or window, which seldom exceed $4\frac{1}{2}$ feet in width, at so small a distance as one-third of that span, or $1\frac{1}{2}$ feet only from the end of a wall.

I shall conclude this article by observing that an exception to the above rule may be admitted in respect to the galleries of countermines, extending under the glacis of a fortress, which are always revetted with brickwork, and arched at top, unless they should be excavated in rock, which is sometimes the case. It will be evident, that in consideration of the pressure of earth on all sides, the abutments or side walls of such galleries need not be made so thick, as would be proper in similar arches constructed above ground.

Observations on the Foundations of Arches, and on the pressure of earth against their abutments.

It is not sufficient that the piers and abutments of arches be properly proportioned, unless all their foundations be perfectly sound.

Whenever the level of the ground differs on two sides of a pier or abutment, this circumstance should also be taken into consideration, on the principle before alluded to, in treating of the abutments of bridges.

A striking proof of the necessity of attending to this subject occurred in Chatham about six years ago, in a storehouse built by Mr. Best the Brewer, after the great fire in Hammond Place. The basement story consisted of five semicircular arches, each nine feet six inches in span, and springing at the height of two feet above the level of the floor. These arches abutted on one side, against the backwall of the building, which was $2\frac{1}{2}$ bricks (or 1 foot $10\frac{1}{2}$ inches) in thickness, being only one-fifth of the span, and which, consequently, ought, according to my view of the subject, to have been insufficient for an abutment; but on this side, the external ground was 2 feet higher than the floor, so that the pressure of earth acted in direct opposition to the thrust of the arches, and no failure took place. On the other side, the front wall of the building, which acted as the second abutment, was 3 feet thick, being very nearly equal to one-third of the span; and this would no doubt have proved sufficient under common circumstances, but here, unfortunately, the floor of the building averaged four feet higher than the pavement of the street, so that this wall was exposed to a double strain, having to do the duty of a revetment as well as of one abutment of a series of arches. This was too much for it, and it was evidently giving way, when the proprietor consulted Colonel D'Arcy, the Commanding Royal Engineer at Chatham,* according to whose opinion the error of the original construction was effectually remedied, by building five internal counterforts against that wall, nearly square in plan, and projecting about 3 feet from the wall, from each of which, as a pier, a quadrant arch was turned in the same direction, to oppose the thrust of the original arches on that side. The upper part of the wall, however, had been so much bulged, that it was pulled down and rebuilt, after those quadrants were added.

After having written so far, I received an account of the recent failure of a warehouse, in Liverpool, of which the particulars shall also be stated, as they bear upon our present subject. †

Fig. 218.

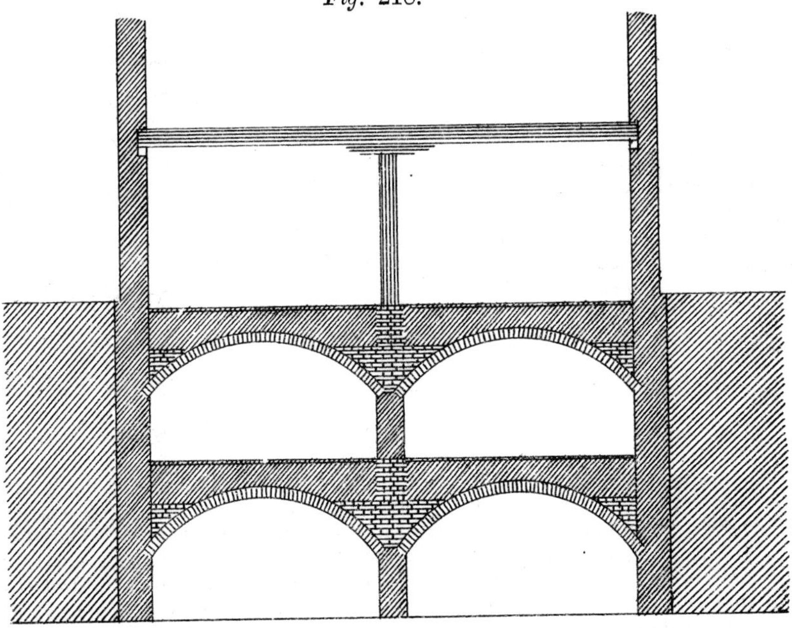

* Now Major General D'Arcy.

* Having seen the fall of a great warehouse at that place reported in the newspapers, I wrote to request Captain Kitson, who is employed in that neighbourhood, to examine and favour me with an account of the construction of the building, which he very obligingly did.

This building consisted of two tiers of vaults, below the level of the ground, over which were no less than five stories with common wooden floors, and a common roof. There were two vaulted apartments in the breadth of the building, the arch of each of which was a segment of about 100°, having a span of 15 feet, and a rise of three and a-half feet, and piers of about four and a-half feet in height.

These arches were supported by a central pier wall of two bricks (or 18 inches,) and by abutments of only $2\frac{1}{2}$ bricks (or 1 foot $10\frac{1}{2}$ inches) in thickness; whereas, according to the rule before laid down, the latter ought not to have been less than five feet thick.

These arches were one brick thick, over which were laid rubbish, and upon this a brick floor, making a total thickness of about 18 inches over the crown of each arch.

Although the abutments were, as has been stated, not half so thick as I should have deemed prudent, yet the building. which was used as a bonded warehouse, and which, consequently, was loaded with very heavy goods, stood for about 10 or 12 months, because the pressure of earth upon each side of those abutments relieved the two vaulted stories from the thrust of the arches, which it effectually counteracted.

At the end of that time, the proprietor very incautiously ordered the ground in rear of the warehouse to be cleared away, for the purpose of building a new storehouse. When the workmen had excavated to the level of the floor over the lower pier of arches on that side, one of them observed the wall to bulge, opposite to the spring of the upper arch on the same side, and in about 5 minutes afterwards the whole warehouse fell down in that direction, with a tremendous crash.*

Of the proper thickness of Common Arches, and on the action of variable weights upon them. Peculiar property of Arches in the face of a Wall.

In respect to the thickness of common arches, they may be only half a brick thick at the crown, when there are inhabited apartments above them, and when the span is small, such as six feet or thereabouts; but if the span be much greater than the above dimension, or if goods or stores of a heavy nature are ever likely to be deposited over them, they should certainly not be less than one brick thick at the crown. Common arches need not, however, exceed one brick and a-half in thickness at the crown in any case, excepting when intended for bombproofs.

In building a series of arches, above which a wooden or stone floor is to be laid, it is proper to fill in the spandrels to about three-fourths of the rise of the arch, above which hollow spaces are left between the summits of adjacent arches, which are sometimes not filled in at all, sometimes partially, and at such intervals only as may be necessary for supporting the stones of the former kind or floor, or the sleepers, upon which the joists of the latter are laid.

---

* Captain Kitson also remarked, that the bricks and mortar used in the walls were of indifferent quality, and that the bond was neglected, all of which defects are said to be very common in the brick buildings about Liverpool; also that the beams of the upper floors, which were supported by central story posts, had only a bearing of about $4\frac{1}{2}$ inches on the walls. All these particulars may have had an unfavourable effect; but it appears to me, that no choice of materials, and no care or skill on the part of the workmen, could possibly have secured such very slender abutments from the risk of giving way sooner or later, for common earth, which was their only security, even without being moved, is always liable to great changes in consequence of alternations of dry and wet weather.

Transverse brick walls, of one brick in thickness, are used for this purpose, which may be laid at intervals of four feet under the former supposition; but at intervals of about seven or eight feet, under the latter supposition. This last is the case at the New British Museum, over the arches of the basement story, where dwarf walls, one brick thick, and at central intervals of $7\frac{1}{2}$ feet apart, support sleepers of $3\frac{1}{2}$ by 4 inches scantling, upon which the flooring joists are laid.

The transverse walls thus used for partially filling in the spandrels of arches, act as counterforts in opposition to the thrust of the upper part of those arches; whilst, at the same time, they press less unfavourably upon the abutments than solid spandrels, having much greater weight, would do.

In considering the nature of arches in the face of a wall the weight of masonry or brickwork over the aperture must not be deemed altogether unfavourable, for it is known that courses of stones or bricks projecting horizontally, so that each gains a little, and converging towards each other until they meet, form an arrangement equivalent to an arch, which has been styled an Egyptian arch, from the circumstance of its having been used in the gallery leading to the chamber in the great Pyramid of Egypt.

Fig. 219 represents this arrangement, and any part of a well built wall

Fig. 219.

of bricks or stones has evidently the same kind of bond. Hence, if we suppose a common arch, A, as in *Fig.* 220, to appear in the face of a wall, it is not the whole weight of masonry above the aperture, but only the irregular portion, B, of much smaller content which presses unfavourably upon the arch and its abutments. Above B, an Egyptian arch is formed by the joints of the stone or brickwork which relieves the common arch, A, beneath it, in the same manner as a common arch takes off the pressure upon a wooden lintel. The supposed Egyptian arch does not appear in any marked manner in the face of the wall, as shown in the figure, but there can be no doubt of such an effect being actually produced.

In high Gothic arches in the face of a wall, there is no downward pressure whatever on the part of the wall above, as may be seen by inspecting some of the pointed windows at Rochester Cathedral.

Fig. 220.

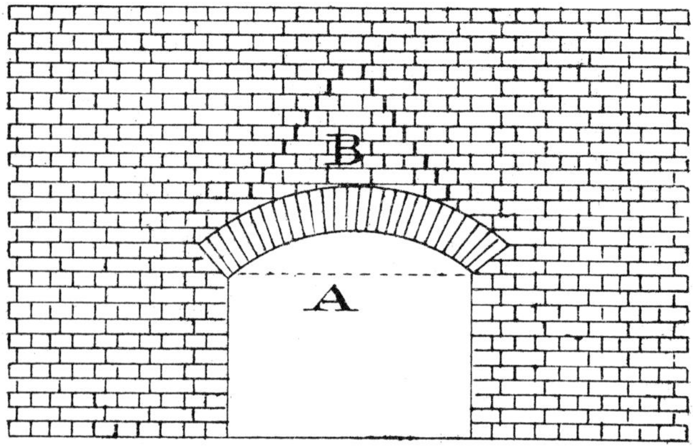

There is another circumstance, not yet discussed, which must necessarily have some influence upon arches, that is, the nature of the weights, which they may be required to support, for it is evident, that an excess of weight, whether general or applied in one particular part, may destroy the equilibrium of an arch, which would remain firm if not so loaded. Hence it is evident, that if an apartment supported by arches is to be used as a store for heavy goods, there should either be a wooden floor over the arch, so constructed as to throw the weight upon the external walls, and thereby relieve the arches, or the latter ought to have thicker abutments than usual, in anticipation of this extra pressure.

After a series of arches shall, however, have been completely consolidated by the setting of the mortar, I am inclined to believe, that any extra weight applied at top will not increase the lateral thrust of the arches in the unfavourable manner that has just been under discussion, but will, on the contrary, act upon the whole of the arches and the abutment walls as one body.

In discussing this subject another point still remains to be considered which may affect vaulted storehouses. If corn, coals, or other such stores be stowed away loose, without being put into sacks, or strong wooden bins, the weight thereof may act not only on the floor, but on the walls, in an injurious manner, producing the same effect on them that loose earth does on the back of a revetment. Hence, abutment walls thus acted upon may be injured or ruined, although well proportioned in other respects, and capable of resisting the thrust of the arches alone, without this additional strain.

Upon the whole, therefore, taking into consideration the various circumstances which have now been explained or discussed in reference to arches, it does not appear advisable to use vaulted chambers in the basement story or stories of a storehouse, lest, by some misapplication of the weight of stores, piled in the rooms above, or by some other circumstance not originally foreseen, such as clearing away the ground on one side, when such vaults are subterraneous, the arches may in time be injured or destroyed, in which case the whole building will be rendered useless. Even if we suppose the architect to have taken every reasonable precaution in his original plan, he cannot be supposed to foresee all the changes that may take place in the application of the several apartments of a great edifice; and the merchant or storekeeper, after the buildings in their charge have stood for many years, will not always think of taking professional advice, in respect to a new arrangement of their stores, and thus may inadvertently cause the ruin of an arched storehouse.

It is true that a series of arches, or of vaulted apartments in the basement story of a building, will afford a very useful preservative against the communication of fire; but considering how very dangerous such vaults, unless made extraordinarily strong, and expensive in proportion, may often prove to the stability of a building, it appears to me much preferable to attain the same object, by means of a fireproof ceiling, not acting in such an unfavourable manner, upon the external walls. This kind of ceiling will afterwards be described.

The same subject continued. General Remarks on Great Arches.

In one of the former articles, having treated of the proper proportional thickness of the piers and abutments of semicircular arches, it remains to make some observations upon those of a different description. The high pointed arches, to be seen in ancient cathedrals, are seldom or never used in modern times, although certainly affording a much finer specimen of ingenuity, and of architectural skill (I shall not say of taste), than the more classical style of building which has since superseded them, as may be seen by comparing the vaulted roofs of some of our ancient cathedrals, with those of St Paul's.

In both, a transverse section would show a nave or principal arch, of greater height and span, than the aisles or smaller arches on each side; and in both, the thrust of the principal arch is communicated to the external walls by arched ribs, but in the more ancient cathederals, all the arches diverge much less from a vertical direction, and their very moderate outward thrust is thrown in a simple and ingenious manner upon external buttresses, attached to the side walls of the building, which walls are light in proportion. In St Paul's, on the contrary, the thickness of the piers and of the abutments or side walls, added together, is more than one-half of the space arched over.

But to dwell on the proportions of the piers of Gothic arches, which are now seldom or never used, would be superfluous. Suffice it here to repeat, that they may be thinner than those of the semicircle, and the actual proportions of the most celebrated arches of this description may be ascertained, by a reference to the numerous ingenious writers who have devoted their time to the illustration of the most celebrated cathedrals, in this and in other countries.

In respect to low arches, such as the flat segment, or the low ellipse, it was also before explained, that their abutments ought to be thicker than those of the semicircle of the same span, whilst the straight arch, or platband, as it is sometimes termed, ought to have the thickest piers of all; but this last kind of arch is seldom used, excepting for very small apertures, such as those of a common door or window. Having, in one of the foregoing articles, laid down one-third of the span, as a proper proportional thickness for the abutments of a semicircular arch, in adopting which, no failure is likely to occur, it strikes me, that one-half of the span is a sufficient thickness for the abutments of a straight arch, and that the thickness of the abutments of the circular segment, less than 180°, and of the low semiellipse, may vary between the aforesaid proportions, of one-half, and of one-third, according as they approach more nearly to the platband or to the semicircle.

Platband, or straight arches, have often been used in stone buildings, in ornamental work, with abutments, whose thickness is much less than one-half of their span, of which many examples are given by Mr. Rondelet. In this case, strong iron ties and bars, intersecting each other in various directions, have been introduced internally into the masonry to bond the whole together, without which such arches would immediately have fallen down, on the removal of the centering used in putting them together.

Upon the whole, I consider that much caution is necessary, in determining the proportions of arches, and of their abutments, when used in the basement or other stories of a common building, for we are absolutely without sufficient records to guide us in the construction of such arches, which are very rare in this country, and I must again repeat, that I should be very sorry to trust to any theory of the subject that I have yet seen. In respect, however, to arches of much greater importance, I mean those of bridges, over great rivers, the same difficulty does not occur. Both in France and in England, but more especially in the former country, elaborate and useful works on this subject have been published, which contain the dimensions of the most celebrated bridges that have been constructed from the time of the Romans. Mr. Gauthey's work, in particular, is invaluable in this respect, since he has stated not only dimensions in detail, but has candidly reported the failures or accidents that took place in many of the bridges in France, whilst in progress. One of the best treatise on this subject, in English, is the article, "Bridge," in the *Edinburgh Encyclopædia*, of which the practical part was written by Mr. Telford.* Of Waterloo Bridge, and other great bridges, executed subsequently to the publication of these works, and consequently not described therein, it is easy to obtain drawings or descrip-tions in London.

Having, therefore, such records before him, an Engineer Officer, who is called upon to superintend the construction of a great bridge, which might possibly be required of him in the colonies, although not in England, can be at no loss in forming a design, so far as the proportions of the several parts are concerned, although he might perhaps find it difficult in those colonies to procure persons qualified to assist him in superintending the execution of the work.

I shall conclude by observing that it has been very common in great bridges to build the spandrels hollow, forming those hollows either as circular or elliptical tunnels; or as small arches of the common form; or with inverted arches at bottom only, but having parallel walls built transversely across, at clear intervals of three or four feet or upwards, with long stones laid over these walls to support the roadway. In one bridge, a very celebrated one over the River Taaf, in Glamorganshire, composed of a single arch in the form of a segment of about 106°, having a span of no less than 140 feet, with a rise of 35 feet,† and which

* The theoretical part is by Mr. Nimmo then only known as a mathematican, now a Civil Engineer, who has for some years been usefully employed in the public service in Ireland.

† This bridge was built by a common Welsh mason in the neighbourhood, who may almost vie with the celebrated Swiss carpenter of Schaffhausen. Until new London Bridge shall be finished, this will remain the largest arch in the British dominions.

was built twice, it having fallen down the first time, there are three unequal cylindrical openings, of nine feet, six feet, and three feet in diameter, across each of the haunches; and it has generally been held out that the previous failure took place owing to the first bridge having had its haunches solid. I have great doubts whether this was the sole cause of the failure, since we have had other arches, of all forms and proportions, filled in solid, without such failures occurring.* But I forbear to enlarge upon this subject, nor shall I exemplify the system of concave spandrels alluded to, of which numerous examples will be found in those writers who have treated professedly of the subject of bridges, which is a study in itself.

Of Cones, such as are used for Tile Kilns, &c.

A building, in the shape of a hollow cone, forms everywhere a species of circular arch, which may be constructed without centering or support of any description, provided that all the joints be made to radiate towards the centre; for, so soon as any one course is carried entirely round the whole circumference, it is impossible that the bricks used in it can be forced inwards.

The best way of building a tile kiln, or other such cone, is to lay all the courses of bricks perpendicular to the sides of the proposed cone, as in *Fig.* 221. A rod of variable length, turning a pivot, must be stretched

Fig. 221.

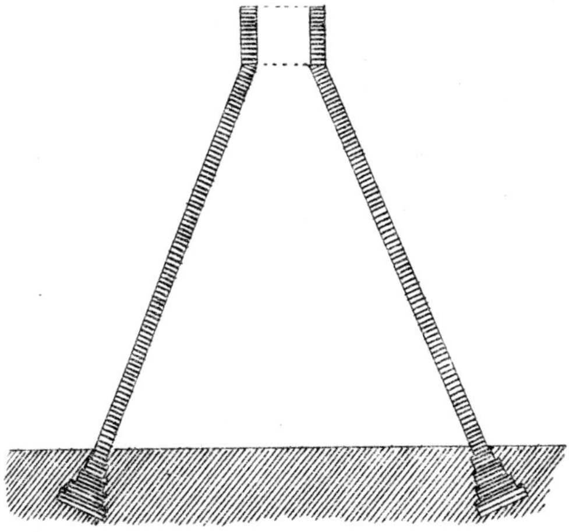

all round from time to time, upon a moveable centre, rising as the work proceeds, in order to regulate the internal outline.

Such is the strength of this form, that the highest kilns are seldom built more than one brick thick, although this dimension would be altogether insufficient for a common wall of the same height.

* I conceive it probable that this accident would not have occured, if the voussoirs of the arch had been thicker throughout.

Of Domes.

When used without further specification, the term dome implies a hemispherical arch, which may be supposed to be generated by the revolution of a quadrant about an upright radius, and of which all the joints, without exception, radiated to the centre. When a dome is of any other form than the above, which is frequently the case, it must be specified by naming the mathematical solid, with which it corresponds in its external form, or the mathematical curve, by which it is generated.

Fig. 222.

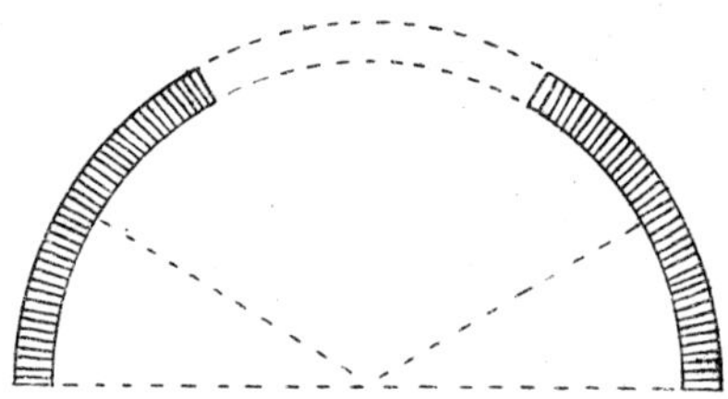

In commencing a dome, as represented in section in *Fig.* 222, the first courses may be built without any centering or support; and as soon as any one course is completed all round, the stones or bricks composing it form a circular arch like that of the cone, which cannot, by any means, fall inwards. After raising the work until the beds of the courses incline by degrees from a horizontal position to an angle of about 30° with the horizon, it then becomes impossible to proceed further without using props or supports to each brick or stone, to prevent it from sliding down the slope of the course below, on which it is to be bedded. But the moment that one whole course is completed, this tendency is again rendered null, because another perfect circular arch is then formed, so that all the props or supports of this course may be removed without risk.

Hence, there is an important difference between the dome and the common arch, usually called a cylindrical or cylindroidal arch,* to distinguish it from the former. The common arch cannot stand at all, without its centering, unless the whole curve be completed; and, when finished, the crown or upper segment tends to overset the haunches or lower segments. The dome, on the contrary, is perfectly strong, and is a complete arch without its upper segment; and thus, as the pressure acts differently, there is less strain upon the haunches and abutments of a dome than on those of a common arch of the same curve, as shown in the vertical mid-section of both.

Hence, a sufficient dome may be constructed, with much thinner materials, than would be proper for a common arch of the same section.

In the Royal Engineer Department, domes can seldom be necessary. There were, however, very handsome stone domes constructed in some of the first Martello towers, built by the British in Minorca, in 1799. Recently, this construction has been set aside, and the annular arch, springing from a small round central pier, and abutting on the external walls of the tower, has been adopted in preference.

St. Paul's Cathedral offers a fine specimen of this kind of work. Internally, there is a hemispherical brick dome 112 feet in diameter, but

* Cylindrical implies a semicircular arch, being a portion of a hollow cylinder. Cylindroidal implies, elliptical and other arches, not being accurate portions of a cylinder, but resembling that form.

only 18 inches, or two bricks thick, built of common bricks, with thorough bonding courses of bricks 18 inches long, and baked for the purpose, at intervals of five feet.*

This dome, open at top, has at bottom a springing course of Portland stone, around which is cut a circular groove, in which is bedded an iron chain, in links of ten feet, run in lead. The external dome of the same church is built of oak, and is 140 feet in diameter, and of an oblong section considerably higher than the hemisphere. It is supported by a brick cone, only two bricks thick, built immediately without the hemispherical brick dome before described; and, at top, the cone supports a massy lantern of Portland stone, said to be 21 feet in diameter, and nearly 64 feet high.

An iron chain, in the form of a hoop, also secures the base of the external dome, and four similar hoops are said to be applied for the same purpose about the brick cone at different heights.

Fig. 223 is a section of the upper part of St. Paul's Church, showing

Fig. 223.

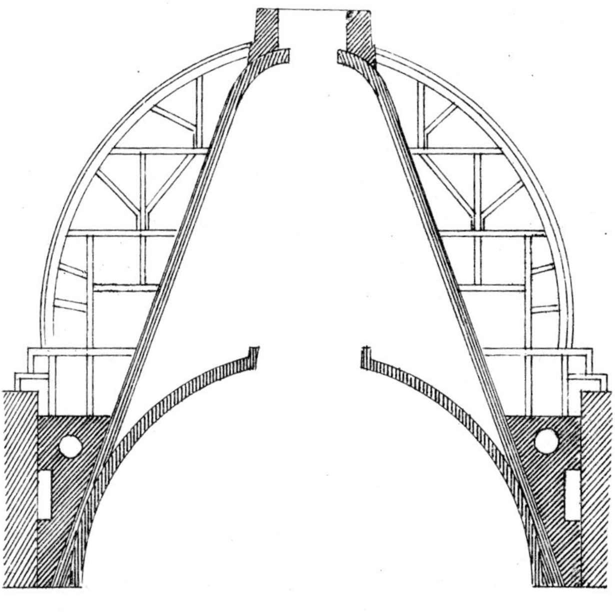

the above construction. The woodwork of the external dome is connected to the cone by iron straps, bedded into stone corbels, let into the brickwork.

The brick cone and the dome are in contact at bottom, where there is a greater thickness than at a higher level, and both are attached to transverse walls at moderate intervals, radiating from the same common

* Hence the general thickness is only about one-seventy-fifth part of its span.

centre, and connected with the external walls of the cathedral. These walls, in which there are bull's-eyes for lightness sake, act as buttresses, and relieve the haunches of the dome from the thrust of the upper part, which, although much less than in that of a common arch, must necessarily have its effect, even in the hemisphere. Towards the bottom the cone is not solid throughout, the upper part of it being supported by piers and arches, the latter of which seem to be of a parabolic form. The external dome consists of a skeleton or framework of oak, covered with planks of the same material, and coated with lead.

The ground plan of the cathedral, being in the form of a cross, the centre of which is immediately beneath the centre of the dome, the various walls thus meeting at right angles, act as very powerful buttresses, upon which no lateral thrust, communicated downwards from any part of the dome, could possibly have the smallest effect.

Mr. Rondelet describes the entire roof of the French Pantheon as being composed of three domes or cupolas, the first or interior of which is a regular hemisphere of about 66 feet $9\frac{1}{2}$ inches span, with a circular opening at top of about 31 feet $4\frac{1}{4}$ inches in diameter. It is built of cut stones, varying from $18\frac{1}{2}$ inches thick at bottom, to $10\frac{1}{2}$ inches thick at top; thus, the thickness at bottom is only about one-forty-third part of the span.

The intermediate dome is a Catenarian curve, having a span of about 70 feet, with a rise of 50 feet, and it has to support considerable weight at top. It has four large openings in its sides to give light, about 37 feet high, by 31 feet wide, which are arched at top, apparently in the form of a parabola.

The outer dome has an external diameter of 78 feet; its height is not stated, but it appears as if it would have been a Gothic arch, very moderately pointed, had it been continued, without forming an opening at top, for the sides of a lantern, which it was intended to support. The thickness of the stonework is about 28 inches at bottom, and 14 inches at top, but a great part of the surface is only half the above thickness, as the dome is laid out internally in piers, supporting three tiers of arched recesses or niches of less substance, and showing like the panels in joiner's work.

I have given these examples of domes, foreign as this subject may be considered to the business of the Royal Engineer Department, to show the very different properties of the dome, as compared with the common arch, it being evident that common arches of the same section as either of the above could not possibly stand.

Niches are small cavities left in massy walls, usually for the purpose of affording places for statues or other ornaments. In brickwork they would be built on the same principle as a half dome, cut vertically through its axis, which they resemble in form, though not in magnitude.

Of light Arches in general, and of those used in the South of France in particular.

It is probable, that the force of mortar, in cementing stones or other materials, is in direct proportion to the number of joints, or, in other words, to the smallness and consequent lightness of the parts so cemented.

Thus, in what is called rough casting, small pebbles or pieces of gravel, are made to adhere to the face of a perpendicular wall, by dashing them upon wet mortar, and by the same means, they might even be suspended from a ceiling, by causing them to adhere to the plaster, in direct opposition to the force of gravity; which, if attempted with larger fragments of the same kind of stone, would be absolutely impracticable. Following up the same principle, the massy voussoirs of Waterloo Bridge can scarcely be supposed to derive any additional stability from the mortar used, which merely serves to create a useful kind of friction, by filling up the joints; whereas, a brick arch of the same span, would derive its chief strength from that source. Hence, also the strength of mortar, in cementing stones, &c., must in some measure depend on the specific gravity of the materials cemented, the lightest material deriving the greatest aid from the mortar, which being in itself a light and weak substance, is liable to be separated from heavier bodies with greater ease, than from light ones.

Hence, if peculiarly light arches are desired for any special purpose, which, of course, must be exempted from the pressure of great weights under the circumstances treated of in the preceding article, it follows that the most effectual mode of forming them is to use very small and light materials, cemented with the most powerful cement than can conveniently be procured.

This principle was acted upon by the Romans, who often used arches, or portions of arches, of small pieces of very light porous stone, apparently laid in abundance of cement upen a wooden centering; and the same practice appcars to have been copied from them by the Arabs. I have observed old arches in the Mediterranean of this description, perfectly sound, in which so little care appears to have been bestowed in the arrangement of the stones, which are usually oblong, that some of them show their largest end in the soffit of the arch.

The same principle was understood by the ingenious architects of the dark and feudal ages, who usually employed arched ribs of building stone of the best quality, whilst they arched over the panels or intermediate parts between those ribs with a smaller and lighter kind of stone; as, for example, in Rochester Cathedral and in Rochester Castle, where the ribs are of a stone equal in quality to the Portland stone; whilst the intermediate parts or panels are frequently composed of chalk, which, in the cathedral, is neatly cut; but, in the castle, consists of rubble work.

In this climate I do not consider that it would be advisable to have the floors of inhabited apartments made of anything but wood. It has, however, been common in France to use pavements of flat tiles, or a coating of cement, for floors, and these have sometimes been supported by very light flat arches, constructed on the principle that was announced in the commencement of this article.

Blondel, in his course of architecture, and after him Rondelet state, that very flat arched floors, composed of bricks, cemented with gypsum or plaster, have been in common use in Roussillon, from time immemorial, and that similar arches have, in some few cases, been executed at Paris, by workmen from that district. M. Rondelet expresses an opinion, that the segment of a circle is a better form for such arches, than the low semiellipse or oval resembling it, although the latter curves appear to have been more generally used.

Both of these writers describe apartments of 18 by 25 feet, at the War Office at Versailles, as having been covered with segment arches of bricks, of which the rise was only one-fifteenth part of the span; and that there were no less than five stories of arches thus constructed: but they do not state the thickness of the abutment walls; and it appears by their figures, that there was a series of such arches.

They are more explicit in describing, the mode of arching the coach houses and stables, of the Marshal de Belle Isle at Bisy near Vernon, of which *Fig.* 224 is a section.

Fig. 224.

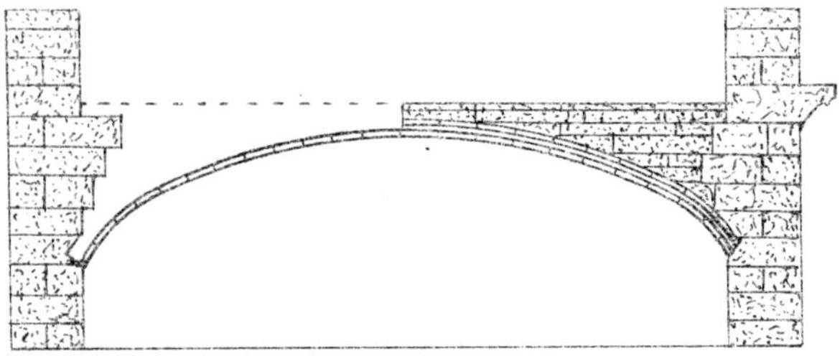

Those arches were of an elliptical form, having a rise of one-fifth of their span which was 32 feet $9\frac{3}{4}$ inches. They were not finished, until a year after the walls and roof had been completed. The walls were built of rubble work, having chains of cut stone, at intervals of about 16 feet.

The thickness of those walls, which served as abutments to the arches, was about 2 feet $8\frac{1}{4}$ inches, thus being equal to about one-twelfth part only of the span.

These arches, were formed of a double thickness of bricks laid flat, with plaster in the joints, and built in succession, one after the other, with a bed of plaster between, and so arranged that all the vertical joints were broken.* The haunches were filled in with small stones, laid also in the same cement. The arches sprang from notches in the walls, above which the regular courses of stone projected inwards, like corbles or gathering courses, thereby locking into the haunches, and giving better bond; to explain which arrangement, only one-half of the arch is represented as being finished, in the figure

Above all, a third course of flat bricks was laid horizontally, forming a pavement.

Before all those arches were finished, M. Rondelet relates that a stone of 4000 or 5000 lbs weight was dropped upon one of them, from a height of four or five feet, which made a large hole through the arch, without doing it any further injury.

From this, as well as from some other experiments recorded by the same author, it appears that arches of small and light materials cemented by gypsum become, as it were, one body, and exert little or no lateral pressure, upon the abutments, excepting at first, because that cement has a tendency to swell in setting.

M. Rondelet seems to think, that similar arches, might be formed with common mortar, although he admits that the parts ought to be

* Rondelet states that there are traces of similar arches of large bricks, laid flat, built by the ancient Romans.

somewhat thicker. I should be very sorry to hazard the experiment, with that material, in any building of importance, although I consider it very possible, that good mortar properly set, might like gypsum, cause an arch of light materials to lose its lateral pressure, if previously supported long enough by the centering. In regard to the wonders, that may be worked with gypsum, any person who has been in Minorca can entertain no doubt upon this subject.

M. Rondelet states, and in this I believe he is right, that in light flat arched floors, such as have been described, it is better to use coved arches, springing from four walls, than a common arch springing from two opposite walls only. He also copies Blondel, in describing some very light arched roofs, formed with similar materials, in the pointed form of Gothic arches ; but I shall not enlarge upon this curious subject, as I do not consider the arches in question applicable to buildings in this country, but it is right that the student of architecture, should have a general notion of the most remarkable varieties, or deviations from our own practice, which have prevailed in other parts of the World. I may conclude however by remarking, that if such arches were desirable, the water cement of this country, usually styled Roman Cement, offers an equally efficient, and more durable composition than gypsum.

Of the use of Pots, or hollow earthen ware Cones, in Arches. A description of light ceilings formed in this manner.

It is said, that there are traces of arched roofs, built by the Romans, with thin bricks laid partly flat and partly on edge, so as to form a series of honeycombs for the sake of lightness. In the church of St. Vital, at Ravenna, built in the fifth century, hollow earthen pots, not unlike common pitchers, are said to have been used in the construction of the dome. And I am informed that the same contrivance, was introduced in Calcutta, within the last 20 years, where apartments of considerable width, were arched over with similar pots, which, having been previously in common use in Syria, were called Syrian floors, or roofs. The same fashion has recently become very general in France, but, according to M. Rondelet, every builder chooses his own model for those hollow bricks. That author states, that some are made hexagonal, others square, or rectangular, as in *Figs*. 225 and 226. They are all made smaller at bottom, to give them the proper arch like form.

Fig. 225.

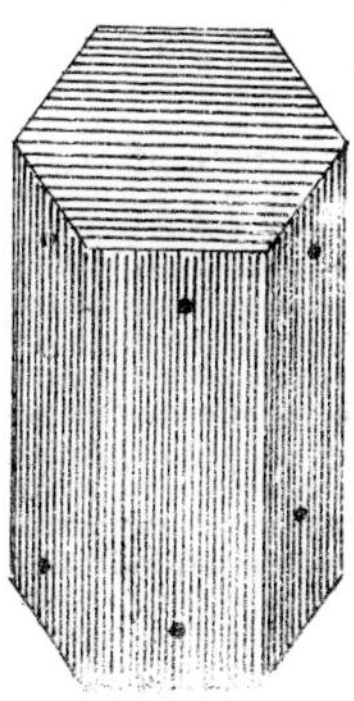

Fig. 226.

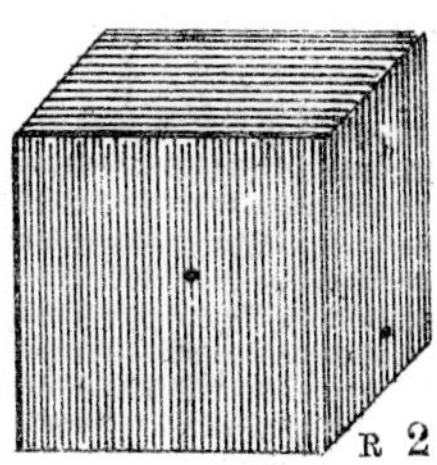

Of late, they have also come into common use in England, but not in private buildings, as appears to be the case in France.

Those which I have seen, and which I believe are the only kind used in this country, and which, according to M. Rondelet, are also common in France, are oblong, square at top, and circular below, and there is a small hole at bottom, so that the lower part of them resembles a common flower pot, In this country they are made of two sizes.

Fig. 227, represents one of these arch pots, which are known by the name of hollow cones at the manufactures. *Fig.* 228, represents the same turned upside down.

Fig. 227.

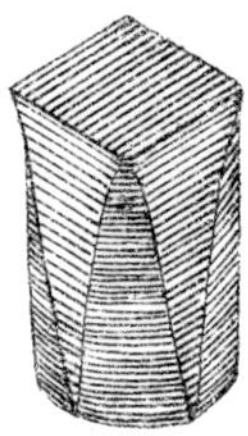

Fig. 228.

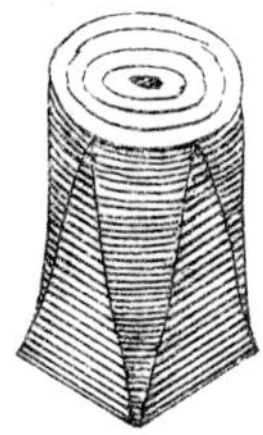

Two only of these pots are at present in my possession, one of each sort, the larger is 8 inches high, and about $4\frac{1}{2}$ inches square at top, and of the same diameter nearly at bottom, and weighs about $4\frac{1}{4}$ lbs.

The smaller kind is $5\frac{3}{4}$ inches high, and about $4\frac{1}{4}$ inches square at top, and of the same diameter nearly at bottom, and weighs about $2\frac{1}{8}$ lbs.

The circular part of both, being the frustum of a cone, appears to taper, and no doubt it must have been intended that the square at top should have been somewhat greater than the diamater at bottom, but it may not perhaps, be easy to make such pots, very accurately ; and whenever they have been used in this country, Roman Cement, has I believe always been employed, which must consolidate the whole into one mass.

I am informed, that Mr. Soane, used arched pots for the Rotunda, or dome of the Bank of England, which appears to be about 65 feet in diameter. In this he may probably have used the larger kind. The same architect, as well as Mr. Smirke and Mr. Nash, have recently employed arch pots, in other buildings of importance, using however the small kind only.

In France it appears that arch pots, have been used to turn complete arches from wall to wall, extending the whole width of an apartment ; and in order to relieve the strain, M. Rondelet recommends inserting iron tie bars, to prevent the walls from being forced out by the thrust of these arches. At Calcutta, the same system of arching over whole apartments was introduced, and as it is a dangerous thing to be too hasty, in adopting arches under new forms, as I have endeavoured to inculcate in the preceding articles, those arch pots light as they were forced out the abutment walls in several instances, and the arches came down.

In England, a safer method of using those arch pots has been introduced, which relieves the wall from the prejudicial thrust, now alluded to. Iron girders are laid from wall to wall, at intervals seldom exceeding six feet, frequently less. The section of these girders is in the form of the capital letter T inverted, and the flanges or projections at bottom, which are sometimes perpendicular, sometimes formed as regular skewbacks, support those light arches which are turned from girder to girder.

Thus the iron girders serve as the piers, whilst the extreme walls serve as the abutments, of a series of those light arches, which, by thus increasing their number, and reducing their span at will, are deprived of all power to injure abutment walls, even of moderate thickness.

Fig. 229 represents in section, an arched floor of this kind, built by direction of Mr. Soane, over the basement story of the new Treasury buildings in which the small hollow cones were used, the span or bearing between the girders appearing to be about six feet, and the rise of each small arch about six inches.

In some of the ceilings of the New Post Office, Mr. Smirke has adopted segment arches, of the same description and of about the same span, but of much greater height, the rise being very nearly in the proportion of one-third of the span.

But the most extensive application of this expedient may be observed in his Majesty's new Palace now building on the site of Buckingham House, under the direction of the eminent architect Mr. Nash.

There, I understand, that all the apartments are to have a ceiling consisting of such arches, which will be supported by iron girders of very great depth* of the inverted T form, above which there will of course be a wooden floor resting on the upper part of the same girders.

In forming certain parts of the roof, a double tier of arches of this description has been constructed, of which *Fig.* 230 is the section. The span of these arches varies from $4\frac{1}{2}$ to 6 feet, their rise is about one-twelfth of the span. The extreme arch pots spring from skew backs of bath stone, resting on the flanges at the bottom of the girders, as represented in the figure.

I am informed that Mr. Nash, proposes to make those parts of the roof entirely flat, which are to be formed by laying over the upper tier of arches, more arch pots to obtain a certain level, then one course of common bricks, and lastly a peculiar kind of cement invented by Lord Stanhope, and composed of Stockholm tar, pounded chalk, and sand mixed together, and applied hot, with slates bedded on the top of all.†

Fig. 229.

* Their depth varies from about 20 inches at each end to about 29 inches in the centre.

† This composition appears to have been adopted in the roof of the casemates at Quebec, before described.

Fig. 230.

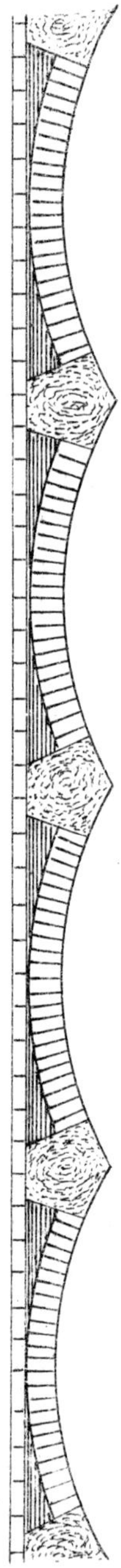

Fig. 231.

This ingenious mode of supporting a series of arches by girders, is not entirely modern. In Belidor's Science des Ingenieurs, he describes the floors of Soldiers' Barracks in France, as having been in the early part of the last century, usually composed of a pavement resting on brick arches, turned between wooden girders, the lower part of each of which was cut to the proper skewback.

In 1798, I recollect the floors of the Soldiers' apartments in the Royal Artillery Barracks at Portsmouth, having been formed in the same manner, as represented in *Fig.* 231.

Of this practice no trace now remains in common use, excepting the trimmer arches, generally formed in front of fire places.

To return to the arch pots before described, although they fit each other at top, they leave irregular interstices at bottom, between the joints where the circles touch each other, which are filled in with cement. The holes at the bottom of these pots, serve as a bond, or as it is technically termed, a key, for the plastering, so that the intrados of an arch of this description may be plastered, without the aid of ceiling joists and laths.

Half pots, divided vertically down the middle, were observed by me, in inspecting those buildings, which were prepared for the purpose of breaking the joints at the intrados of those light arches. *Fig.* 232 exemplifies the above particulars, being the plan of part of the intrados of an arch of this kind.

Fig. 232.

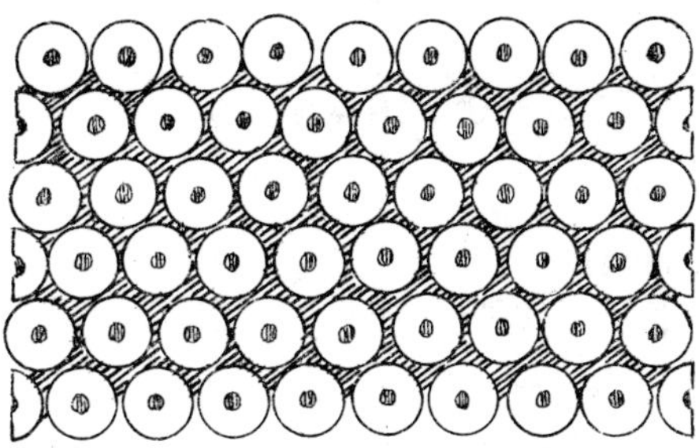

Arches of this description resting on iron girders, compose the kind of fire proof ceiling, before alluded to, which is not injurious to the external walls of a building, like a series of common brick arches, such as were before described. I shall conclude this article by observing that the light arches now under consideration, appear to have strength enough, either to support a pavement of flat stones or of paving bricks or tiles, or to combine with the iron girders, in supporting light joists to carry a common wooden floor, which is most suitable for inhabited apartments in this climate.

After I had written so far, I received from Mr. Sydney Smirke* an account of a manufactory at Salford near Manchester, of a very peculiar construction, which affords a useful lesson in regard to the subjects now

* To whom I am indebted for much other valuable information, he being of the same profession, and a younger brother of the eminent architect of that name.

under consideration, the more especially, as part of this manufactory stood, whilst a portion of it fell down, after being completed.

The building was from 80 to 100 feet long, and 40 feet wide. The floors were supported by very flat segment brick arches, nine inches thick, and having a rise of nine inches. At one end of the building three of those arches had a span of eleven feet nine inches; all the others had a span of nine feet only. These arches rested on iron girders as piers, and on the end walls of the building as abutments, all the walls of which were three bricks, (or two feet three inches) thick.

The iron girders extended transversely across the whole width of the building in two lengths, being supported in the centre by hollow iron pillars, six inches in diameter, and at the ends by stone templets let into the side walls. These pillars had a flange for the under side of the beams to rest upon, and another flange above the top of the beams, for receiving the base of similar pillars in the floor above. Those two ends of each iron beam, which met in the centre of the building, had semicircular sockets embracing the neck of the pillars, with flanges at the ends of those sockets, for receiving connecting bolts and nuts.

The iron beams were of the form of an inverted T. Their depth which, according to custom was greatest at the centre, varied from $16\frac{1}{2}$ to $11\frac{1}{2}$ inches, including the bottom flange, which was $5\frac{1}{4}$ inches in width, and $1\frac{1}{2}$ inch in thickness. The thickness of the stem or vertical part of the beam varied from seven-eighths of an inch at top, to $1\frac{1}{4}$ inch at bottom, where it met the said flange, as in *Figs.* 233 and 234, which show the central and the end sections of it.

Fig. 233.

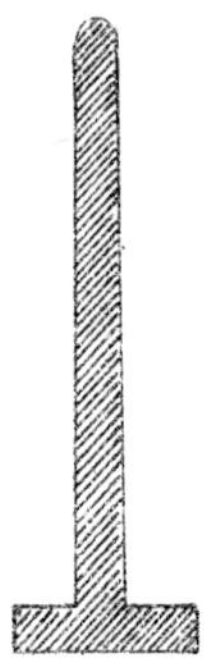

Fig. 234.

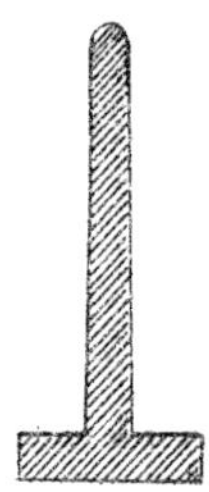

The bottom flanges, by supporting the extreme bricks, served as skewbacks of the arches before described. Iron tie rods extending from beam to beam, and from the extreme beams to the end walls, were inserted through each, in order to connect all the iron beams and those two walls of the building together; and thereby to counteract the external thrust of the arches. These ties were placed at intervals of about 4 feet apart, and penetrated the centre of each beam, on the level of the crown of the flat arches nearly, so that they also ran through the arches themselves, as shown in *Fig.* 235, which is a section of those arches, taken in the longitudinal direction of the building. I am not informed in what manner these rods were connected with the end walls, but it may be presumed that they ran entirely through the brickwork, and were secured by nuts, screwed to the outside of circular plates, of about nine inches or a foot in diameter, through the centre of which they also penetrated, this being a very common method of securing walls against a similar thrust or strain, by means of rods of this description. The manufactory was no less than five stories high, and by this construction was rendered completely fireproof, there being a level pavement over the brick arches, as shown in

Fig. 235.

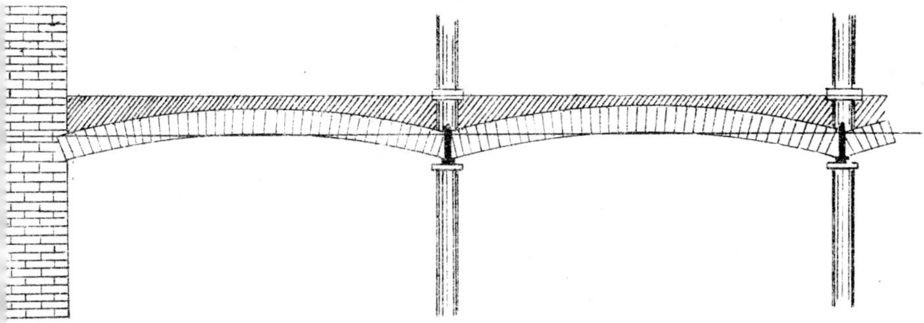

the last mentioned figure. Unfortunately, the iron girders appear to have been too weak, for those three arches which had a greater span than the others, and which, consequently, bore a greater share of the weight of the flooring, &c., fell down from top to bottom of the whole building, whilst the abutment wall on that side as well as the whole of the smaller arches stood. Hence the breaking of the beams must necessarily have been the cause of this accident, which occurred in August, 1824.

The above instructive example proves the very great utility of securing flat arches by iron tie rods or bars, for, without these, there can scarcely be a doubt but that the whole building would have fallen to the ground, as the abutment walls must have been altogether inadequate to resist the thrust of such a series of flat arches in so many successive tiers; for I am not informed that they were built with any cement superior to common mortar. In respect to tie rods, Mr. Sydney Smirke informs me that they are not only used in France, as mentioned by M. Rondelet, but that they are also very commonly used in vaults and in arcades, in Italy, and more especially in the neighbourhood of Naples, where they are applied about one-third way above the spring of semicircular and other arches; so that they are apparent to the eye, whereas, in very flat segment arches, they may be partially or entirely concealed.

I ought to have mentioned that tie rods have been used by Mr. Nash for connecting several of the great iron girders of the floors of the King's new Palace, before described.

Supplementary Remarks on Arch Bricks, or Bricks moulded to the Arch like form.

Since the article on this subject was lithographed, Mr. Nicholson, of Rochester, one of the principal contractors for the magnificent new works in Sheerness Dockyard,* informs me that in building the elliptical arches over the Mast Pond in that yard, he used bricks of the common size nearly, but moulded to the arch like form, agreeably to

* In partnership with Messrs. Jolliffe and Sir Edward Banks. Mr. Nicholson, whilst in partnership with Mr. Baker, of Rochester, also built the Ordnance Barracks at this station, of which the brickwork fell more especially under his superintendence.

the curvature of the proposed ellipses. The body of each of those arches is built with bricks of this description, and is two bricks, or 18 inches thick, whilst the extremities of those arches, which appear in the end view of the building, are of granite, and of a very handsome elevation.

Mr. Nicholson also informs me that his father, who carried on the same business, had some unusually long coping bricks made for coping the walls of the Infantry Barracks at Chatham, which may still be observed on inspecting those buildings. This was done by the desire of the Duke of Richmond, then Master-General of the Ordnance, during whose administration the affairs of barracks belonged to the Ordnance Department, as they do now.

Since the article before alluded to was written, I have also observed coping bricks, moulded on purpose, for the parapet walls of small bridges, convex at top, and flat at bottom, and of the length of 9, or of $13\frac{1}{2}$ inches, according to circumstances, but of moderate width and thickness.

Mr. Nicholson, who is also a brick manufacturer on a very extensive scale, informs me that large bricks for peculiar purposes, or arch bricks having a splay or summering to any given curve, may be made for great arches, at a moderate extra expense, which, in public works of importance, it would generally be worth while to incur; but that for the general purposes of building, nothing can be more convenient than the common standard brick of this country.

I have pleasure in correcting some remarks in my first article on this subject, in deference to the opinion of Mr. Nicholson, whose experience and skill as a builder are well known to the public.

Having now concluded the subject of brick arches, I shall next treat of chimneys, first describing the methods in common use, and afterwards the new and improved system of Mr. Hiort.

General Remarks on the Chimneys of Buildings.

The term chimney is general, including not only the fireplace of an apartment, but also the funnel or vent, for carrying off the smoke from thence above the roof of the building.

The fireplace has jambs on each side, and a mantle over the opening. Immediately below it is the hearthstone, and in front of that is the slab, which also is usually of stone or of marble.

The funnel, or opening above the fireplace, contracts by degrees as it ascends, until it assumes a uniform section of small dimensions, called the flue. The sides of the funnel, where this takes place, are called the wings, and this contraction is termed the gathering of the wings. The narrow part, where the flue commences, is the throat of the chimney.

In buildings of one or more stories, if the plan of each floor be alike, or nearly so, which is very common, all the fireplaces of corresponding apartments in the several stories will stand one over the other, or nearly so, each being situated opposite to the centre of its respective apartment, and each being attached to the same wall, which, in a street, is almost always a party wall between two adjoining houses. In an insulated building of moderate size, it is almost always an end or flank wall. In a very large building fireplaces may also be attached to cross walls.

The gathering of the funnel of each chimney takes place within a moderate distance, not exceeding two feet or thereabouts above its own fireplace; after which the flue is made to turn either to the right or left, in order to clear the position of the fireplace of the apartment immediately above it.

Fireplaces, and even their flues, usually require a greater depth, measuring from front to rear, than the thickness of a well-proportioned wall will admit. Hence, they are almost always made to project inwards, into the apartments which they are intended to warm.

The whole breadth of this projection, as seen within an apartment, including the fireplace, and the brickwork on each side of it, is called the chimney breast, which is usually of equal width from the foundation of the house upwards.

The chimney breast in the basement story has only one flue or opening in the upper part of it, being the flue of its own fireplace; in the next story, as you ascend, the chimney breast contains two flues; in the next, three, and so on; and thus the wall becomes more and more honey-combed, in proportion as it rises.

The whole series of chimneys of several apartments thus arranged, constitute a stack of chimneys, and that part in which the several flues combined rise above the roof, is called the chimney shaft.

In the lower apartments of a building the flues must necessarily be curved as they ascend; in the attic story, and sometimes in the one beneath it, this is not a matter of necessity, but it is generally esteemed better to give even to these a moderate curve, in order that rain or sleet may not beat vertically down upon the fire.

Fig. 236 represents a stack

Fig. 236

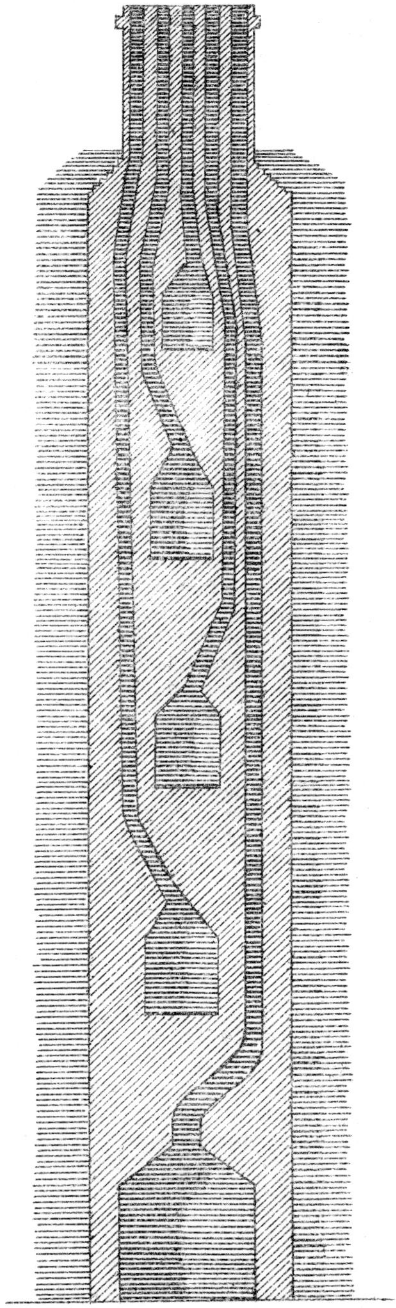

of chimneys in a building of five stories, carried up in the mode now usually adopted.

The flues of the two lowest stories of the building ascend, one on the extreme right, the other on the extreme left, of the whole stack, whilst the flue of the uppermost story is in the centre of the shaft, there being an odd number. If the number were even, no one flue would be exactly in the centre, but the flues of the two uppermost apartments would be the nearest to it. As a general rule, the flues of any lower apartment or pair of apartments are nearer to the extreme ends of the stack than those of another apartment or pair of apartments on a higher level, the latter being always towards the centre.

In carrying up a stack of chimneys, the partition wall between any two flues is technically termed a with, which term is also sometimes applied to the extreme or boundary walls of the whole.

A stack of chimneys may be single or double. It is single at the flank walls of an insulated house. It is always double in a street or row of houses, or in a building of any considerable extent, such as a barrack; for there two sets of chimneys, built back to back, against the same party wall, are carried up in the same stack, and issue forth above the roof in the same common shaft.

Formely both fireplaces and their funnels were made of enormous size, as may still be remarked in some old baronial halls, or in old fashioned farm houses; and, excepting the kitchen and one or two other apartments, no others were provided with fireplaces. Such old fashioned fireplaces always smoked in certain winds, being built contrary to those sounder principles which came to be understood after the middle of the last century, and which were then developed by several writers, of whom Dr. Franklin and Count Rumford are the most celebrated, but who were both anticipated in this subject by Mr. Robert Clavering, a builder of London, in an essay on chimneys, published in 1779. This judicious author has laid down the same rules nearly that were afterwards more fully developed by the others.

Having explaned the general nature of chimneys and defined the technical terms which apply to their component parts, I shall proceed in the next article to illustrate in detail the principles now generally received and acted upon in the modern practice of chimney building.

Of the rules now usually followed, in respect to the construction of Fireplaces, Chimneys, &c.

According to the present practice, every apartment of a building has its own fireplace, although constant fires are not required in all, even in the depth of winter.

It is essential that every fireplace shall have its own distinct flue, terminating above the roof of the building, for if the flue of any one fireplace lead into that of another, one or both of them will be sure to smoke.

The air heated by the fire, being thereby rarefied and rendered lighter, rises up the flue and draws the smoke up with it; whilst colder air is continually rushing into its place from below. Hence, the throat of a chimney should be small, in order that no air may pass up without coming into close contact with the fire; and it is also desirable that the flue should not be larger than is necessary for conveying up the smoke and heated air. If the funnel be too large, a chimney will be sure to smoke in certain winds. The continual renewal of the air is, of course, necessary whenever fires are burning, which, in common dwelling-houses, is generally effected by means of the crevices in the woodwork about the doors and windows, which seldom fit so close, even when shut, but that some air is continually entering; at other times the doors and windows themselves must occasionally be opened, for such crevices cannot always supply sufficient air to a large company, particularly in a well-lighted apartment; but, as the sudden opening of doors and windows, and especially of the latter, after rooms has become very hot, is prejudicial to health, other means of ventilation have often been resorted to in public buildings, in order to prevent the necessity of incurring that risk.

The position of doors and windows has also some effect. They should never, if possible, be situated in the same wall with the fireplace, for in that case, suddenly opening either one or the other, but more especially a door, creates a side current of air, that will drive the smoke into the room, and prevent it from ascending by the flue.

It is essential that all the walls about a stack of chimneys, and the withs or partitions between them, shall be built with the best mortar, and that the flues shall be well pargetted, and all perfectly air tight; otherwise, as bricks and common mortar are porous, cold air and damp may penetrate from the outside, or from the flue of one chimney in which there is no fire into the flue of another chimney having a fire lighted in it, and by cooling or condensing the heated air in the latter may cause it to smoke.

Moreover, much depends upon the height of a flue, the shortest being in all cases the most liable to smoke, for which it is generally supposed that there is no remedy but raising the shaft. Hence, if all other circumstances be alike, the upper apartments of a house are more liable to smoke than the lower ones.

There is another circumstance which often has a very prejudicial effect upon chimneys—that is, their being commanded by higher buildings or objects; which, when the wind blows in certain directions, often occasions violent eddies, that force the smoke back into the apartment. For this also, it is generally supposed that there is no effectual remedy but by raising the shaft, or by using some peculiar chimney pot—turn-cap or cowl—to turn off those eddies. Such are the known facts or principles generally received about the causes and remedies of smoky chimneys.

Of the dimensions proper for the Flues of Chimneys.

The flues of chimneys having hitherto been built of a square or rectangular form, the size of a brick must be considered in their proportions. It is acknowledged by the most experienced builders that nine inches, or one brick square, is a sufficient vent for carrying off the smoke of a common fireplace, but 9 inches by $13\frac{1}{2}$ (or 1 brick by $1\frac{1}{2}$ brick) has been more generally used, because this dimension admits of a chimney-sweep boy ascending, for which purpose the former aperture is too small.

It is also acknowledged that a flue of $13\frac{1}{2}$ inches, or $1\frac{1}{2}$ brick square, is a sufficient aperture for carrying off the smoke of the largest kitchen fire.*

The thickness of withs is seldom more than half a brick, or 4½ inches, including the pargetting, but it appears to me desirable that the partition which divides two sets of flues, built back to back, as well as the extreme or boundary withs of a stack of chimneys, should all be nine inches, or one brick thick. This dimension is, however, to the best of my information, very rarely adopted, excepting in public buildings, or in the private houses of opulent families.

Of the dimensions proper for common Fireplaces.

Having discussed the size of flues, that of fireplaces remains to be considered.

It sometimes happens in the colonies, seldom or never in England, that, instead of using a regular metal grate or stove, iron bars only are provided, which are fitted into brickwork or stone, so arranged as to resemble a common grate.

In either case a rectangular aperture is left in the original brickwork, of sufficient size to admit of any arrangement that is in view; and the fixing of a regular metal grate, or stove, or the building of a fireplace with bars and brickwork, within the said aperture, is an after fitment; and so, also, is the fixing of the ornamental chimney piece in front of it.

Count Rumford, whose writings on this subject deservedly attracted much attention, recommended that the sides of a fireplace, within the jambs of the original aperture, should be coved, as it is techically termed, so as to form an angle of 135° with the back of the same fireplace; and he also recommended, that the throat of a chimney immediately over the fireplace should be contracted to a depth not exceeding four inches.

Fig. 237 represents in plan, and *Fig.* 238 in section, the large aperture

Fig. 238.

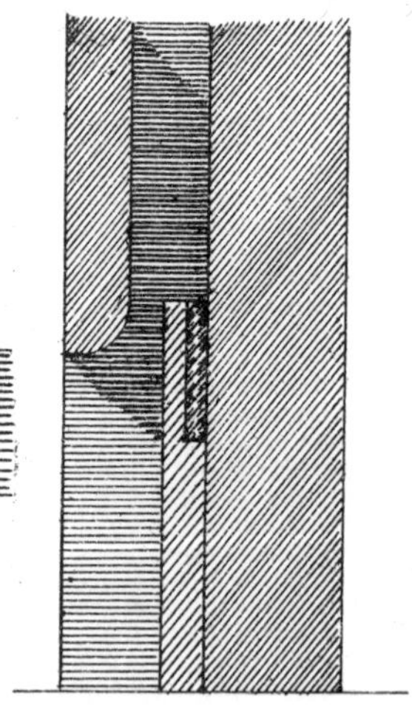

Fig. 237.

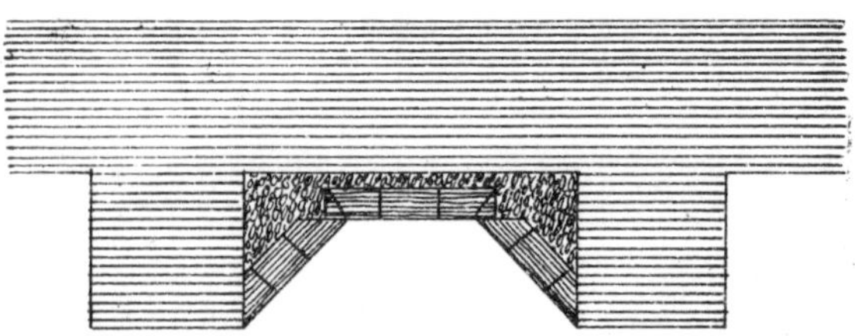

---

* Some recommend 18 inches, or two bricks square, in preference, but I believe that the former dimension is quite sufficient.

of a fireplace, contracted according to the Count's plan, or Rumfordized, as it was usually termed.

The opening of four inches being insufficient to allow a chimney-sweep boy to ascend, it became necessary to place vertically a moveable flat stone, or large flat tile, at the centre of the back of the fireplace, which was taken down whenever the chimney was to be swept, and replaced afterwards.

The distinction between a grate and a stove is well known. The former rises only about half the height of the original aperture, and terminates on that level, so that all the upper part of the fireplace requires to be fitted with brickwork, by means of which the covings and back are formed; and so much of the original square aperture as may appear necessary is thus filled in. The stove, on the contrary, completely occupies the whole of the original aperture, to which it is either exactly suited, or, if too small to fill the whole space, the deficiencies are made good either by brickwork or by iron plates; and the irregularity of this part of the work is afterwards concealed by the ornamental chimney-piece. In all stoves, for the reason before stated, the back plate is moveable at top, and may be pushed from its proper and usual position to a greater width, in order to admit of the chimney being swept. I need scarcely explain that Register stoves are those made with still greater nicety than common stoves, having a contrivance for varying the aperture, not only for the above purpose, but also to suit the state of the fire, and in proportion to the greater or less quantity of smoke that may be ascending.

In planning a building an architect may either adapt the opening of each fireplace to the known dimensions of the most approved grates and stoves in common use, or he may choose his own dimensions, and have grates or stoves afterwards made by order to suit them. If he should wish to leave the fixing of the pattern to the taste of the proprietor of the building, the safest way is to make the aperture a little larger than the known dimensions of those articles in common use; if too large the surplus space can easily be filled in; if too small, the original chimney jambs must be cut, which may injure the bond of the brickwork.

In the Ordnance Department, now including the affairs of barracks, the present pattern of grates and stoves are as follows:—

1st. The officer's mess room stove is 3 feet 6 inches square, and 11 inches in depth.

2nd. The officers' barrack room grate is 2 feet 10 inches wide, 1 foot 10 inches high, and 11 inches deep.

3rd. The grates for servants' apartments, in the basement and attic stories of officers' barracks, are feet inches long, feet inches high, and inches deep.

4th. The mess kitchen range is 5 feet 9 inches wide, 3 feet high, and 18 inches deep.

5th. The private soldiers' large barrack room grate is 4 feet wide, by 2 feet 6 inches high, and 1 foot 4 inches deep.

6th. For civil officers' houses, the pattern stove is 3 feet wide, 3 feet 3 inches high, and 11 inches deep.

7th. For civil officers' houses, the pattern kitchen range is 4 feet 6 inches wide, 1 foot 9 inches high, and 18 inches deep, with an oven 2 feet square, and 18 inches high.

I am informed that all the stoves for dwelling houses made in the

Carron Iron Works are of one uniform size, being 3 feet 2 inches high, and 3 feet in width, and that most of the stoves made in London are of that dimension, or nearly so.

Generally, speaking, therefore, it is altogether unnecessary, excepting in a kitchen, to make the aperture in the original brickwork, for any fireplace, greater than about 3 feet 6 inches, either in width or in height. The height should not in any case be less than about 3 feet, in order to allow room for a sufficient ash-hole below the fire; and if proportion be studied, a fire place narrower than the same dimension of 3 feet would not be pleasing to the eye; but the width may be reduced to two feet, or thereabouts, in attics, where symmetry is not regarded, this space, or even less, affording room enough for a sufficient fire.

In the fireplaces now under discussion, the depth left in the original brickwork must depend on the nature of the flue, and is always somewhat greater than what is absolutely necessary to receive a stove. Suffice it here to say, that it is never less than 13½ inches, nor greater than 18 inches, being made either equal to one brick and a-half, or to two bricks, according to circumstances.

It is to be remarked, that fireplaces are never made larger in very extensive apartments than in small ones, for it has been found more advisable, when rooms exceed a certain size, to have two fireplaces, and stoves of moderate dimensions, in preference to one fireplace, and a stove of greater magnitude. Accordingly, in ball rooms, and in great dinner rooms, and in military mess rooms, there are usually two fireplaces, and in the Elgin marble room, in the new British Museum, there are no less than four.

If an architect should wish to have chimney pieces larger than usual, and in some degree proportioned to the size of a very spacious apartment, it is evident that, without increasing the size of his fireplaces, he may give to the ornamental work about them any dimensions that he pleases; in like manner as a small picture may be set in a large frame.

Of the dimensions proper for the Fireplaces of kitchens.

It appears from observations in this neighbourhood, that an aperture in the original brickwork of from six to eight feet in width, is sufficient for the kitchen fireplace, even of a well-frequented inn or mess-room; the only one wider than the above being that in the Officers' Mess-house in the Ordnance Barracks at this station, which is ten feet wide; but that dimension seems more than necessary, and a part of it has accordingly been blocked up.

Within the aperture there is always a grate in the centre for roasting, besides which there is usually an oven, or a hot plate, and a boiler. From the latter a pipe sometimes leads to a steaming apparatus, which, when used, is generally used at a little distance from the fireplace; and the hot-plate or oven is also sometimes at a distance from it, in which case the latter has generally its own fireplace, in the form of a small furnace, with a metal flue of about four inches in diameter, leading into the kitchen chimney; and here, it may be observed, that there is no objection to such communication of the small flue, under those peculiar circumstances, into the greater, although inadmissable in regard to the chimneys of inhabited apartments.

The large fireplaces of kitchens now alluded to are arched in the usual manner, with segment arches, over an iron turning bar, and the opening usually varies in height from about four feet six inches to five feet six inches, or thereabouts, scarcely ever exceeding the last named dimension. The depth of the kitchen fireplace, sometimes, but very seldom, exceeds two feet. It is more usually less than that dimension. Dampers, as they are termed, or sliding iron plates, are used for the purpose of shutting, occasionally, the flues of the ovens or other side flues, opening into the great flue, which rises immediately over the roasting fire, or open grate, placed in the centre of a kitchen fireplace.

I treat now of the fireplaces of kitchens situated under the inhabited apartments of a dwelling-house. In respect to the cooking-houses of barracks, and of great hospitals, or of colleges, &c., where the wants of several hundred persons are to be attended to, some peculiar apparatus and arrangements for cooking on a great scale, with economy and efficiency, have usually been adopted, which it would be superfluous for me to attempt to describe at present. Suffice it to say, that the architect who is intrusted with the planning of such cooking places ought, of course, to see and inquire into the merits and defects of those of similar establishments, previously constructed and actually in use.

Rules for determining the dimensions of Chimney Breasts, &c.

It was before stated that the dimensions of the flues of common chimneys are generally 1 by 1½ brick in the clear; the latter dimension is usually in depth, or according to the transverse section of the wall; not in breadth, or according to the longitudinal section or elevation of the wall to which the stack of chimneys is attached. Hence, in regulating the width of a chimney breast, it is proper to allow for the width of the fire-place of the attic story, to which must be added the width of the flues of all the other stories, together with the thickness of the withs between and about the several flues.

For example, in a building of five stories, including the basement, let us suppose the fireplace of the attic story to be made 2 feet 4½ inches wide in the clear, and that all the flues are 9 by 13½ inches (or 1 by 1½ brick) in the clear, excepting that of the kitchen chimney, which should be 13½ inches (or 1½ brick) square in the clear; also, that the withs are all 4½ inches (or ½ brick) thick, excepting the extreme withs at the flanks of the chimney breast, which are one brick thick.

The width of the chimney breast required in that apartment, and, consequently, in those below it, will be as follows:—

	feet.	inches.
Width of the attic fireplace....................	2	4½
3 flues of apartments below, at 9 inches ...	2	3
1 flue of the kitchen chimney..................	1	1½
4 withs between flues, at 4½ inches	1	6
2 extreme withs, at 9 inches	1	6
Total	8	9

By curtailing the attic fireplace to 18 inches in width, which is enough for heating a room, though not a pleasing proportion to the eye, and by curtailing the extreme withs to 4½ inches, which I do not recommend, the width of a chimney breast of a five story house may be reduced to 7 feet $1\frac{1}{2}$ inches, which arrangement is very common in private houses built on speculation.

Having determined the proportion of the chimney breast in reference to the attic fireplace, it will be obvious that the fireplace below that may be made wider by nine inches, or by the breadth of one flue, and that a similar increase may be made in succession to the width of any other lower fireplace if required, still preserving the chimney breast of the same dimensions throughout. But, as was before observed, no common fireplace need be made wider than about three feet, or say three feet six inches, at the utmost.

The kitchen fireplace, only, is an exception; but under the supposition of the chimney breast being 8 feet 9 inches wide, there is room for a fireplace of 5 feet 9 inches in width in the clear, with jambs, each two bricks thick, which would be sufficient under ordinary circumstances. But if we suppose the kitchen fireplace to be eight feet wide, which, as was before remarked, is not uncommon in this neighbourhood, then the chimney breast in that floor would require to be no less than 11 feet wide, but it would of course be superfluous to carry up this great width higher than the level of the ceiling of the basement story.

By the Building Act, which was enacted for the prevention of fire in London, the back wall of every chimney must be, in the basement story, not less than $1\frac{1}{2}$ bricks thick; in all other stories, not less than one brick thick; and these thicknesses must be continued to the height of one foot above the mantle, after which they are allowed to be diminished by half a brick.

Fig. 239.

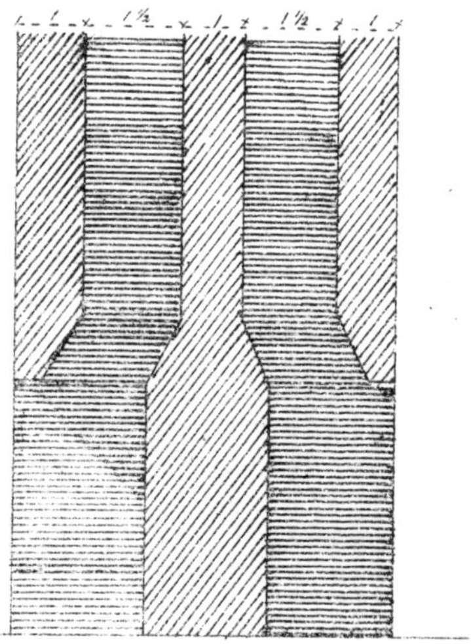

It appears to me, however, on considering the provisions of the Building Act,* that it fixes the minimum dimensions and precautions, consistent with safety, in order not to compel private individuals to adopt too expensive a style of building. I would therefore recommend, in preference, in respect to public buildings, that all chimney breasts, and all the extreme withs at the flanks of the same, as well as all those withs which rise above and in continuation of chimney breasts, whether of a double or single stack of chimneys, shall nowhere be less than nine inches.

Fig. 239 represents the section of two fireplaces, built back to back, and attached to a party wall, supposed to be two bricks thick, in which both the chimney breasts and the partition with between the two sets of flues are each one brick thick; but as it is proper to advance the grate as far into the room as

* The particulars of the Building Act are to be found in all the builders' price books published in London.

possible, the chimney breast is only made half a brick thick over the turning bar, and increases to one brick at the height of about nine inches over the mantle ; and, for the same reason, the chimney back, over which the partition with rises, is made two bricks thick opposite to each fireplace.

Fig. 240 is the plan or horizontal section of the same chimney, taken on the level of the fireplaces ; whilst *Fig.* 241 is the horizontal section

Fig. 240.

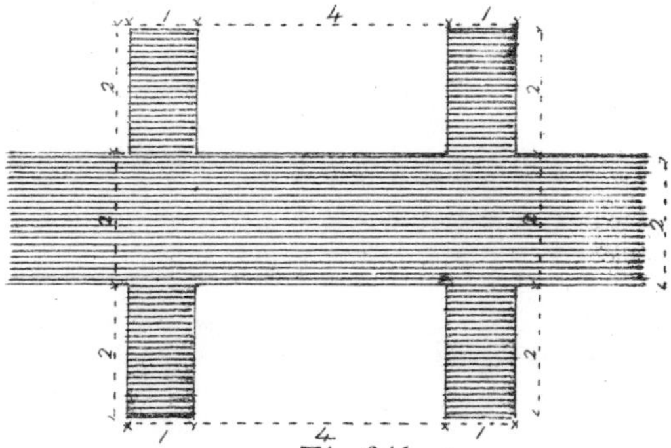

Fig 241

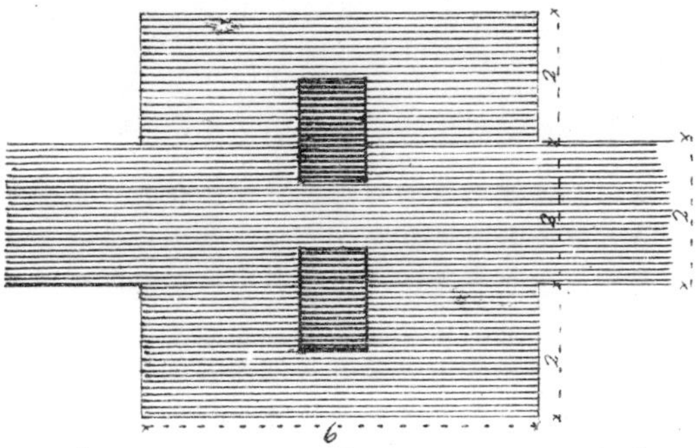

of the same, taken about a foot above the mantles of those fireplaces, and of course passing through the flues.

In all those figures the dimensions are marked in bricks' lengths, and it will be evident, on comparing the whole together, that the projection of chimney breasts into their respective rooms, on each side of the party wall, cannot be less than 18 inches, or two bricks, in order to obtain the standard thickness of one brick for each of the chimney breasts, and also for the intermediate with parallel to them, which divides the two sets of flues.

In a single stack of chimneys, connected with the flank wall of a building, a projection into the room of 1½ brick will be sufficient, when the wall is supposed to be two bricks thick, as shown in *Fig.* 242, and, generally speaking, the projection of a chimney breast may be half a brick less in a single than in a double stack.

From what has been said, the width of the chimney breast of the same stack may, or may not, be greater in the basement story than in that of the story immediately above it. From thence the width of the chimney breast will remain the same, until it clears the ceiling of the attic story. Above that level, and immediately below the roof, a further reduction of width takes place, and the chimney shaft commences, which contains space sufficient for all the flues, and their withs only, and here, as I said before, the withs may be reduced to half a brick, without exception.

Fig. 242.

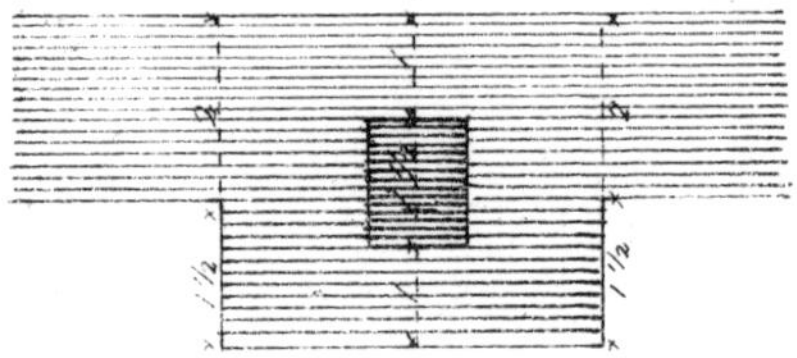

Hence, the length of the chimney shaft of the supposed five story building, which agrees with the breadth of the chimney breasts below, after the latter has been reduced, will be as follows :—

	feet.	inches.
4 flues, at 9 inches each	3	0
1 kitchen flue	1	$1\frac{1}{2}$
6 withes, at $4\frac{1}{2}$ inches each	2	3
Total feet..................	6	$4\frac{1}{2}$

The depth or thickness of the same chimney shaft, measured transversely and supposed to belong to a double stack, will be as follows :—

	feet.	inches.
2 sets of flues, at 1 foot $1\frac{1}{2}$ inch.........	2	3
3 withs, at $4\frac{1}{2}$ inches	1	$1\frac{1}{2}$
Total feet..................	3	$4\frac{1}{2}$

Fig. 243 represents the plan, or rather the horizontal section, of the

Fig. 243.

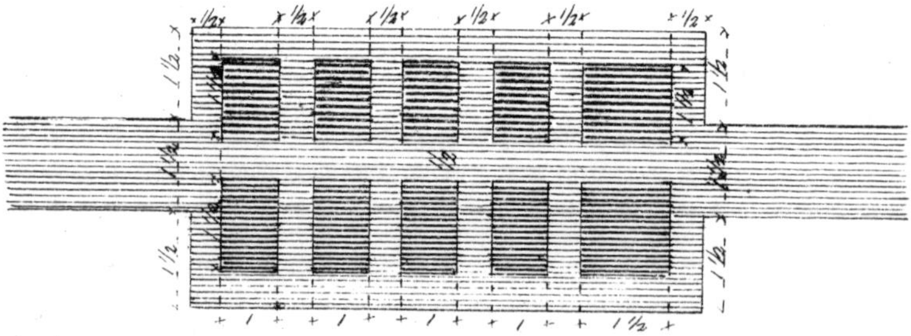

chimney shaft now under discussion, the dimensions being marked in bricks' lengths, as before.

Fig. 244 represents the plan or horizontal section of the same stack, taken on the level of the floor of the attick story of the supposed build-

Fig. 244.

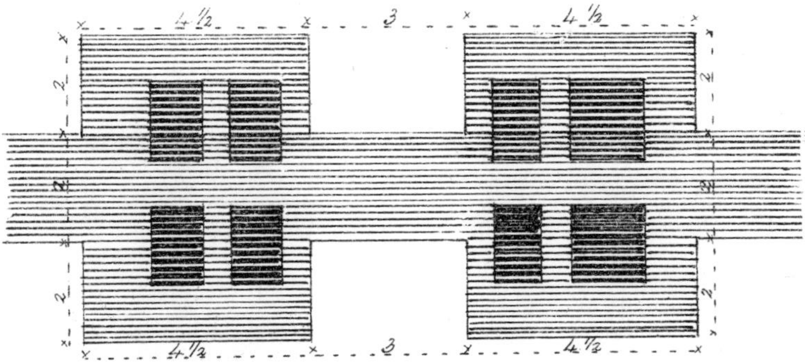

ing, at which level the boundary and partition withs are not less than one brick in thickness.

Fig. 245 represents the plan or horizontal section of the same stack,

Fig. 245.

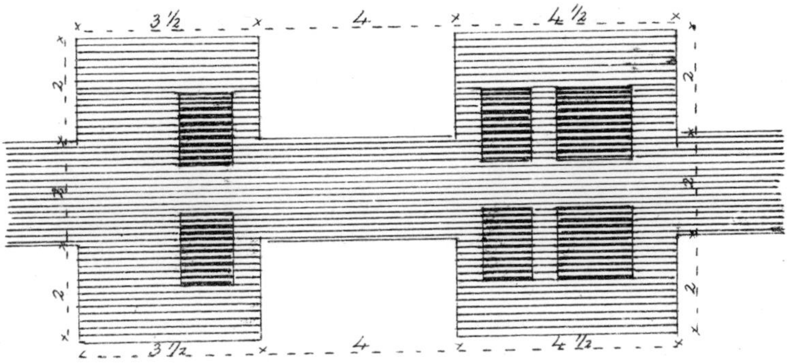

taken on the level of the floor below the attic story, which, in a house of five stories, is usually termed the second floor.

Fig. 246 represents the plan or horizontal section of the same stack,

Fig. 246.

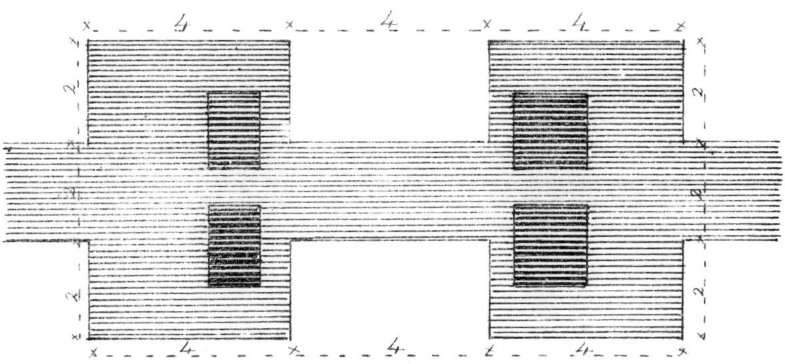

taken on the level of the next lower floor, which is usually termed the first floor.

Fig. 247 represents the plan or horizontal section of the same stack,

Fig. 247.

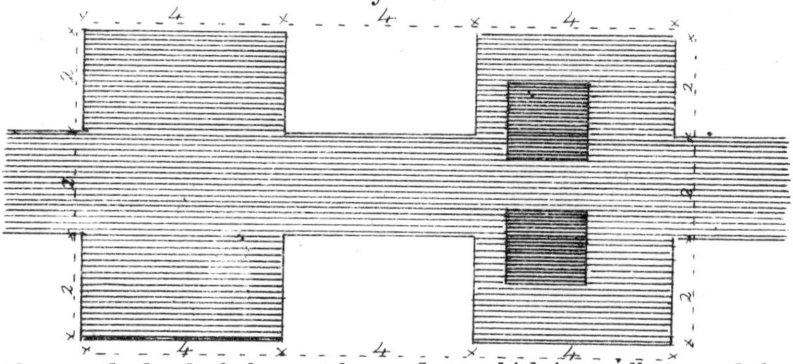

taken on the level of the next lower floor, which is usually termed the ground floor.

Fig. 248 represents the plan or horizontal sectiou of the base of the

Fig. 248.

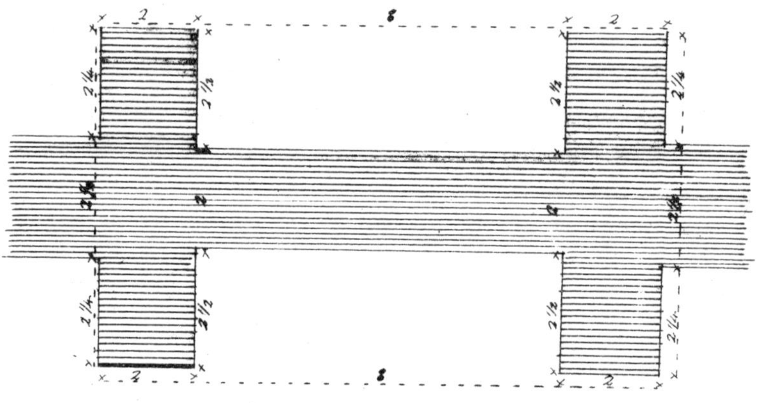

same stack, taken on the level of the fireplace of the basement story.

In apartments, generally, the fireplace is supposed to be two bricks deep, and the projection of the chimney breasts the same; but, if more depth be desired in the kitchen fireplace, let that dimension be increased to $2\frac{1}{2}$ bricks in depth, which will be quite enough, and then there will be an offset, on the face of the chimney breast, on the level of the floor of the story immediately above the basement.

All offsets, or diminutions, when such occur in the chimney breasts of a building, should be immediately above the ceiling of a room, and thereby concealed from view when the building is finished, as they would offend the eye if seen within an apartment.

In reference to the above examples, it may be remarked that $4\frac{1}{2}$-inch withs, without exception, are most proper for the upper part of the shaft of a chimney, after it clears the attic ceiling, and is about to penetrate the roof, as there is no advantage, but rather the contrary, in having it larger than is absolutely necessary. But in all the inhabited apartments of a house, the thickness of nine inches is evidently most proper for the chimney back, but more especially for the front and flanks of the chimney breast, because the battens for the laths and plastering, by which those parts are to be covered, require to be secured by iron hold-fasts driven into the joints of the brickwork; but instead of iron, those battens are often attached to wooden plugs, used for the same purpose, although such ought certainly not to be employed about any part of a chimney. Now, in a nine inch with, there is sufficient space to fix those hold-fasts firmly, without penetrating more than half the thickness of the wall. In a $4\frac{1}{2}$-inch with, on the contrary, the battens cannot be sufficiently secured without a risk of ruining the bond of the brickwork, nor are such withs in this case altogether exempt from danger in respect to fire, especially if wooden plugs be used.

At or near the top of a chimney shaft there are two or three salient courses projecting a little way beyond the sides, which make a kind of finish to the shaft. At this part of the work plain tiles (flat thin tiles) are often used, which give a better base to this projection than common bricks could do in so very thin a wall. The finishing course of a chimney is also sometimes worked with plain tiles, in order to obtain a gentle slope at the extreme top.

Of the common method of executing the brickwork of Chimneys; with remarks on the propriety of using cement instead of common mortar, and pargetting.

In commencing the brickwork of a stack of chimneys, on the level of the foundation, the projection of the chimney breast ought to have its footings, in the same manner as any other part of the walls. These should include not only the jambs, but the hearth.

In building the jambs of a fireplace, the same rules are followed as in common brick walls, as nearly as possible.

The gathering courses, by which the throat of a chimney is gradually contracted, are made each to overlap or project beyond that which is immediately below it, by any small dimensions, not usually exceeding $1\frac{1}{2}$ inch, or thereabouts, until the desired change of width is effected.

Fig. 249 represents the longitudinal section of a fireplace, and part of its funnel.

The jambs are first carried up vertically, in the usual manner. Above the top of the fireplace the construction of the throat of the chimney is effected by projecting courses. The flue is then carried up vertically for a few courses, after which is represented a turning to the right, by gathering or projecting courses on one side, and by receding courses in steps on the other. In the projecting or overhanging courses, the angles of the bricks are always cut away during the execution of the work, instead of leaving them square, as in the figure, which has been done for the sake of clearness. In the receding courses there is no cutting, but small triangular fragments are often laid on the steps to make a smooth surface.

When several $4\frac{1}{2}$-inch withs are all made to turn to the right or left,

Fig. 249.

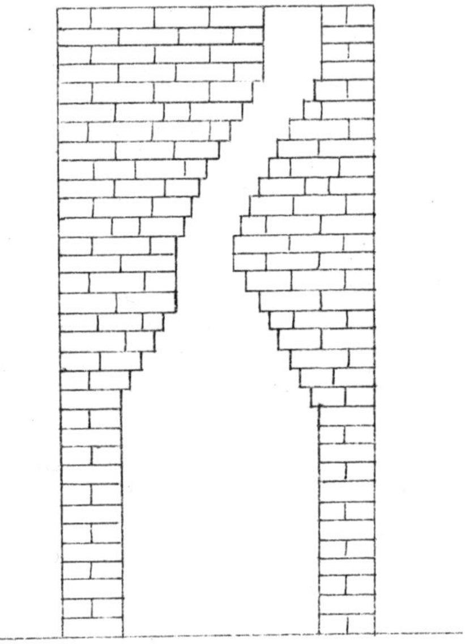

at once, as shown in *Fig.* 250, unless the projection or step in each course be very small indeed, the work cannot be strong, taking the cuttings into consideration, which are represented in this figure.

Some builders fill the mass of the chimney breasts of the lower stories with rubbish inclosed in a mere casing of brick.* This trifling saving of materials is to be reprobated, as it may be the cause of a stack of chimneys being blown down in a gale of wind, which not unfrequently occurs in ill-built houses, to the risk both of lives and of property.

Considering the great number of changes which continually take place in the direction of flues by those small projections or steps, which are not any aliquot part of a brick, like those used in the footings of walls, it will be evident that in building a stack of chimneys whole bricks cannot be used, as in a common wall. In almost every course, bricks cut a little shorter, or small bats must be used; and, therefore, no system of regular bond can be laid down applying to all those varieties; and much must be left to the judgment of the bricklayer, who will take care to break the joints as well as he can, which may always be done by an intelligent person. Indeed, I have seen a wall eight feet high,† built entirely with brickbats, in which, notwithstanding the smallness of those generally used, scarcely any two joints were unbroken.

Fig. 250.

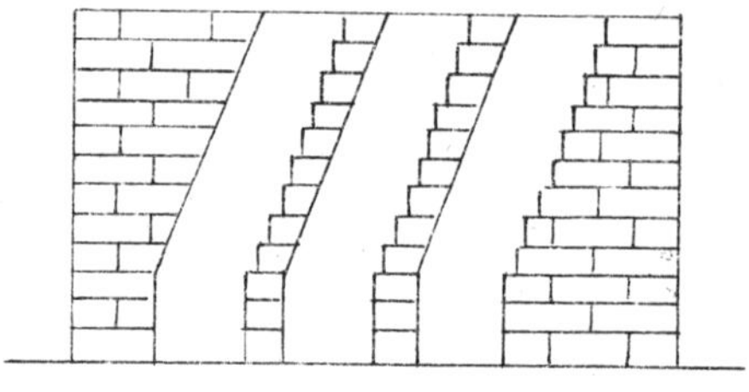

Since the bond of chimneys must necessarily be imperfect in many, if not in most parts, of a great stack, and since it is desirable that every part should be air tight, it appears proper that the work should be executed either with cement, or with the very best description of

* These cavities, filled with rubbish, are technically called pockets.

† Built at St. Mary's Creek by my direction: to be blown up with gunpowder for experiment.

mortar, mixed with tile dust to prevent it from settling; and, in respect to the withs, these also should be plastered with cement, or with a superior kind of mortar, instead of the common pargetting, in which horse or cow dung is an ingredient.

I have recently learned that the last-mentioned composition is not always used in buildings of importance. Mr. Clavering, whose book I had not seen when the article of pargetting was written, recommended, nearly 50 years ago, that the mortar of chimneys should be mixed with coal ashes, to give it the properties of cement, and that the pargetting should be composed of two parts of stone lime (or Dorking lime), one part of fine drift sand, and one part of coal ashes, or of brick dust, for the first coat, and of fine plaster of Paris, mixed with stale small beer, for the second.

In the new King's Mews Barracks, the pargetting of the flues consisted of one part of Dorking lime, one part of brick dust, and two parts of river sand.

Of the Brick Bond proper for the shafts of Chimneys, when the boundary withs are one brick thick.

It was observed that the boundary or extreme withs of a stack of chimneys should all be nine inches thick, until the shaft is on the point of rising above the roof. Our present article, therefore, applies to that part of the stack which is above the ceiling of the attic story but below the roof. We will first suppose the shaft under discussion to belong to a single stack of chimneys, containing three common flues of 1 by $1\frac{1}{2}$ brick, and one kitchen flue of $1\frac{1}{2}$ brick square.

The external withs being built English bond, let the bricks of the intermediate transverse withs be each let into the stretching courses of the long sides by a depth equal to one quarter of a brick. The inside course of stretchers of each of the said long sides will, therefore, be broken by the intermixture of closers and half bricks, as shown in *Fig.* 251, which represents one course of the supposed shaft.

Fig. 251.

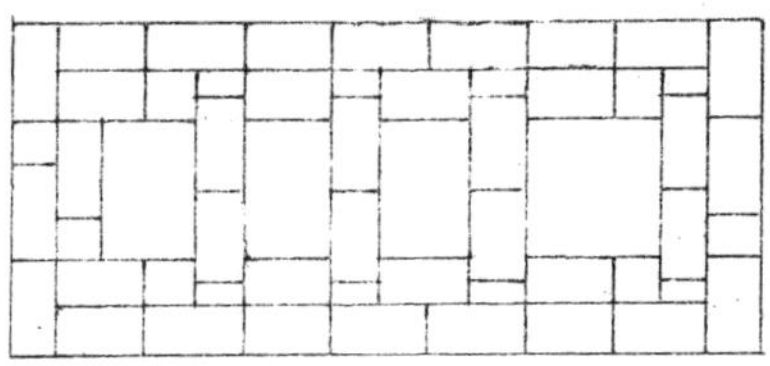

The next course in succession being a heading course in respect to the long sides of the shaft is represented in *Fig.* 252.

Fig. 252.

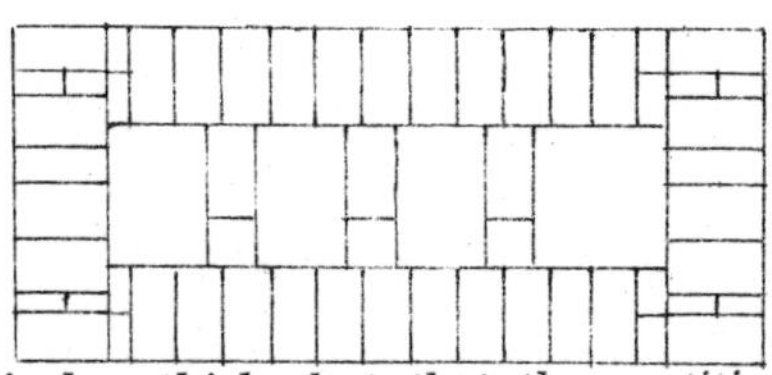

The arrangements shown in this and the former figure, alternating with each other in every two courses, will produce as good bond as can be desired.

Secondly, when the shaft belongs to a double stack. In this case we will suppose the extreme, or boundary withs, still to continue nine inches thick, but that the partition with between the two sets of flues has been reduced to half a brick, or $4\frac{1}{2}$ inches only, by gathering in the two long sides of the shaft, just above the ceiling of the attic story. We will also suppose that three common flues and one kitchen flue are contained in each portion of the double stack, making eight flues in all.

Figs. 253 and 254 represent the bond that may be used in the alternate courses of such a stack.

Fig. 253.

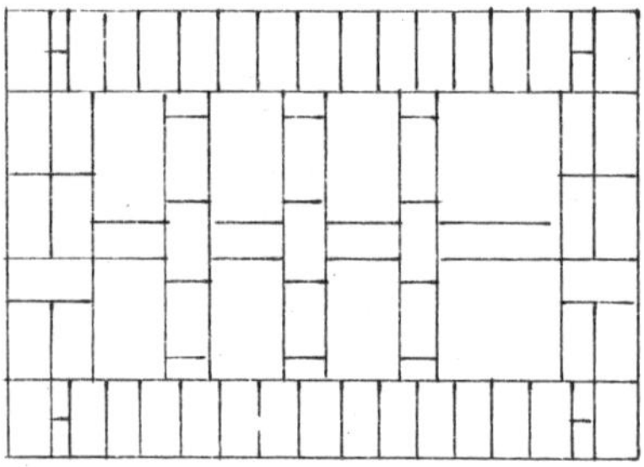

Fig. 254.

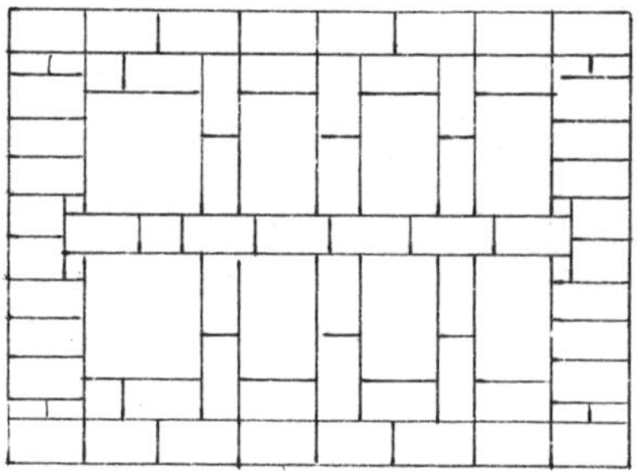

In reference to the second of those figures, another method occurs, of bonding the two extreme bricks of the partition with into the heading courses of the nine-inch withs at the ends of the shaft, which is by cutting those two bricks in the form of a salient angle, and cutting out a re-entering angle in those bricks of the said heading courses which are to meet them. This arrangement, represented in *Fig.* 255, may be substituted in alternate courses in lieu of that shown in *Fig.* 254. A salient angle of this kind is technically called a bull's-nose, and the re-entering angle into which it is fitted is called a bird's mouth.*

Fig. 255.

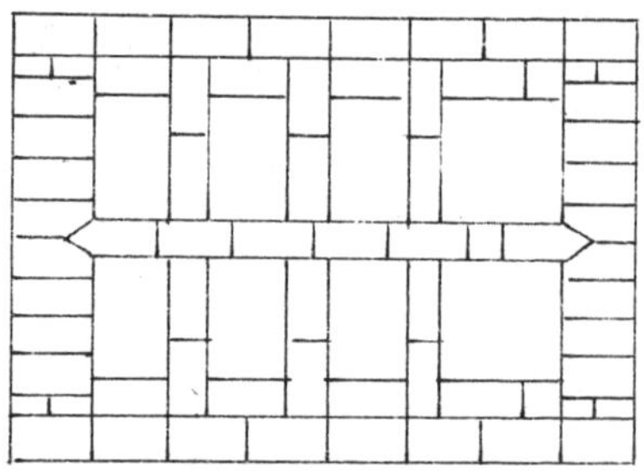

Of the Brick Bond, proper for the shafts of Chimneys, having all their withs half brick thick.

Just before they rise above the roof, the shafts of chimeys are usually diminished to four and a half-inch withs all round, as was before explained.

In illustrating the bond proper for this part of the work, we will first suppose the shaft to belong to a single stack, and to contain the common flues and one kitchen flue, as in the preceeding article.

* These words are also used as verbs and participles by mechanics. In carpentry, for instance, a piece of timber is often said to be bird's mouth.

Under this supposition, Flemish bond for the long sides of the shaft, with real and sham headers in alternate courses, offers a very simple method of combining all parts of the work in an advantageous manner.

This arrangement is exemplified in *Figs.* 256 and 257, which represent

Fig. 256.

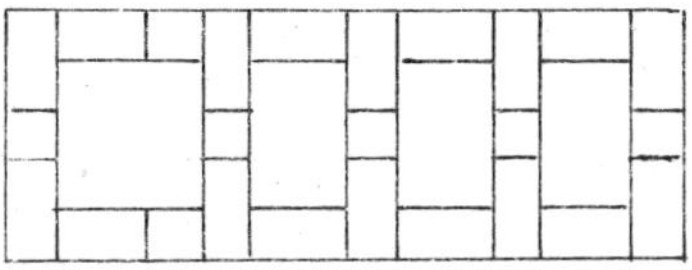

Fig. 257.

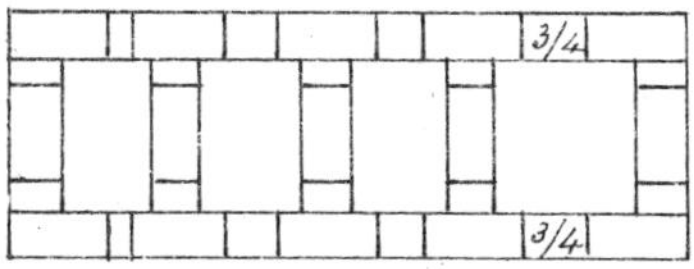

two successive courses of the chimney shaft under discussion, which are supposed to alternate in this manner as high as the salient courses, usually added towards the top, which were described in a former article.

It will be observed, that in the second of those figures, the transverse withs are formed by one whole brick and two closers, equal in length to one and a-half brick in all. An equally good bond may, therefore, be obtained for those withs, by the occasional use of two three-quarter bats, and the same remark will apply to other examples of transverse withs that will afterwards be given in treating of the bond of chimney shafts.

Another method of building chimney shafts, of the nature now under consideration, is to use all stretching bond in the external withs; but as this arrangement would leave the internal or transverse withs without any bond at all, excepting the bricks be mitred or cut like king closers, at the junctions of the transverse withs, they are accordingly so cut, in each alternate course, whenever stretching bond is used.

This arrangement is exemplified in *Figs.* 258 and 259 which represent

Fig. 258.

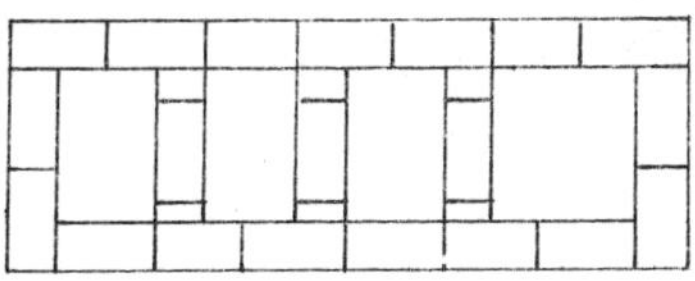

Fig. 259.

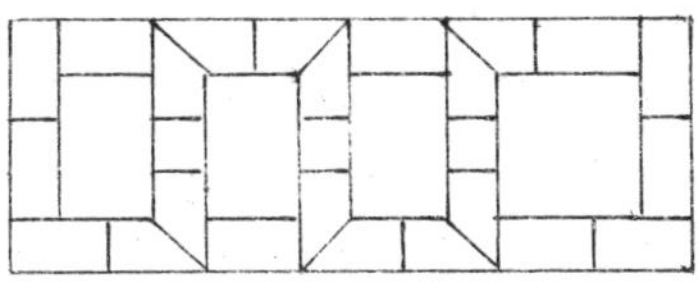

in plan two successive courses of a chimney shaft of the same description as the former, but built all stretching bond externally.

Fig. 260 is the elevation of some courses of the same.

Fig. 260.

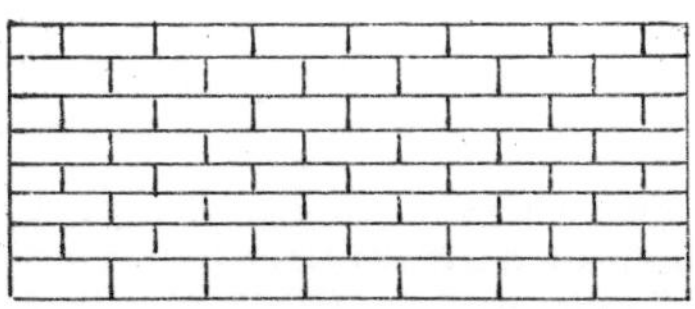

We will next suppose the shaft to belong to a double stack of chimneys, having on each side of it three common flues and one kitchen flue (eight in all), as in one of our suppositions.

In this case also, Flemish bond offers a simple and efficient method of arranging the bricks properly, as shown in *Figs.* 261 and 262, which represent in plan two successive courses of a shaft of the description now under consideration.

Fig. 261.

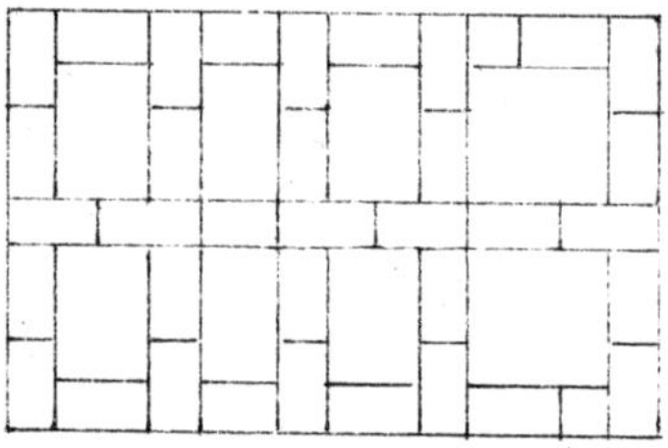

Fig. 262.

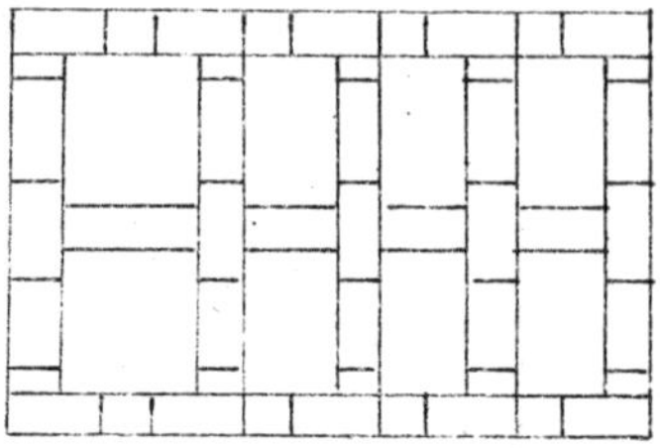

It may be remarked, that instead of using all sham headers in alternate courses, in the long side of a chimney shaft, three-quarter bats might occasionally be used in those courses, and equally good bond obtained. I shall not give a figure to illustrate this arrangement, as it may easily be understood, by means of your small brick models.

The shaft of a double stack of chimneys, of the nature now under consideration, may likewise be built, all stretching bond externally, by mitring the transverse withs in alternate courses, in the manner before explained, in treating of the same bond, as applied to a single stack of chimneys.

This arrangement is exemplified in *Figs.* 263 and 264, which represent two successive courses of the brickwork of such a shaft.

Fig. 263.

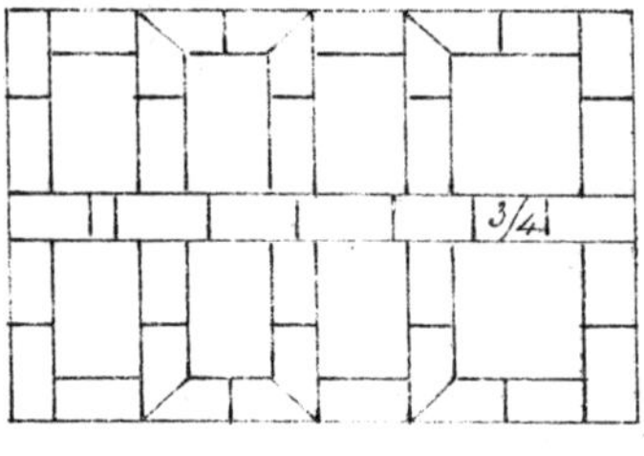

Fig. 264.

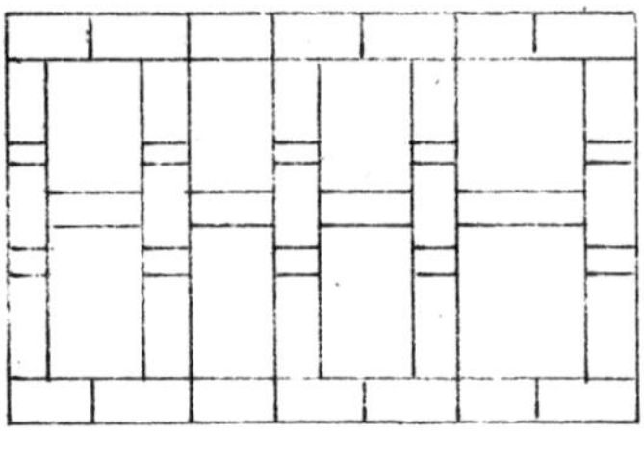

It may be remarked, that so long as a kitchen flue one and a half-brick square forms part of a stack of chimneys, it is absolutely necessary that the depth of the other flues, which are oblong, shall be greater than their width, otherwise there would be an unseemly irregularity in the front of the chimney breast. In all the foregoing examples, I have, therefore, followed this arrangement, and do not think it worth while to give examples of any other.

Varieties in regard to the Arrangement and Construction of Chimneys.

The distinction between single and double houses is well known, the latter, which are the most common, having two sets of apartments in the width of the building.

In double houses the roof is sometimes formed with a single span, having its ridge extending longitudinally over the middle of the building. In this case, as there must necessarily be two stacks of chimneys to every double house, it is usual to combine them in one common shaft rising from the centre of the roof, in order to avoid the necessity of raising two shafts on different sides of the ridge, and lower down the slope, and consequently of greater apparent height, and of a more awkward appearance.

To effect this object, an arch is used connecting the two stacks, and as the flues of chimneys are light in themselves and are always bonded into the walls to which they are attached, such arches are not attended with any great strain upon their piers.

Fig. 265 represents this arrangement which is so very common that

Fig. 265.

examples of it may be seen everywhere, for those arches often appear externally*, but are always concealed in the interior of the building, as the recesses on each side of the chimney breasts are often converted

* For example, at the store keeper's house in the Gun Wharf.

into cupboards or closets; and this would always be done in preference to showing such arches internally.

When the roof of a double house has any other form than the above, each stack of chimneys is carried up independently.

In large houses there are frequently, in the upper stories, small apartments, such as dressing rooms, &c., having insulated fireplaces, not rising in the usual manner from the great stack of chimneys in the basement story.

Sometimes such apartments have their fireplaces in one corner. Those angle fireplaces are supported either by turning a small arch from wall to wall, or what is equally common, they may spring from a piece of stone, of a sound tenacious quality (such as Yorkshire stone), inserted into both walls. The flue of such a fireplace is usually carried up at the same time that the walls are built, and the internal angle is then cut off to suit the back of the flue, the whole of the brickwork being bonded properly together.

When the fireplaces now under consideration are situated on one side of the small apartments to which they belong, instead of being formed at an angle, they generally project inwards in the usual manner, but without having any support from the lower floors. In this case, therefore, the jambs are sometimes supported by stone corbels, let into, and projecting from, the brickwork of the wall; sometimes the support is afforded by gathering courses of bricks projecting one over the other, until the proper dimension is obtained, as shown in *Fig*. 266.

Fig. 266.

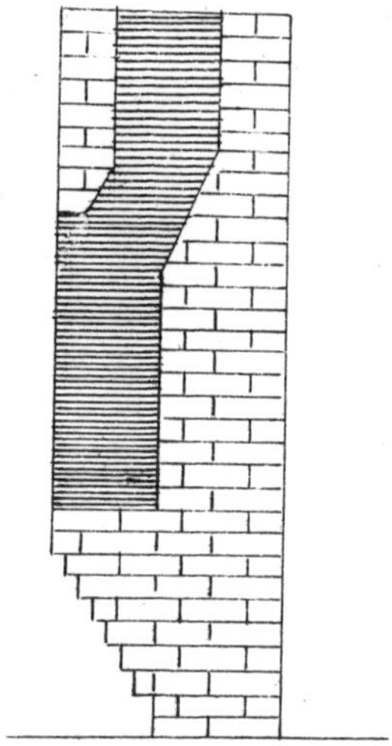

Sometimes the back of the chimney of an upper apartment is made to project, entirely beyond one of the external walls of a building, in which case the same system of stone corbels or brick gathering courses is resorted to as before, and there they are usually apparent to the eye; whereas those corbels, &c., are never allowed to appear internally in a finished building. This last arrangement of a flue, to project externally from the upper part of the wall of a house, is prohibited in London by the building act, nor is it to be recommended; but is here noticed merely as a variety, and is repesented in *Fig*. 267, in which a stone corbel is supposed to be used.

Fig. 267.

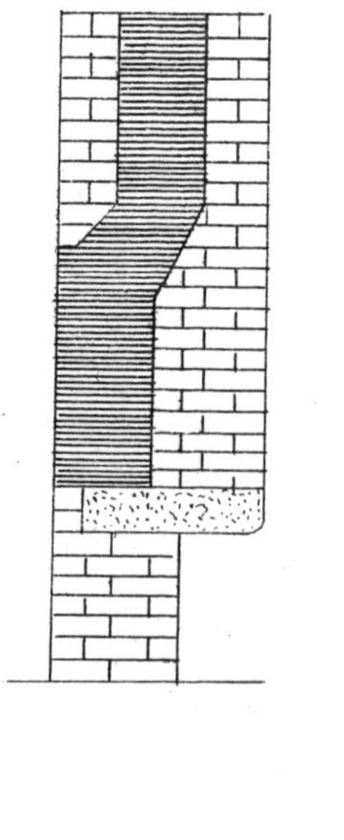

Instead of stone corbels or gathering courses, the fireplace of a small apartment in an upper story is often supported by an arch in the story below, thrown over for that purpose, between the principal stack of chimneys and the nearest wall that can conveniently be used as a second abutment.

This arrangement, which is very common, is represented in *Fig*. 268.

If those arches appear in a room over a recess, between the chimney breast and wall, they are

Fig. 268

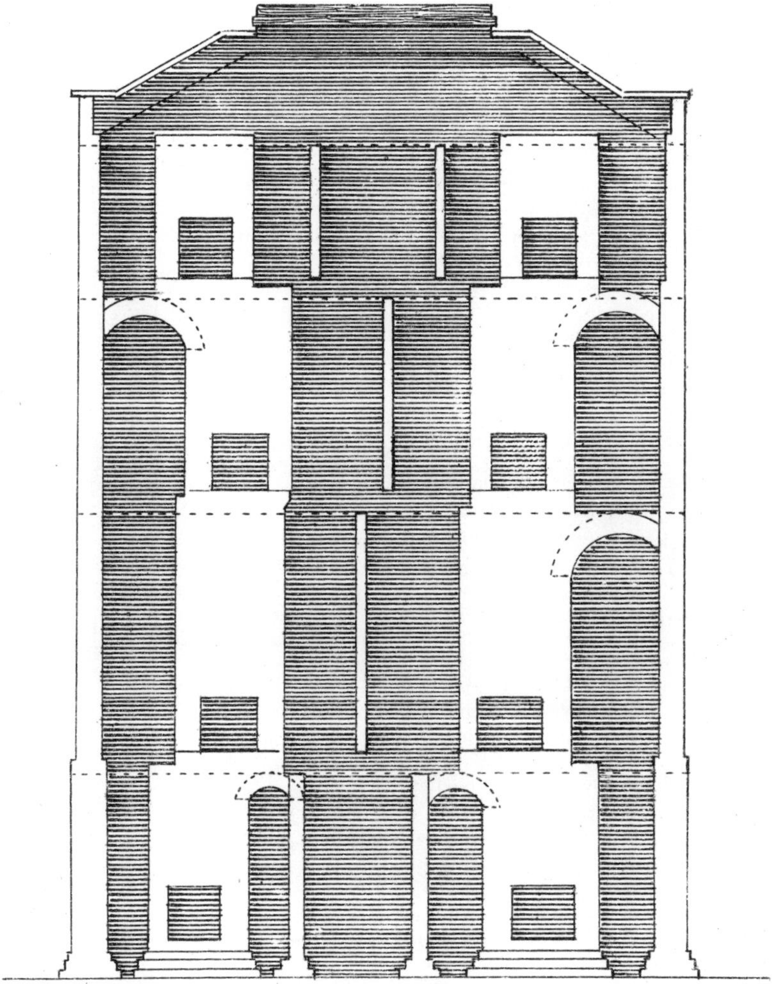

usually rendered ornamental, but such recesses are sometimes concealed by fitting them up for book-cases, cupboards, or wardrobes. When one of two recesses is arched for use the other is usually arched for symmetry, in finishing the room, but not with brickwork.

Sometimes the flues of chimneys are gradually made to incline backwards, and are conducted into the centre of the thickness of a party wall, or of the flank wall of a substantial building, instead of being carried up vertically in a stack. By this arrangement the chimney breasts of each successive apartment may be made no broader than is necessary for containing one fireplace; and the chimney breasts of the attic story usually terminate a little above the level of the ceiling of that floor. This arrangement also, is supposed to have been adopted in *Fig.* 268, which shows the internal elevation of a set of chimneys of this description.

Fig. 269 is the section of the fire-place and flue of the attic story of the same building, showing the setting back of the flue, and the abrupt termination of the chimney breast, as before explained.

Fig. 269.

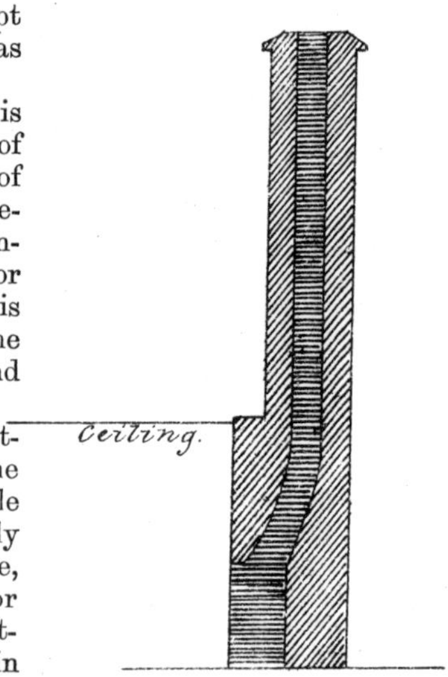

The kitchen of a lofty building is sometimes an insulated apartment of one story only, in the back yard of the house. In this case the flue is generally conducted by means of a leaning arch, or by resting on wooden or iron beams, on which the brickwork is built until it meets a back wall of the house, to which it is attached, and thus it is made to clear the roof.

In buildings having very low apartments, not affording room for all the flues to change their direction by gentle curves, they are often made nearly horizontal in some part of their course, and in this case the withs of one or more flues are often formed, as a matter of necessity, with long stones in those particular portions.

Fig. 270.

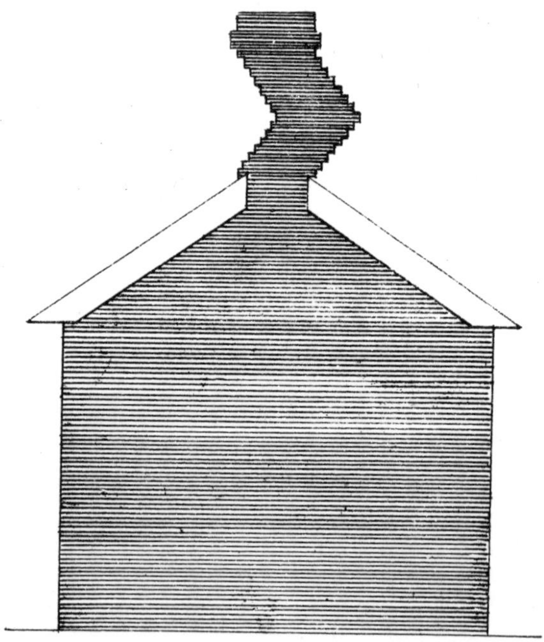

When chimney shafts, and, more especially, those of single flues, are raised to a more than usual height, it is customary to secure them by iron bars or guys, which are, of course, more effective on three sides than on one or two only.

The shafts of chimneys about London are not always carried up vertically above the roof, in the manner that has been described. In many buildings, but usually of an inferior description, they have been made to assume a fantastic appearance, by giving to the flues a curved form, by means of gathering courses projecting first to one side and then to another, as in *Fig.* 270.

Of Chimney-pots, Cowls, or Turncaps.

In the foregoing articles I have fully described the system of chimney building now in common use.

Formerly, instead of carrying up all the flues in one and the same common large rectangular shaft, rising uniformly to a certain height above the roof of the building, it appears to have been the custom to build a distinct shaft for each flue, as may be seen in an old house in St. Margaret's, Rochester, where there are three distinct square shafts, rising about six feet above the flank wall of the building, and placed in such a manner that one diagonal of each agrees with the centre of the wall. Hence the wind has a free vent between each of those shafts, and is turned off by the angles which they present to it.

Formerly, also, it was the custom to build chimney shafts of very curious ornamental brickwork, resembling sometimes independent columns, sometimes a cluster of columns combined together, of which specimens may be seen in Rochester, and in Gillingham, and in an old farm house near New Brompton, now in a dilapidated condition, and all built of woodwork, and of lath and plaster, except its very curious stack of chimneys. This old practice was probably copied from the Venetians, who generally make their chimney shafts appear as ornamental columns.

The same has been done at the Bank of England.

The great shafts of our modern chimneys, in which many flues are combined, oppose a considerable and direct obstsuction to the wind, instead of turning it off, like the small and insulated shafts of former times. Accordingly, it has become the custom, probably from finding the necessity of it, to erect round chimney-pots of earthenware over each flue, with an opening of some inches between every two adjacent ones.*

Recently, square chimney pots of stone, or of cement resembling it, and also affording a free passage to the wind, have become common, and are more ornamental than the former. They both taper by a gradual diminution towards the top.

Cowls, or turncaps, of metal, have also been frequently used, which, having vanes attached to them, and turning on a spindle or pivot, always oppose their back to the wind, and leave the opening for the smoke on the leeward side. These have been made of the most fantastic forms that can well be imagined, but an obvious inconvenience results from using them anywhere excepting over an insulated flue. When placed over a chimney shaft containing many flues, there is not

---

* These pots are often square at bottom, but this part is usually concealed from view.

room to have turncaps for all; hence, it is usual to give a turncap to one or two flues only out of a great number, which overtop the others, and give to the whole arrangement a very whimsical appearance. For the same reason, turncaps of unequal height have sometimes been used in adjacent flues.

Chimney pots, &c., are fixed to the top of a shaft by bedding them, as it were, in a slope made round them with brickbats, fragments of tiles, &c., and mortar,

In respect to the merits of metal cowls or turncaps, the best way is to enquire of those persons who have used them; not to take the opinion of those who deal in them. This rule suggested itself forcibly to me, from the circumstance of having recently enquired into the effect of one of a very ingenious and promising description. On going to the only house in this neighbourhood where it had been erected, I found that the proprietor had recently pulled it down, as useless, after giving it a trial of several months.

Remarks on the difficulties attending the subject of Smoky Chimneys. Account of some expedients for relieving this nuisance.

The subject of chimneys is a very difficult one, and attended with anomalies not always reconcilable to those rules which experience has generally proved to hold good.

For example, most of the chimneys in the Ordnance Barracks at this station smoke a little occasionally, although built on the highest ground in the neighbourhood. Perhaps their being a little larger than the usual dimension may be one cause.* But what is remarkable, some of the chimneys of the towers of the barracks smoke more than other chimneys that are lower; and some of the upper rooms, although having much shorter flues, smoke less than lower apartments, whose flues terminate in the same shaft. In Fort Pitt, the chimney of one low building that had been uninhabitable in winter, was cured of smoking by altering the fireplace, raising the flue a little, and putting on a turncap. In the same fort a kitchen, which smoked so violently at times as to fill every apartment in the house above it with smoke, was cured by a violent gale of wind blowing down all the chimney pots and cowls of the shaft in which the flue of this chimney had been carried up.

In the Garrison Library at this place,† a one-story building commanded by higher objeets, in which both chimneys were constructed exactly alike, one smoked, whilst the other did not, which must have proceeded from a very slight difference of situation, for the whole distance between the two is less than 40 feet. A back window having been cut last summer in the room which smoked, for another purpose, and without any expectation of curing the evil, considerable benefit was found to have resulted from this alteration, on the approach of winter.

In chimneys belonging to the same shaft, the wind will, at times, force the smoke from the top of one flue down another, and thus will fill a room with smoke in which there is no fire. The only remedy for this evil, which, however, very rarely occurs, is to shut up the flue of

---

* I believe that they are one and a-quarter by one and a-half brick wide or $11\frac{1}{4}$ by $13\frac{1}{2}$ inches. The flues of the chimneys in the infantry barracks are larger and more troublesome.

† A building erected for a library and reading room by subscription of the officers of the Royal Navy and Army, and of the Civil Officers of both services at this station.

that chimney in which there is no fire, by a chimney board, or otherwise.

An expedient, which I have known to be of some use in lessening the inconvenience of smoke, consists in introducing a small metal tube at the back of the fireplace, open at the top, which must point upwards and rise above the level of the mantle piece. This tube must have no communication with the air of the room at its other end, which must be fed with cold air from without, or from some cool apartment, in which there is no fireplace. When the chimney is built against the external wall of a house, this pipe is made with an elbow, part of which may go right through the wall, as shown in section in *Fig.* 271.

Fig. 271.

The air in the upper part of the tube being heated by the fire is, of course, rarified, and being fed with cold air from without, rushes up the chimney in a continual current, and with such force as tends to prevent smoking in some degree.

Another expedient which I have known to effect an improvement in some of the chimneys at the Ordnance Barracks, and also in the chimney at Fort Pitt, before alluded to, consisted in first reducing the throat of the chimney to a smaller depth, on Count Romford's principle; and besides this narrow vertical opening, making another vent for the smoke about six inches square, a little above the back of the grate, which aperture, being cut through brickwork or stone, communicates with the flue about or a little above the height of the mantle piece.

Fig. 272.

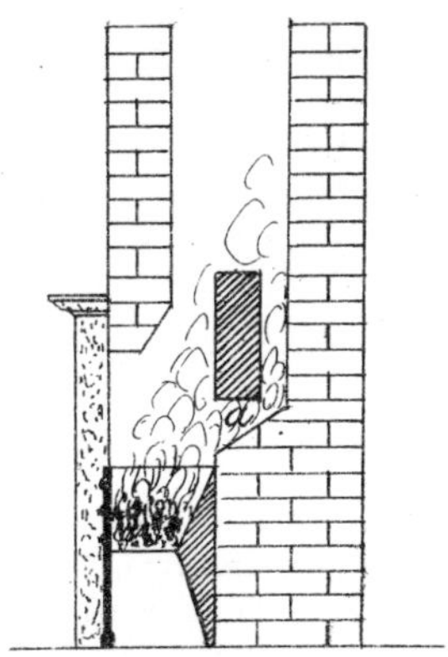

Fig. 272 represents this arrangement in section, in which the small aperture above the back of the fire place is marked by the letter *a*.

The benefit derived from this horizontal vent, to which the smoke is very strongly attracted, appears to have some analogy with a most beneficial improvement, recently introduced in the smithery of His Majesty's Dockyard at this place, by Mr. Parkins, the present master builder of the yard. Formerly, every fireplace, or hearth, as it is styled, in a smithery, had a very large chimney for carrying off the smoke, consisting of a brick flue with a great projecting hood at the lower part of it, suspended by ironwork, which, on account of its size, was necessarily about 12 feet above the hearth, to prevent it from interfering with the crane by which anchors and other heavy pieces of iron are moved to and from the fire. Owing to the very great intervening space between the hearth and the hood above it, the greater part of the smoke, instead of finding its way up the flue, was dispersed about the the whole smithery, as shown in *Fig.* 273.

The alteration made by Mr. Parkins, by which he cured that nuisance, consisted in building a vertical flue behind each hearth, or a pair of vertical flues between those hearths, which are placed back to back. A square opening, about one foot above the hearth, communicates with and forms the commencement or lower part of each flue, and thus the smoke enters into the flue horizontly, and immediately changing its direction rises vertically up to the chimney top. It is curious to observe the flame, and the smoke proceeding from it, which one might naturally suppose would take a vertical direction and fill the whole smithery, both rushing as it were out of the natural course, to enter the horizontal opening.

This arrangement is shown in *Fig.* 274, which represents a fire made

Fig. 273.

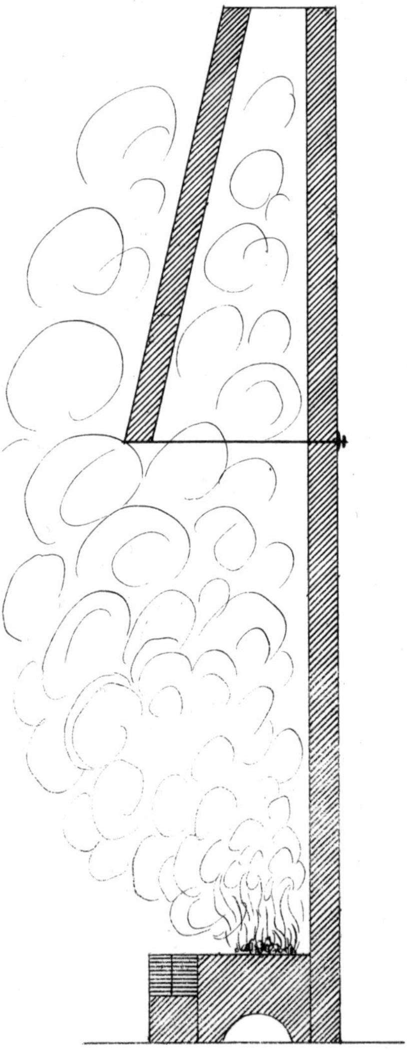

Fig. 274.

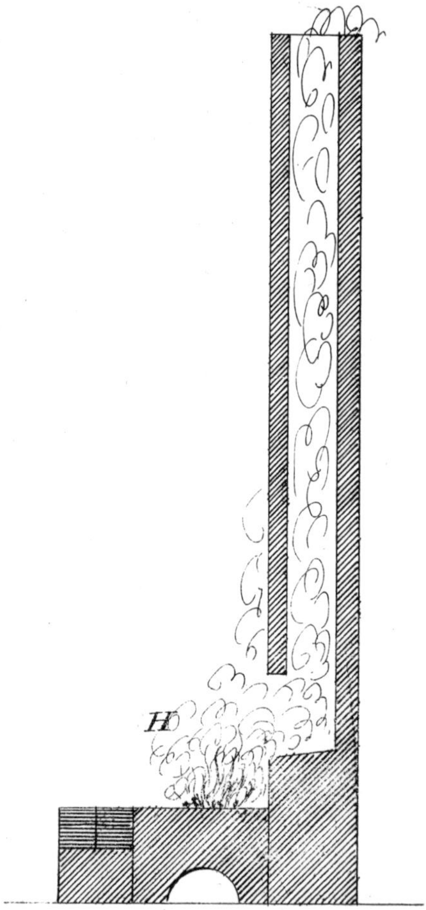

on the hearth, *H*, the smoke from which enters horizontally into the adjoining flue, and from thence ascends vertically to the top of the building. Still some of the smoke escapes, which is unavoidable in a large smithery, but not in sufficient quantity to annoy the workmen, as heretofore.* This example seems to prove that a fireplace might be thrown entirely forward into a room without occasioning smoke, provided that there were a horizontal opening a little above the flame, leading into a vertical flue. Indeed, Mr. Parkins informed me that he took the idea of the above improvement from what were called Staffordshire grates, which were made of a hemispherical form, and which usually projected into the room much further than the front of the flue.

Mr. Hiort's Improved System of Chimneys described.

Mr. Clavering, the intelligent writer on chimneys, whom I have before quoted, recommended in 1779, not only that all flues should be most carefully plastered with superior mortar, but that they should be of the circular form throughout, and that all changes of direction should be gradual, and that they should be swept by a circular brush, to be worked by ropes from the top and bottom,† instead of employing chimney sweep boys.

The same system of circular flues has also been more recently recommended, in the brief notice of practical architecture contained in the interesting article on "Civil Architecture," in the Edinburgh Encyclopædia, the able writer of which‡ has, however, enlarged more upon the higher branches of that important subject.

These suggestions, however, were never generally acted upon in this country until Mr. Hiort, of His Majesty's Board of Works, turned his mind to the subject, who has recommended and taken a patent for what may be considered an entire new system of chimney building; for although he has adopted the circular flue, with its own distinct independent shaft, and other arrangements not unknown before, yet the combination of the whole is original, and highly ingenious, and, at the same time, sufficiently simple.

All the chimneys of His Majesty's New Palace, now building under the direction of Mr. Nash, have been constructed according to Mr. Hiort's system; which has also been adopted by Mr. Smirke, in Lord Manver's new house, in Portman Square, and in the upper apartments of the New Post Office, the lower ones having been finished before the subject was brought to his notice.

In Mr. Hiort's chimneys there is a circular flue of uniform dimensions throughout its whole height, commencing without any gathering from the turning bar or mantle, and terminating at the roof.

These flues may be of three several diameters, viz., 10 inches, 12 inches, and 14 inches, which last dimension he considers sufficient for kitchen fireplaces, the two former for the fireplaces of other apartments.

The flues are built with quadrant bricks, made for the purpose, four of which complete the circle, or one course; but it requires two courses,

* The Blacksmiths of Chatham Dockyard wrote a letter to Mr. Parkins, expressing in strong language their gratitude for this improvement, which they declared had not merely conduced to their comfort, but might be the means of prolonging their lives.

† A very perfect apparatus for sweeping chimneys on this principle has since been invented by that ingenious mechanic, Mr. Smart, of Southwark.

‡ Mr. Telford.

or eight in number, to complete a set. They are kiln burnt, of the best brick earth, and glazed inside.*

Of those eight quadrant bricks, two pairs are alike in the first course, and one pair in the second course. Hence, there are only five patterns used in the two courses, although eight bricks are required in that height.

The peculiar bricks made for Mr. Hiort's chimneys are all numbered; those for the 10-inch flue are marked I, I, II, II, which completes one course, and III, IV, IV, V, which completes the second course. Those for the 12-inch flue are numbered in a similar manner, from VI to X, inclusive, and those for the 12-inch flue are numbered from XI to XV, inclusive.

Figs. 275 and 276 represent in plan two successive courses of Mr.

Fig. 275.

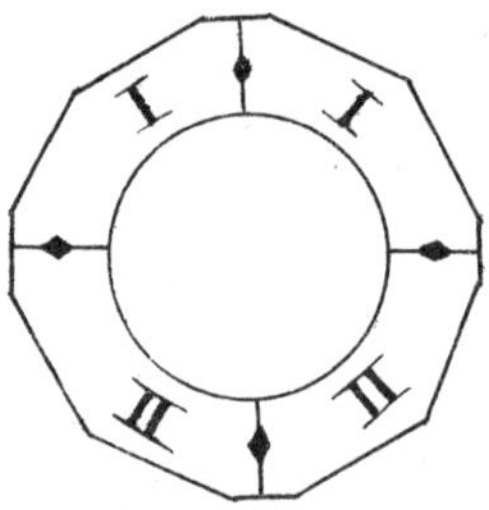

Fig. 276.

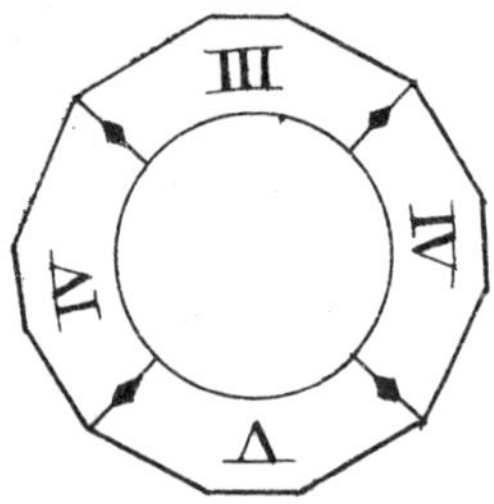

Hiort's chimney bricks, of the kind calculated for a 10-inch flue. They are perfectly circular inside, but polygonal outside, as shown in those figures. At the end of each there is a notch, which serves as a key for the mortar in the vertical joints.

The quadrant bricks, when placed so as to complete the circle, are not of equal height throughout, but of the wedge-like form. Hence, in building those flues, if the thin side of one course be laid on the thick side of another, the flue will ascend vertically; but if the thin side of one course be laid upon the thin side of another the flue may be made to diverge gradually to the right or left, as is always necessary in a stack of chimneys, and thus any change of direction that can possibly be required, may be effected, at the same time that the perpendicular section through any part of the flue will always be a perfect circle.

Fig. 277 is the section of part of one of Mr. Hiort's flues, exemplifying the particulars that have been described.

The extreme height of each course is nearly 4½ inches, and the least height 1½ inches. Hence, in raising a flue vertically, the height of two courses is six inches, the average height of one course being three inches, including the mortar, which however adds very little, if made according to Mr. Hiorts direction, who wishes it to be as fine as the putty used for ornamental brickwork.

In building those circular flues, Mr. Hiort carries each of them up within a square flue, and in a stack of chimneys of this description the withs of the square flues, which thus inclose the former and separate them from each other, may be built partly with common bricks and

* The manufactory of these bricks may be seen at Mr. Baker's brickfield at Darland, near the village of Luton, in this neighbourhood.

Fig. 277.

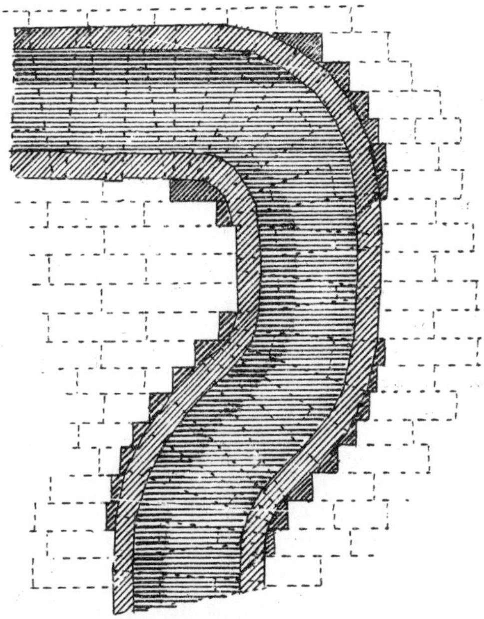

partly with narrow bricks of the same length and height as a common brick, but only of half the width.

By the use of those narrow bricks which Mr. Hiort calls *club bricks*, a single stack of 10-inch flues may be made within an 18-inch wall, for the quadrant bricks and club bricks, being each two inches wide occupy no more thickness in the front or back than a half-brick with, built of common bricks would do.

At the angles of those narrow withs, Mr. Hiort recommends the use of another peculiar kind of bricks, called *coin bricks*, which have a small return like the capital letter L.

By this arrangement, represented in *Fig.* 278, intervals are left all round the circular flues, between the outside of them and the sides of the wall in which they are carried up, and the transverse withs of communication; and although the quadrant bricks touch those walls or withs in certain parts, they are so made, that small open spaces are left at intervals from the bottom upwards; and when there is a stack of chimneys, of which the several flues are carried up near to each other, Mr. Hiort recommends that the transverse withs shall not go right across the wall, from side to side, in every course, but that here and there, communications shall be left, as shown in the same figure, which is taken from Mr. Hiorts book, and which shows the intermixture of common bricks, quadrant bricks, club bricks, and coin bricks.*

The vacant spaces thus left round all those circular flues, are completely closed both at top and bottom, so that they have no direct

* The latter are marked *a*, *a*, *a*, *a*.

communication with the air of the apartment, nor with the external air from above, nor with that of the circular flues, which they surround.

These spaces are, by Mr. Hiort, termed warm air chambers, as he states that a fire in any one flue of a stack of his circular chimneys will warm the whole of those chambers; and the quiescent air therein contained is not liable afterwards to have its temperature changed by cold or wet weather. This is another principle not absolutely new, for it has frequently been applied to prevent cold and damp from penetrating through the external walls of a building, as will afterwards be explained.

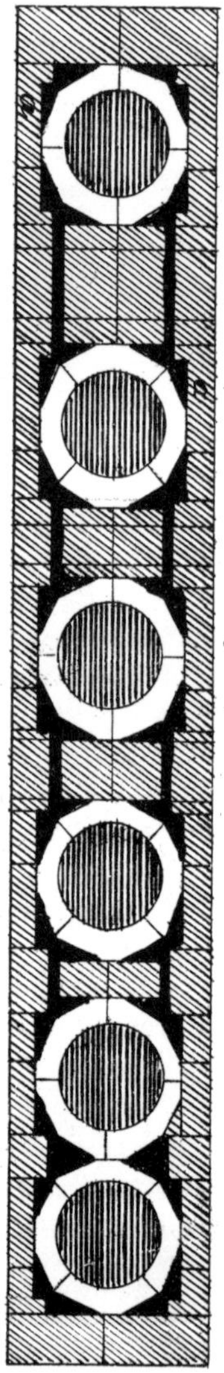

Fig. 278.

In building any of those three sizes of circular flues, Mr. Hiort recommends the bricklayer to provide himself with a wooden mould, equal and similar to one entire course of the proposed flue, upon one side of which are marked the number of quadrant bricks of one course, and on the other the numbers of the quadrant bricks of the next course, that ought to be applied over it as shown in *Figs.* 275 and 276. After having commenced his flue, if the bricklayer wishes to diverge to the right or left, and is in doubt what sized quadrant bricks he should use, and in what position he applies his mould to the course last finished, which enables him to work without hesitation; for the only precaution necessary is not to use the same numbers in two successive courses, for which reason, in consulting his mould, he must turn it upside down when necessary.

Mr. Hiort recommends that all the flues shall be built with very fine stone lime, or Dorking lime mortar, mixed with smiths' ashes sifted; and that the quadrant bricks shall be wetted at top and bottom, and at the ends, but on no account to be soaked in water before they are used. He recommends the rectangular withs, inclosing these flues, also to be built with fine stone lime mortar, but in this part of the work he dispensed with the ashes.

The bricks used in the square withs, which separate the circular flues are laid in the usual manner, forming steps or gathering courses, at all the turnings. Mr. Hiort remarks, that if the circular flues should in any part be made to assume a horizontal direction, or nearly so, then flat stones, extending across the wall, should be used in the inclosing withs, in preference to bricks, which practice is usual under the same circumstances in common flues.

The fireplaces for Mr. Hiort's Chimney may be of the height and width suited for common grates, but of less depth. The general depth of the aperture is equal to the semidiameter of the flue, + 4 or $4\frac{1}{2}$ inches, in addition to which there is a semicircular groove or recess in the back equal to the said semi-diameter.

Thus, for example, if the fireplace is to have a ten-inch flue, to five, the semi-diameter, add four inches, which gives nine inches for the general depth of the fireplace, to which add five for the semicircular recess at the back, and the extreme depth is 14 inches.

Figs. 279 and 280 represent two successive courses of a fireplace, built according to this rule.

Fig. 279.

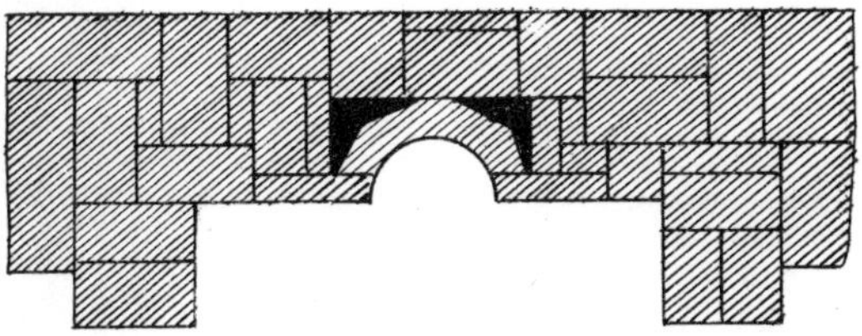

Fig. 280.

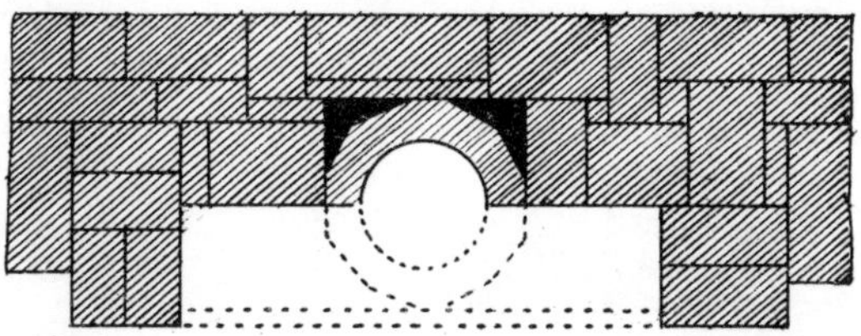

In each alternate course, two inches are cut off from the ends of the quadrant bricks, in order that the joints may be broken there by introducing club bricks, as stretchers at each coin.

In one of those figures, dotted lines show the position of the turning bar, and of the finished or perfect circular flue which commences abruptly at that level.

Excepting this circle, the whole of the rectangular aperture of the fireplace is to be entirely closed at top, which may either be done by common bricks, laid flat across from the turning bar to the back wall, as in *Fig.* 281, or by coving the same spaces, by means of quadrant

Fig. 281.

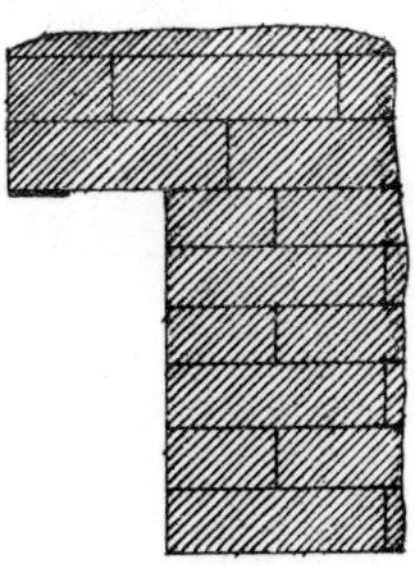

Fig. 282.

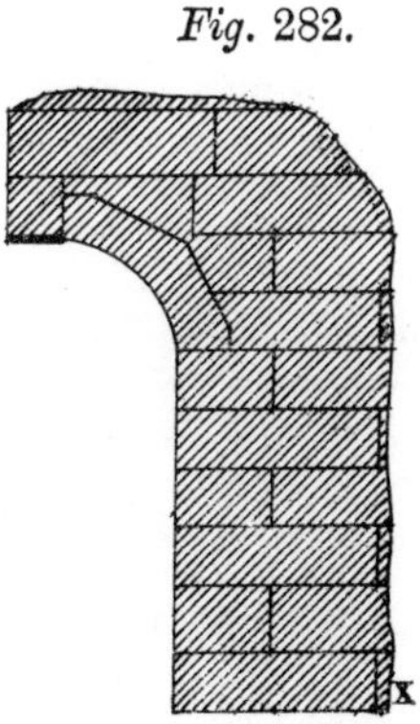

bricks placed vertically and forming a portion of an arch, springing from the back of the fireplace, and abutting against the course of bricks placed on the turning bar, as in *Fig.* 282.

A grate or stove of any pattern may be used with those fireplaces, by altering the back of the ironwork. The grate part, containing the fuel, must have bars in the back as well as in front. Above this level a plate, shelving upwards, must be added, to meet and fit into the semicircular groove, at the back of the fireplace. At the hearth Mr. Hiort recommends a slab being placed parallel to the above plate, and filling in like manner, the bottom part of the same groove.

Fig. 283.

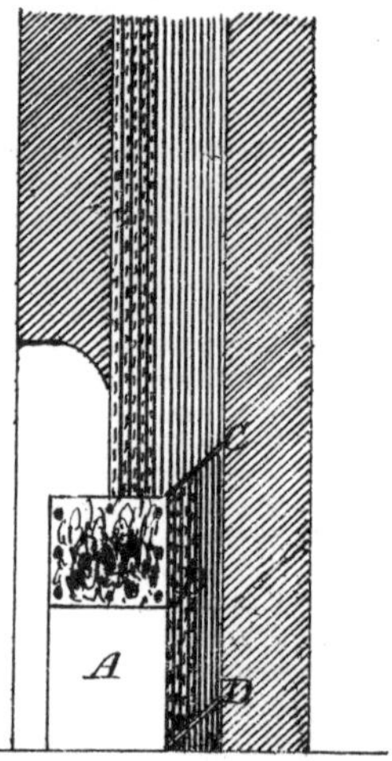

Fig. 283 represents these arrangements, being a vertical section through the centre of the fireplace. The ashhole, *A*, below the grate, communicates with the semicircular groove, *B*, in rear of it, thus admitting a free passage of air behind, as well as in front, to feed the flame. *C* is the shelving plate above the back of the grate. *D* is a similar plate or stone below resting on the hearth.

A party of bricklayers whom I inspected at work, building a double stack of chimneys on Mr. Hiort's plan, near the top of a house, informed me that these circular flues can be carried up as quickly as any other part of the work, and that, after a little practice, they can work without the mould, which, however, they kept near them. I observed that they did not take so much pains as Mr. Hiort recommends in mixing common bricks, club bricks, and coin bricks, in the very careful manner before shown in *Fig.* 278, nor did they endeavour to effect the complete communication of all the quiescent air chambers throughout; nor is it to be expected that bricklayers in general will work with so much nicety, unless there were men selected for this particular branch, and no other. But I observed that the insides of all the circular flues were exceedingly well finished, and all the turnings made with great care. In some parts the bricklayers inserted iron hoops from one side of the wall to the other, with the ends turned back; and at intervals they used thorough courses of Yorkshire stone, through which the necessary apertures were cut, by direction of the architect, in order to bond the work properly; for although the form of these flues is strong, yet all the parts are so minute that, unless the workmanship and mortar or cement used be of the best quality, the bond is scarcely equal to that of a common stack of chimneys on the old system.

Fig. 284.

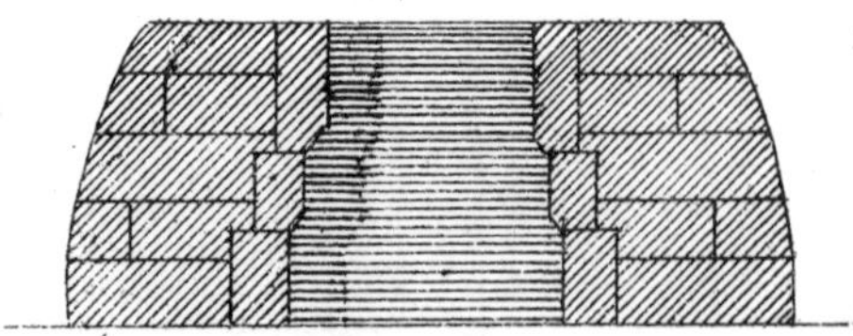

Just before his flues clear the roof, Mr. Hiort reduces the 12 and 14-inch flues to a diameter of 10 inches only, so that the upper part of all the three is precisely alike. This diminution is effected by making use of trumpet-mouthed bricks, as he terms them, tapering from the greater to the less dimension, as shown in *Fig*. 284, which represents a 12-inch flue, thus changing into a 10-inch one.

Above the level of the roof Mr. Hiort uses a distinct shaft for every several flue; and in preparing for those shafts, he directs that the central intervals from flue to flue shall be made exactly equal to 1 foot $10\frac{1}{2}$ inches. In this state the flues need not be carried up higher than 18 inches above the gutter, or to the underside of the coping, with which level the bottom of the stone base or plinth of the chimney shaft is to agree.

The chimney shaft consists of a cast iron cylinder, in continuation of the ordinary circular brick flue, and fitting close down upon the latter by means of a flange two inches wide.

This is enclosed in a hexagon, the sides of which are two inches thick, and which may either be built with club bricks gauged or with Bath stones having upright joggle joints, cramped and united with Roman cement. Externally each shaft has the appearance of a small pedestal, having a stone base or plinth at bottom, and a cap at the top. The air chambers are continued in the intervals between the cyilnder and sides, as far as the bottom of the cap, which is rabbitted and fitted to the top of the iron.

The stone base is square at bottom and slopes at top on each side, from the outside upwards, until it meets the upright side of the hexagonal shaft. The stone cap projects equally beyond those sides, and the top of it also slopes upwards from the outside towards the centre. The particulars, which have been explained, will be illustrated more clearly by means of the following figures.

Fig. 285 represents in plan the upper part of two circular flues, over

Fig. 285.

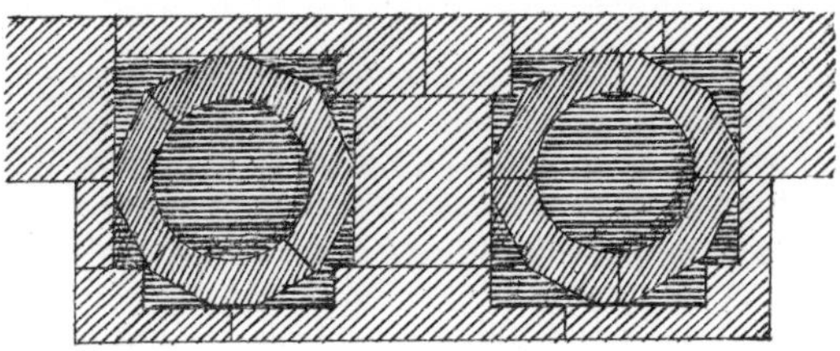

which the above shafts are about to be applied, the interval, from centre to centre, being here reduced to the standard of 1 foot $10\frac{1}{2}$ inches.

Fig. 286 represents in plan the two shafts, showing the internal iron

Fig. 286.

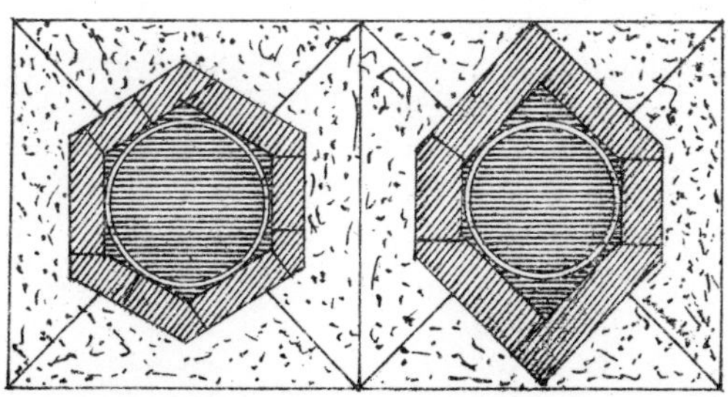

cylinders and the continuation of the air chambers between them, and the stone or brick hexagon. In one of those shafts the extreme outline is a regular hexagon; in the other an oblong hexagon, which latter form may be adopted to give a little more extent of base, if the shafts should be higher than usual.

Fig. 287 represents the external elevation, and *Fig.* 288 the vertical section, of the same chimney shafts.

Fig. 287. *Fig.* 288.

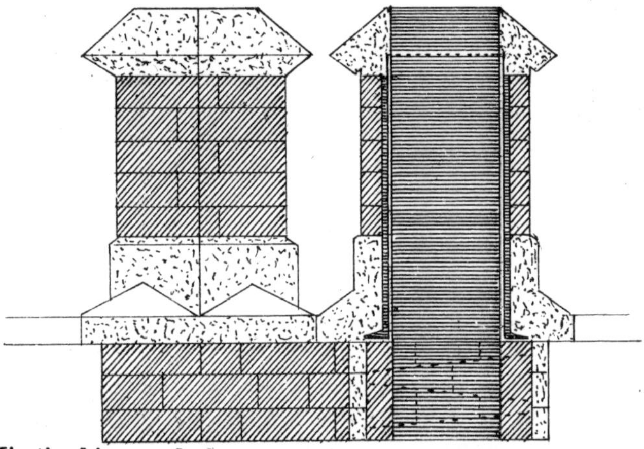

Mr. Hiort's chimney shafts are not more than three feet high, including the base and cap, and have a neat and rather ornamental appearance. He considers it quite unnecessary to raise them above the ridge of the roof of a building, provided that they are higher than the coping of the parapet wall. Hence in a valley between two ridges, they may be entirely concealed; and on the external slopes of a roof they may be raised in steps one above the other, so that those only which are at or near to the ridge rise their full height above that level.

Fig. 289 represents the top of a house, with five of Mr. Hiorts chim-

Fig. 289.

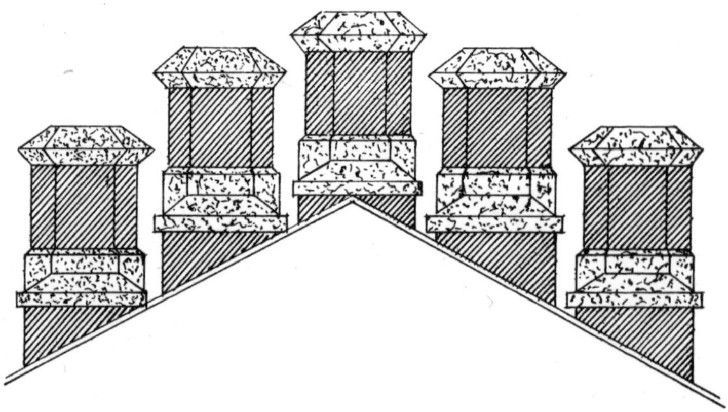

ney shafts, arranged in the manner that has just been described, that is, all independent of each other, and no higher above the slope of the roof than will just admit of the entire base or plinth of each shaft clearing it.

In connecting lead with those chimneys, Mr. Hiort directs that the flashing or part entering into the brickwork and from thence turned down, shall be prepared and placed at the time the brickwork is building, instead of opening the joints of the finished brickwork and inserting it afterwards, according to the common practice, which he considers injurious to the bond of the work.

Having described Mr. Hiort's system, I shall now briefly state the superior advantages which he alledges that it possesses.

1st. He maintains that the circular form is that which ascending smoke naturally assumes, and that by filling the whole flue it leaves no room for eddies of cold air to descend, as in the angles of square flues, and no large superfluous space, as at the gathering of the wings of flues of the common construction.

2ndly. That the gradual windings prevent that accumulation of soot, which takes place in the abrupt and angular turnings of common flues, and that the square withs surrounding his circular flues, and forming a double inclosure, prevent the latter from being injured by plugs, which are often prejudicial to the single withs of flues of the common construction, and sometimes even dangerous in respect to fire.

3rdly. That the warm air chambers inclosing all his circular flues protect them from the external air, and prevent the rarefied air ascending therein from being condensed in cold or wet weather; whereas flues of the old construction are continually exposed to the unfavourable action of external air penetrating through the pores of the bricks, and the common pargetting, which is of a very perishable nature in itself, and is deranged by the chimney sweeps ascending and descending.

4thly. That the polygonal form and superior slope of the chimney shafts, with the intervals between every flue, turn off the wind and allow it to pass freely; and that in blowing between and over those shafts, the wind does not close so immediately upon the other side as when circular pots are used.* Hence, there is always a space to leeward for the smoke to escape; whilst on the contrary, in the common system of large shafts, combining a great number of flues, an extraordinary obstruction is first created by the great mass of brickwork, which increases the fury of the blast, and produces gusts of wind, forcing down the smoke, in spite of the chimney pots, cowls, or turncaps used in hopes of remedying this evil.

Mr. Hiort asserts that such are the advantages of his system, that even a partial alteration on his plan has effected a cure in old chimneys that were never before free from smoke.† I cannot as yet vouch for any particular fact of this nature from my own personal knowledge, but I must confess that his principles appear to my mind more than plausible; and as some of the most eminent architects have already adopted his system, and as very many old chimneys have recently been altered according to the same rules, the merits of his invention will soon cease to be a matter of doubt.

---

* In describing this part of his system, Mr. Hiort mentions, with applause, the old arrangement of independent shafts for each flue, which formerly prevailed in England.

† The partial alterations alluded to consist in putting up a small hexagonal shaft at top, and in altering from thence downwards as far as the attic ceiling, or from the fireplace of any apartment upwards, as high as the ceiling of that one room only.

If Mr. Hiort should be fortunate enough to make good his promises to the public, and if he should actually prove to have discovered a specific remedy for one of the greatest annoyances in domestic life, the only question hereafter will be, in building a house, whether the proprietor chooses to have the common rectangular flues hitherto in use at a moderate expense, or more perfect ones at a greater; for the new chimneys described in this article will undoubtedly be rather more expensive than the old system, which they are intended to supersede.

For further particulars I refer to Mr. Hiort's treatise on chimneys recently published.

Another query naturally suggests itself in the event of the soundness of M. Hiort's principles being hereafter fully established by experience, whether cast iron pipes might not be substituted in lieu of the circular flues composed of quadrant bricks.

Having written so far upon this subject, I received the supplement to Mr. Hiort's Treatise, more recently published.

In this he states that for kitchen chimneys it will be necessary, not to commence his circular flue with the air chambers inclosing it abruptly at the turning bar, but at so much higher a level as will admit the usual apparatus of a smoke jack. Hence, in a kitchen the fireplace and lower part of the flue immediately above the fire, may be of the old construction, whilst the remainder of it is executed according to the new principle. He also admits that, in very large kitchen fireplaces, it may possibly be necessary to have more than one of his circular flues.

He further cautions persons who contemplate an alteration of their chimneys according to his principle, not to commence in the winter or wet season of the year, as it is essential that the whole of the new work shall be perfectly dry before any beneficial result can be obtained.

Of other modes of Heating and Ventilating Apartments, that have been proposed.

It is well known that in Germany, and throughout all the North of Europe, apartments are heated by close stoves, and pipes rising within the room itself, instead of open fireplaces, placed beneath the flue of a chimney. Such stoves are frequently used in shops in winter in this country, and occasionally in the halls of gentlemen's houses, but seldom or never in the apartments occupied by the family, for although they are certainly more economical of fuel, there is a strong feeling against them in this country. They do not convey the pleasurable sensation attached to the favorite expression of "an Englishman's fire side"; and some persons imagine that they are unwholesome. It is difficult, however, seriously to believe that all those nations are so much in error, as is generally supposed.*

Recently, three several methods of heating apartments have been introduced in this country.

1st. The method said to have been invented in 1792 by Mr. Strutt, and afterwards applied to the heating of the Infirmary at Derby, which is described by Mr. Silvester, in a book entitled "The Philosophy of Domestic Economy."

---

* Dr. Franklin says, that the German settlers in Pensylvania, who fitted up their houses with stoves had the greatest advantage over the English settlers, who used open fireplaces, in point of economy and comfort.

The latter gentleman, a civil engineer by profession, has since applied it extensively to other buildings, as for example to the London Hospital, Whitechapel, and to the new county gaol at Maidstone, and he has more recently applied the same principle to the fitment of the ships, employed in the last Arctic Expedition, on which occasion his arrangements gave the greatest satisfaction to Captain Parry, and to the officers and men under his command.

The apparatus used at the London Hospital consists in introducing the external air by a shaft and small subterraneous gallery, or brick culvert into the basement story of the building, from whence, after being heated, it is distributed by a hot air flue into the apartments above. A vane or turncap fixed at the top of a small hollow pillar, causes air to descend down the shaft on the same principle as the windsail used on boardship, from whence it is led to the hot air chamber in the basement story. This chamber, which is of brick, incloses within it a large close iron vessel, called a cockle, under which the fire is made, and by heating which the air in the chamber is heated. But there is no communication between the interior of the cockle and the hot air chamber; hence, no smoke can penetrate from the fire into the latter; that is carried away by first descending through certain spaces, left at the bottom of the cockle, from whence it is conveyed into a common brick flue. The fire-place is a close stove, such as is commonly used for producing great heat for other purposes.

The hot air flues are of laths and plaster, and the openings from thence into the apartments above are of the hopper shape, in order to throw the current of air up towards the ceiling of those apartments.

The physicians at the London Hospital consider this apparatus to have been favourable to the health of the patients; but it is not always used, excepting in very cold or damp weather, as the building was fitted with common fireplaces beforehand, which also are still used, but which were not found adequate to the size of the wards.

At Maidstone Gaol this apparatus is now also seldom used; so that the best place for inspecting it will probably be at the New British Museum, where several furnaces for supplying heated air have been constructed in different parts of that extensive edifice, partly by Mr. Silvester, and partly of Mr. Haden, of Trowbridge, in Wiltshire.

The hot air stoves in both cases are placed in the basement story, under the groined arches. Both use cockles, but of different kinds; Mr. Silvester's is of a square form, but of greater height than extent of base; and having its top composed of two intersecting semicircular arches coved at the angles. It is made of cast iron three-sixteenths of an inch in thickness, and having its parts rivetted together like the boiler of a steam engine. Parallel to, and surrounding the iron cockle, at a little distance from it is a quadrilateral inclosure of brickwork, one brick thick, also coved at top, and completely honeycombed by numerous apertures, into which are inserted square iron or earthenware tubes, extending inwards to within three-quarters of an inch of the surface of the cockle.

This honeycombed inclosure is placed immediately within the hot air chamber, built of brickwork, which incloses all, and which is arched over at top, excepting the hot air funnel or flue, the area of which is said to be nine superficial feet for each hot air stove in the Museum.

The space between the honeycombed brickwork with its tubes, and the external walls of the hot air chamber, is divided into two distinct parts, about half way up the cockle; by this arrangement the cold air entering from below is forced to pass inwards through the lower tiers of tubes, until it is brought nearly in contact with the lower surface of the cockle; and in ascending it is forced to pass through the upper tiers of tubes, and thereby close to the upper surface of the cockle, before it can find its way into the hot air chamber and its flue.

Fig. 290.

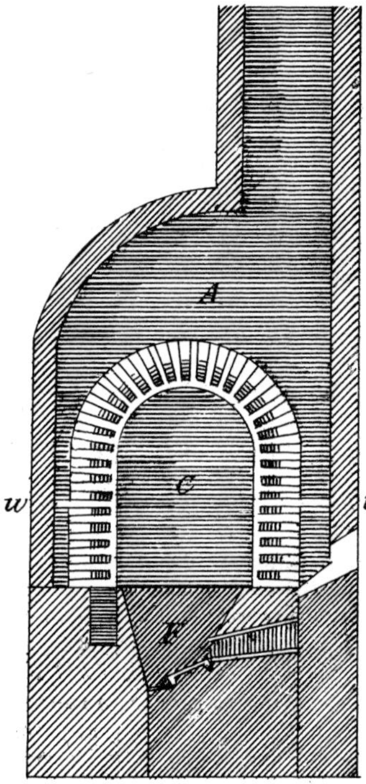

Fig. 290 is the section of one of Mr. Silvester's hot air stoves.* A fire at *F* heats tho cast iron cokcle, *C*, which is inclosed within the hot air chamber, *A*, between the walls of which chamber (*w w*), and the sides of the cockle, may be seen the honeycombed brickwork and the tubes before described.

Mr. Haden's cockles have also a quadrilateral base, like Mr. Silvester's, but are much flatter, so that in section they are nearly oven-shaped. I am informed that Mr. Haden also surrounds his cockle with tubes similar to those before described, and for the same purpose; but of this I cannot speak with certainty, for I have not seen his apparatus laid open, nor any drawing of it. He is said so far to have improved upon Mr. Silvester's system, that his apparatus is in itself cheaper, and admits of coke being used instead of coal, which is a great advantage.

In one of the apartments of the east wing of the new British Museum there are two square holes in the floor, at some distance from the walls, by means of which the heated air enters from below. These are covered by small hollow pilasters, about three or four feet high, and of an ornamental appearance.

In other parts of the same wing the hot air enters by narrow oblong apertures close to the floor, and cut about four or five inches deep into the walls. Similar apertures allow it to ascend from this level into the upper apartments, which have rectangular openings in the walls a little above the floor, with square flues from thence to lead off the heated air to the top of the building. This affords the means of comple ventilation.

Another method of heating air has been introduced by the Marquis de Chabannes, who originally came to this country as a French emigrant. It is described in a "Treatise on Conducting Air by Forced Ventilation," &c., published by him in 1818. His method was tried at

* This figure has been copied from Plate II. of Mr. Silvester's publication, before quoted.

Covent Garden Theatre and at Lloyd's Coffee House, and having also been previously tried at a military opthalmic hospital, in London; the Military Lunatic Asylum, at Fort Clarence, in this neighbourhood, has recently been fitted up in the same manner, under his superintendence.

His system consists in passing a number of small close iron pipes through a furnace, which are fed by means of a cold air flue or culvert, communicating with the external air. These pipes being set up vertically, open at top into an hot air reservoir, placed immediately above the fire, from whence one large iron pipe leads the hot air into the apartment which is to be heated. The smoke of the fire is led off by a common flue.

The same apparatus on a smaller scale may be applied to the heating of one apartment by means of the fireplace of another. Instead of a common stove or grate the Marquis uses for this purpose a peculiar stove, with small vertical iron pipes at the back of it, communicating with a cold air flue below, and a reservoir and hot air flue above, the latter being formed by an iron pipe or pipes, which may of course be led into any apartment in the house. In the grate or stove by means of which this apparatus is put into activity, a common fire is lighted, so that the apartment in which it stands is heated in the usual manner. The fireplace itself has nothing extraordinary in its appearance, and the smoke is carried off by the common chimney, to which this apparatus is fitted.

Whenever heated air is used, some provision must of course be made for carrying it off, and keeping up a constant change. Mr. Silvester recommends, in his treatise, that in every apartment into which air is forced from the hot air stoves, there shall be a foul air flue leading to the roof of the building of equal area to the hot air flue entering from below; and, in like manner, as he proposes a turncap to force air down the shaft and cold air culvert, he proposes a similar turncap at the top of the foul air flue, with this difference, that the former is always to open to windward to receive air, whilst the latter is to open to leeward to admit of its escape. By this means he conceives that even in summer, when his stoves are not heated, a beneficial ventilation will be produced throughout the whole of a building thus fitted, because, as he observes, the air in a subterraneous culvert is cooler in summer and warmer in winter than the atmospheric air.

The Marquis de Chabannes, in particular, is not satisfied with providing apertures for that purpose. He attaches considerable importance to what he calls forced ventilation, for obtaining which he describes two methods.

First, by mechanical means, which he considers an inferior method to the second, which consists in drawing off the foul air by the agency of fire.

For this purpose, whilst his heated air is made to enter near the lower part of a room, there are pipes from the upper part of the same leading into a second furnace above, where a fire is lighted, and thus these pipes are also heated, and the air within them is rarified. Consequently, an upward current is produced, continually extracting the foul air from the apartment, whilst the purer air entering from below, and heated by means of the lower furnace, supplies its place.

At Fort Clarence the chimney of the furnace itself, which heats the apartments, is the agent for rarefying the air within the foul air pipe, which is led into the upper part of that chimney. At Covent Garden Theatre, when fitted up on the Marquis's principle, the gas chandelier afforded so much heat as to effect a part of the forced ventilation, which he considered necessary, whilst a part of it was effected by means of a furnace in the upper part of the building.

This apparatus the inventor has denominated by the quaint title of his Patent Calorifere Fumivore Furnace. I am not aware whether his apparatus yet remains in use at Covent Garden Theatre, or at Lloyd's Coffee House, but the reader who wishes to know more of it may inspect the thing itself at Fort Clarence.

The London Hospital does not, perhaps, afford the most favourable specimen of Mr. Silvester's apparatus, which was an addition not contemplated in the original plan of that building. There, also, heated air is applied to the purpose of forced ventilation; for, adjacent to the end of every large ward, there is a small apartment serving as a water closet, in the wall of which rises the flue of a fireplace situated in the basement story. An air pipe opening into each of these flues, and turning up for a few feet, extracts the foul air from those closets, and, of course, also from the wards, which are separated from them only by a door, purposely made with open joints, so that, when shut, it does not fit close to the floor. Contrary to the arrangement of the Marquis de Chabannes, before described, the openings for heated air are nearest to the ceiling of each ward of the London Hospital, whilst the escape for the foul air is on the level of the floor.

It should have been mentioned before that Mr. Silvester recommends in his Treatise that every foul air flue shall have two openings, one near the floor, the other near the ceiling; the former to be used in winter when the stove is in action, the latter in summer when the air circulates without being heated.

The third method of heating apartments which remains to be noticed is by means of steam, which system has been largely treated of in a recent publication of Mr. Tredgold, on the principles of warming and ventilating public buildings, &c., in which he gives it a decided preference over all others; and at the same time lays down rules for the supply of fresh air by ventilation. He alleges in favour of steam that as it cannot produce a temperature exceeding 212°, it is incapable of burning the air, and thereby rendering it unwholesome, which is an objection that has often been urged against the use of air heated by fire in close iron vessels, or in stoves and pipes of that metal;* and he states that in the manufacturing districts, where both methods are in common use, the manufactories heated by steam are more healthy than those heated by stoves. He also mentions several magnificent hot-houses where steam has been used to a great extent to produce the temperature necessary for the cultivation of large tropical plants.

On consulting practical men, who have seen both methods tried, they do not appear generally to agree with Mr. Tredgold in giving a preference to steam over hot air stoves. I have been informed that a steam apparatus originally in use at the Penitentiary at Millbank, for warming that extensive building, has been removed, and hot air substituted in

* Mr. Silvester observes that the cockle of a hot air stove on his principle should never be heated to more than 280°, as an excess of heat would injure the apparatus. The intense heat of the Marquis de Chabanne's air pipes, which are inserted in the middle of a furnace, appears an objection to his system.

lieu of it, but on account of what particular reason I could not learn. I have also been told by an eminent architect that there is a great objection to the use of steam, as applied to the warming of gentlemens' houses, in the necessity of the various pipes which, in such buildings, may be of great extent, being always kept perfectly watertight, which is attended with great trouble and expense; and the whole apparatus requires more than usual care and skill in the management of it, for should any leak occur, the condensed steam must injure the building and its furniture. I have not yet been able personally to inspect any

Fig. 291.

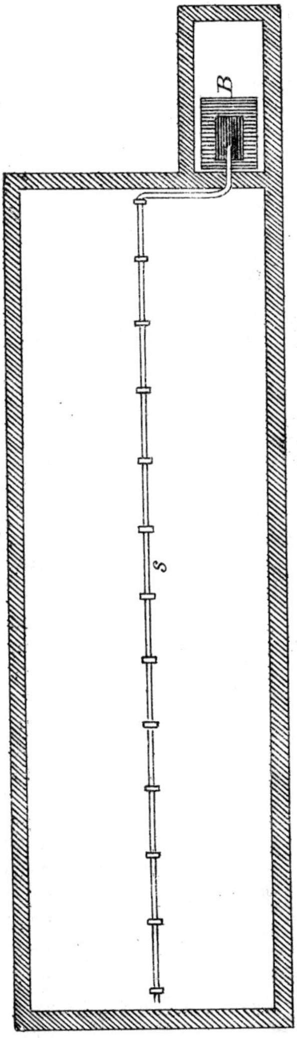

Fig. 292.

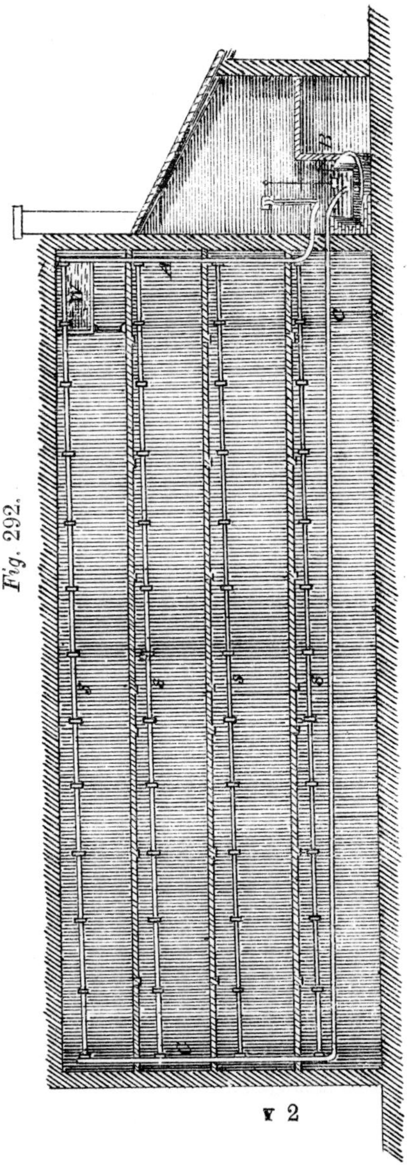

public building heated by steam,* to compare it with the other methods before described, there being none, to my knowledge, heated in this manner in the neighbourhood of the metropolis; and, therefore, with due deference to the opinion of that able writer, and considering the objections urged by others, I am in doubt whether Mr. Tredgold may not have overrated the superior advantages of this peculiar method of warming apartments.

Figs. 291 and 292, copied from Mr. Tredgold's publication, are a plan and longitudinal section of a silk mill at Watford, Herts, warmed by steam, by means of a wrought iron boiler, *B*, placed in an adjoining shed, and containing 38 cubic feet. From this the main steam pipe, *A*, is led vertically upwards, after entering the adjacent end of the building, and has branch pipes, *s*, *s*, *s*, *s*, leading horizontally from it along the whole extent of the ceiling of each room. At the other end of the building, which is 106 feet 4 inches long, by 33 feet wide, is a descending pipe, *C*, which returns back to the boiler the water of condensation. The upper story, eight feet high, is warmed by a pipe of three inches in diameter; the next, eight feet eight inches high, by a four-inch pipe; the next, nine feet high, also by a four-inch pipe; the lower story, nine feet high, by a five-inch pipe. A cistern at *W* supplies the building with water. This apparatus was substituted in 1817, in place of 13 iron stoves, with pipes passing out at the windows and through the roof, and the change is said to have conduced to the benefit of the pro-proprietors and the comfort of the children employed.

In drawing towards the conclusion of this subject, I may remark, that I have not yet seen enough of the various expedients above mentioned for heating and ventilating apartments; nor am I provided with sufficient data to enable me to form an opinion whether any of them are possessed of such advantages, as to render it advisable to adopt them for buildings in general, in preference to the common fireplaces and chimneys, which constitute the favourite arrangement of the people of this country. I think it possible, that hereafter some definite opinion may be formed by an impartial person. At present a great proportion of the community are as strongly biased against them as the inventors are sanguine in their favour.† For my part, I am at present rather doubtful so far as common apartments are concerned; but heated air seems to have its advantages, as applied to churches, large public libraries, like that of the British Museum, and to those infirmaries, in which the wards are of great extent, as in the London Hospital; and it may also deserve the preference over common fireplaces in certain large manufactories, or in those workshops and warehouses in which an equal temperature is desirable for the preservation of delicate and valuable articles.

Heated air may also be applied with great advantage to the drying of linen in a great hospital, or other public institution. An apparatus of this kind is used at the London Hospital,‡ and the same has recently been recommended for the General Hospital at Fort Pitt, by Lieut. Colonel Burgoyne, according to a design of Mr. Silvester, who has

---

* Mr. Jeffrey, a banker in Chatham, has the lower apartments of his house heated with steam, with which he is perfectly satisfied.

† The County of Lancaster Office in London was at first warmed with hot air on Mr. Silvester's plan, but the clerks making the strongest objection to it, common fireplaces have since been used.

‡ Where I was informed that it gave great satisfacion. Not so the washing apparatus, which has been less frequently used.

offered to supply that part of the work which falls within his own province. The linen is suspended on frames, called horses, which slide upon iron railways in and out of the drying closet. This is a small apartment, proposed to be about 10 feet square and 10 feet high. It has an iron floor resting on small brick piers and arches forming flues beneath. A furnace heats those flues and consequently the whole of the iron floor, and has a flue or chimney leading from under that floor to the roof of the building. The closet itself is to be heated by a hot air stove with a cockle, and tubes on the principle before descibed, but entirely of iron work; the hot air chamber and its walls only being of brick. From the closet rises what would be termed a foul air flue, in a common apartment, but here it is called a vapour flue, to carry off the steam of the clothes whilst being dried. It opens near the level of the floor and from thence ascends to the top of the building. The hot air stove also has its own distinct flue for carrying off the smoke according to custom. The horses, when pushed in or drawn out to their full extent, shut up the vertical openings, in which they move in such a manner as to prevent the entrance of cold air.

To conclude, as the investigation is important, and may be attended with amusement as well as information, I would recommend the student of Practical Architecture, not only to peruse the authors upon this subject but to visit and inspect the various methods of warming apartments and to make inquiries on the spot; for it is not only difficult to understand any of the systems recommended by the descriptions contained in books, but it is almost impossible to form any just opinion of the comparative merit of those systems from such descriptions alone; for all the authors alluded to are equally sanguine as to the superior advantages of their own favourite plans.

In designing an edifice the architect must necessarily make up his mind beforehand as to what system of heated air if any he will adopt; after deciding upon which he will consult the engineer or mechanist who is to provide stoves, furnaces &c,; and, according to his demand, he will prepare the cold air culverts, the spaces for the stoves or furnaces in the basement story, and all the foul air flues that may be necessary.

In the Ordnance Hospitals at Woolwich and Chatham, there are openings for the purpose of ventilation, about nine inches square, cut through both walls of each ward, just below the ceiling. On the outside of the building, each of these apertures has a thin square stone fitted to it, which is perforated by a number of small round holes. Inside is a wooden shutter, which when open admits the escape of foul air, or the entrance of atmospherical air. The same system of ventilation has been adopted in the new King's Mews Barracks in London, and in the new Naval Hospital now building at this station.

Of the connection between the Brickwork and the Woodwork of Buildings, with a General description of Chain Timbers and Wall Plates.

It is not my intention at present to enter at large into the subject of the woodwork of buildings, which properly belongs to carpentry, but merely give a general notion of it, so far as it is connected with the brickwork, whilst in progress.

The nature of the woodwork, frequently used under the foundations of buildings, has already been explained.

The principal woodwork connected with the walls of buildings whilst in progress, consists of chain timbers, and wall plates, which are always laid longitudinally, and are inserted wholly or partially into the brickwork.

Chain timbers imply those which are completely buried in the wall, so that no part of them is seen after the brickwork is raised to a higher level.

The ingenious and judicious arrangement of chain timbers adopted by Mr. Baker, for the purpose of securing the foundations of his new buildings in Lancaster Place, London, has already been described. In the same buildings he has introduced two lines of chain timbers in every story, from the foundations upwards, one immediately below and the other above each tier of windows, the latter of which answer the double purpose of chain timbers and of lintels. These longitudinal chain timbers, inserted into the middle of the walls, are connected with transverse timbers of the same description, introduced into all the party walls, the continuity of which is only broken by the stacks of chimneys, where it would of course be improper to use timber of any description.

The use of such timbers is to bond the brickwork together, and to prevent those partial settlements to which new brickwork is always liable, for the reasons that were before explained, in treating of foundations; also to prevent the thrust of the flat arches of windows from forcing out the end piers, and to prevent the cross or party walls from separating until the mortar sets, after which those injurious tendencies towards partial settlements or separations cease to operate.

The use of chain timbers buried in the centre of the walls of buildings, and which was, to the best of my information, not before adopted systematically, or to any extent, is undoubtedly a great improvement upon the common system of bond timbers, which will next be described; for it is evident that those chain timbers are from their position incapable of being burned out, and in the event of their rotting, a small groove not in any great degree injurious to stability is left in the centre of the wall, which does not destroy the connection of the brickwork.

The more common arrangement is to use no other timbers for this purpose than the wall plates, as they are termed, the principal use of which is to support the framing of floors, and of roofs, and to prevent the heavy girders of the one or the trusses of the other from straining the brickwork more in some parts than in others.

In substantial buildings, chain timbers and wall plates are never made less than about $5\frac{1}{2}$ inches thick, to correspond with two courses of bricks in height. Sometimes they are made equal in height to three courses of bricks or about $8\frac{1}{2}$ inches.

Wall plates being placed flush with the inside of the brickwork, and continued longitudinally the whole extent of the walls, produce a less effectual bond than central chain timbers, and are accessible to fire; and if they should decay in process of time, to which timbers in this situation are very liable, a groove is left on one side of the wall, which weakens it much. In cross walls the wall plates are sometimes placed close to each other, and cut up the whole thickness of the brickwork. For example in the cross walls of a handsome public building, recently erected, which were two bricks thick, I have seen double wall plates laid in this manner each five and one-half inches high, and nine inches wide, thus cutting the brickwork completely in two, so that if the timber should decay, those walls must give way, or become disfigured by unseemly fractures so as to require their being pulled down, and in the event of fire, there is a chance of the same injurious effect taking place.

Fig. 293.

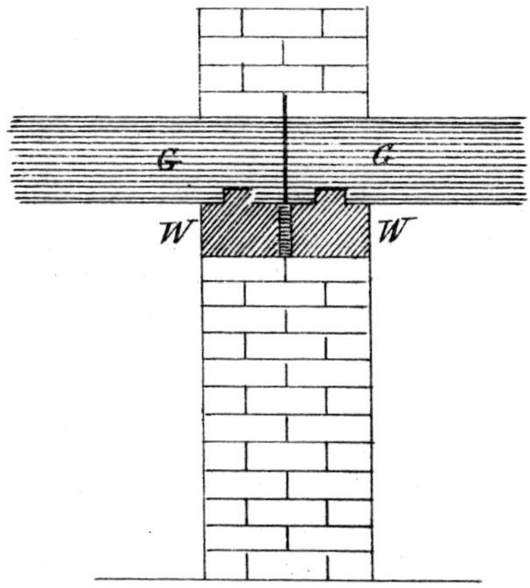

Fig. 293 represents this arrangement; *WW* are the wall plates, introduced into a two-brick wall, and supporting the girders, *G G*, which butt against each other in the centre of the wall, and thus the continuity of the brickwork is completely destroyed by the wall plates, but of course only in a slight degree by the girders, which are at intervals of about ten feet apart; but even of this I do not approve, as the meeting of girders from contrary sides of a wall might easily be avoided.

To return to Mr. Baker's improvement, instead of inserting his wall plates into the walls, in the manner that has been described, he has supported them by four projecting courses of brickwork, with stone corbels at intervals. The use of the latter, which pass entirely through the wall, is to confine the wall plates in their proper places, by means of vertical iron pins, inserted into both, and fixed in the stones with lead. The scantling of his wall plate is about four inches in depth by four and one-half inches in width.

Fig. 294.

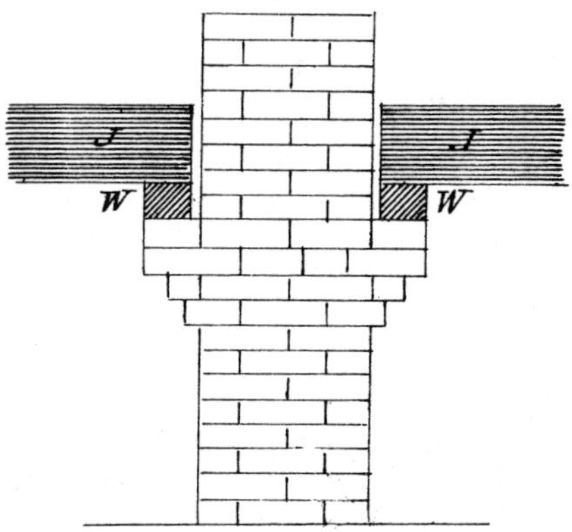

Fig. 294 represents this arrangement, being the section of a portion of a party wall; *W W* are the wall plates of the floors of two adjacent houses supported by four gathering or projecting courses of bricks, of which the two lower courses each project $1\frac{1}{4}$ inch, whilst the third projects $2\frac{1}{2}$ inches, making 5 inches in all. The fourth or uppermost course of bricks—which, in co-operation with the stone corbels, supports the wall plate—does not project beyond the third.

J J are flooring joists in those apartments where single floors are used; and it will be observed that no part of the woodwook enters into the wall.

In those apartments in which he also used larger timbers to support his joists, he rested their ends on the stone corbels before described, but not considering five inches a sufficient bearing, he let the ends of those timbers into rectangular holes in the wall, covered with small brick arches and having a free passage of air all round, and he took care not to allow the ends of any of those timbers to meet, as in the foregoing example.

Unless cast iron girders be introduced into public buildings, which have certainly superior advantages when circumstances admit of their being used, nothing can be better or more judicious than the above arrangement, of which, although probably not any one detail is new, yet I am not aware of the whole having been combined into a regular system previously to the construction of those houses in Lancaster Place.

Mr. Smirke, the architect, besides using chain timbers, such as have been described, has introduced a very judicious system of additional bonding in the walls of the new County Court House, at Maidstone, now building under his direction, which has handsome stone fronts, backed with a mixture of bricks and of rubble stone quarried on the spot. At certain intervals he discontinues the rubble, and builds three or four courses entirely with bricks laid in Roman cement, which from its property of setting immediately connects those courses into one compact mass, probably not by any means equal to timber in the first instance, but free from its disadvantages, and answering the same purpose to a certain degree.

Of the bond timbers commonly used in walls. Remarks on the disadvantages of the common practice of inserting an excess of Woodwork into the walls of buildings in this country. Brief notice of the improved system recently adopted by some eminent architects.

Besides wall plates for supporting the floors and roofs, and which, as I have said before, are scarcely ever less than 5½ inches in height, it has been usual in this country to lay slighter bond timbers, 4 inches wide, and not exceeding 2½ inches, or one course of bricks in height, longitudinally throughout the whole inside of the walls of buildings, at intervals of from 2 to 4 feet apart; and in the apertures left for doors and windows, similar bonds are used transversely, corresponding in length with the thickness of the walls, which from their being so much shorter, are called wooden bricks. In external walls, where the doors and windows have a reveal, these last do not extend to the outside, but only as far as the reveal, or so far as is afterwards covered by the woodwork of the finished door or window.

The bond timbers that have just been described, and which are the only woodwork of a building technically known by that name, although chain timbers and the common wall plates, from their greater strength better deserve that title, are used for the purpose of securing the materials for the interior finishing of apartments in this country, which formerly consisted of wainscotting or woodwork throughout, as may often be seen in old buildings—for example, in the Deanery at Rochester; but in the present times, apartments are only lined with woodwork at bottom, above which laths and plaster are used, both being fixed to battens, or vertical pieces of wood, of small scantling which are nailed to the bond timbers.

The courses of wooden bricks inserted into the sides of the apertures for doors and windows are in like manner used for the purpose of securing the linings and other woodwork of the said doors and windows.

The common arrangement of bond timbers in an apartment is as follows:—

The lowest course is laid about six inches above the level of the floor, and is technically called the skirting bond.

Between this and the known position of the wall plate which is to support the floor above, the remaining courses of bond timbers are usually laid at equi-distant intervals; and they extend all round the walls of the apartment, on the same level throughout, excepting in chimney breasts, or opposite to flues, where they are discontinued at the distance of at least 9 inches from the openings of those flues.

The short transverse timbers, or wooden bricks, used in the openings of doors and windows, are fixed one at top and another at bottom of the aperture. It is usual to employ twice as many transverse bond timbers of this description as there are longitudinal ones in the walls, unless the latter are placed nearer to each other than usual, and they are so arranged that a part of the transverse ones shall meet the ends of the others, not on the same level, but so as to rest upon them.

Fig. 295, which is the internal elevation of one wall of a small apartment, in which there is a window, explains this arrangement.

Fig. 295.

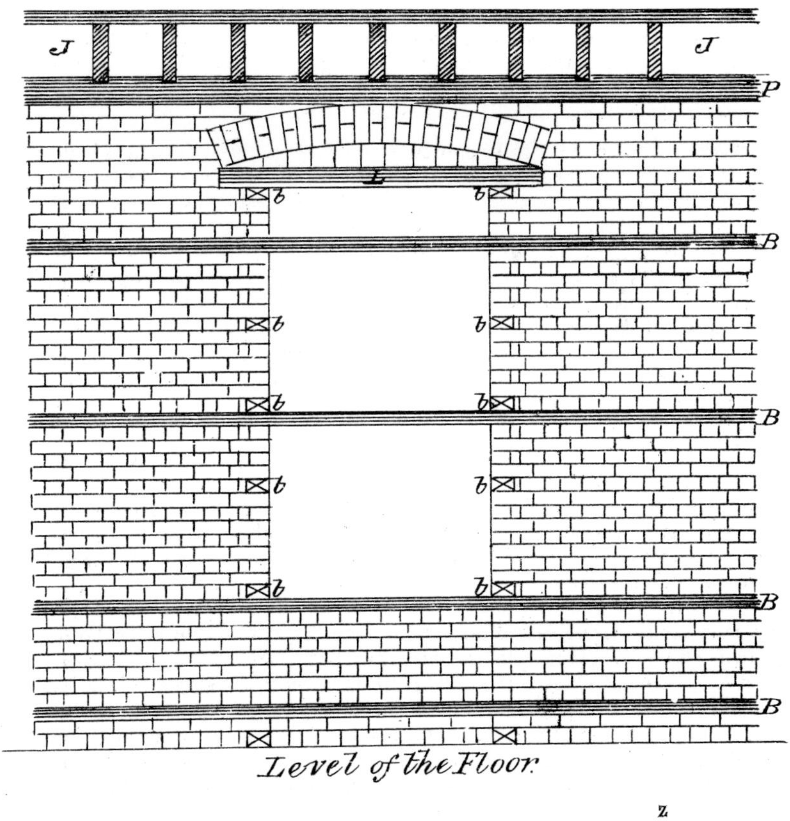

B B B B, &c., &c., are four longitudinal bond timbers, inserted into the walls at nearly equi-distant intervals between the floor of the supposed apartment and the wall plate *P*, which supports the joists *J J*, &c., and the floor of the apartment above.

b b, &c., &c., are the short transverse bond timbers, or wooden bricks on each side of the window, to which the frame and linings are afterwards fixed.

L is the lintel over the window, which was treated of in a former article, and which in one or more pieces covers the whole aperture, excepting the small ornamental straight arch outside.

In London, and in other cities, it is also customary to support shop fronts by wooden posts and beams, equal in one direction to the whole thickness of the brickwork. The former are called story posts; the latter, which are merely lintels of more than usual span between the bearings, are called bressummers.

In the common style of building, the joists of floors are usually let into the walls, at central intervals of about 1 foot apart, which require holes seldom exceeding 3 or 4 inches in width, but from two to four courses of bricks high, and the ends of binding joists, which are more substantial than the above, and of girders, which are still larger than either, are also let into rectangular holes in the brickwork.

Upon the whole, taking into consideration the manner in which the brick walls of common buildings are cut into by wall plates, bond timbers, and lintels, as well as by the rectangular holes for the timbers of floors, which have just been mentioned, it will appear that such walls may be said to be completely honey-combed, even according to the present style of constructing common dwelling-houses and shops in the metropolis. Formerly, a still more objectionable system prevailed there, for until the law entitled the Building Act came into force, it was the custom for the timbers of one house to be connected with those of the adjoining ones, from one end of a street to the other, and the same system still prevails more or less in all provincial towns, to which, unfortunately, the provisions of the law alluded to do not extend. The roofs, likewise, were often formed with eaves, that is to say, projecting over and beyond the external walls of the building, and it was also customary in the better kind of houses to form an ornamental cornice, entirely of wood, towards the summit of, and extending beyond, the face of the front wall. And in former times a pernicious practice also prevailed of inserting wooden girders into the breasts of chimneys, of which there are examples in many old buildings in England, and which has eventually been the cause of a number of houses being burnt to the ground, after apparent security for a number of years. Wooden turning bars were also used for chimneys instead of iron, and thus the buildings that have been described literally appeared as if made on purpose to burn.

When I commenced the compilation of this work, I knew little or nothing of practical architecture, and was therefore much surprised, on enquiring into the details of building in this country, to which I had never before paid any attention, to discover that the walls of houses in general were intersected with continued lines of woodwork, in such a manner that if a fire should take place or dry rot enter, not only must the woodwork perish, but if the walls do not also fall down at the time, they are so much injured that they must necessarily be pulled down afterwards. I was shocked by such a system, which, however excusable in a private individual of limited income, who is often obliged to go to work, not in the best, but in the cheapest way, appeared to me to be highly objectionable, and, indeed, altogether to be reprobated for public buildings; and under this impression it was my intention to have written strongly against it, having in early life, whilst stationed in the Mediterranean, been accustomed to a style of building which either sets fire entirely at defiance, or makes it impossible that the burning of the woodwork in one house shall communicate the fire to the next. On making further observations and enquiries, I was happy to find that two of our most eminent architects, Mr. Nash and Mr. Smirke, had already introduced a more judicious style, which I have no doubt will soon be adopted for all new buildings of importance; so that instead of laying down my own ideas as speculations not yet tried in this country, I shall only have to describe their methods, which, however, are not yet adopted, nor even known so generally as they ought to be. In the buildings of the architects alluded to, some have fire-proof ceilings of arch pots, or of curved iron plates, or even of long thick slates. Iron girders are invariably used, and no timbers are inserted into the walls in the form of wall plates or bond timbers, so that the woodwork used, which in this climate is indispensable to comfort, is not liable to rot, and if the whole of it should be burnt out, the walls braced firmly together by the iron girders, and being composed of solid brickwork, not honey-combed in the usual manner, will remain uninjured after the conflagration.

But in this part of my treatise, which is purposely confined to brickwork, it would be superfluous to enlarge upon these matters in detail, which more properly belong to carpentry. I shall, therefore, conclude this article by briefly noticing some expedients that bear upon the subject now under discussion.

Instead of continued wall plates, which, if incombustible and incorruptible, would certainly afford the firmest support to floors, templets or shorter pieces, 3 or 4 feet long, of oak or stone, are sometimes used to support girders, which answer the purpose of spreading the weight over a greater space of wall, instead of concentrating it over one point, which is prejudicial to stability. Iron templets may also be used for the same purpose; and, indeed, in a new fireproof storehouse in Sheerness dockyard, the cast iron girders and binders which support the stone floors rest on iron wall plates.

Instead of continued bond timbers for battening, oak bricks have sometimes been used, which are of the size of a brick, and may be built into the stretching courses of English bond, alternating with common bricks; and those courses in which the oak bricks are introduced may be at the same intervals apart, in respect to height, as common bond timbers when used.

Sometimes the battening of apartments is fixed to plugs, or small wooden wedges of about 2 inches wide, and half an inch in extreme thickness, which are driven into the walls at horizontal intervals of about 1 foot, and at vertical intervals of from 18 inches to 2 feet apart, the joint into which they are driven being previously cleared of mortar for the purpose by a chisel. In this arrangement, which I have recently seen in some new buildings of importance,* neither bond timbers nor

* His Royal Highness the Duke of Clarence's house, St. James's, designed by Mr. Nash, and Lord Manvers' house, Portman Square, designed by Mr. Smirke.

wooden bricks are used for any part af the walls, excepting for the apertures, that is, for the jambs of doors and windows.

Iron holdfasts, pointed like spikes, are also used to secure battening instead of woodwork. These are driven into the wall alternately on contrary sides of the batten, and are provided with nail holes for fixing to the wood, which they thus retain in its position. Iron holdfasts should always be used in battening of chimney breasts, into which no timber ought on any account to be inserted.

Of securing the ground floors of brick buildings against damp. The peculiar precautions necessary for the Foundations and Floors of Powder Magazines.

The bricks used in this country, and the common mortar in which they are usually laid, are both of a porous quality, in which wet will not merely penetrate, but will even ascend to a higher level, by a process similar to capillary attraction. This may be observed in brick buildings in almost every part of the kingdom, the lower courses of brickwork being usually damp for at least two feet above where it is in contact with earth; and in those cases where a wooden floor is supported by dwarf brick walls, nothing is more common than that such a floor shall decay in a few years, especially if the external earth should be higher than the floor, or on the same level with it.

This cause has ruined the floors of several of our powder magazines, whilst all the rest of the woodwork remaind good.

In order to prevent damp from ascending, and producing the prejudical effect just described, the brickwork immediately above the level of the earth should be carefully laid in cement for several courses. Dwarf brick walls, when used for supporting the joists of a ground floor, should also be built with cement; and the external ground must always be lower than the internal, as is the case in the area walls of well built houses. But when it is not convenient or necessary to make a regular area, a narrow vacant space may be left all round the walls of the building externally, the opposite side of which has its own small retaining wall, and which may be covered with Yorkshire paving stones at top, with iron gratings at intervals, to admit the free entrance and circulation of air. This expedient has been adopted by Mr. Smirke at the new British Museum, and at St. John's new church at Chatham.

The narrow space alluded to need not necessarily be covered, so far as the object of keeping the walls dry is concerned, but it would lead to many inconveniences to leave it open.

In a powder magazine it is necessary to make air holes, not only in the lower part of the external walls, according to custom, but also across the dwarf walls, under the level of the ground floor, in order to admit of thorough ventilation below that floor, from one side of the building to the other. The air holes in the external walls may be covered with an iron grating, or with an iron plate pierced with holes to prevent animals from entering; and it is desirable that the space below the boarding of the floor should be 3 feet high, that is, high enough to admit a man stooping, in order occasionally to examine the state of the floor from below. In this kind of building also—in which, above all others, dryness is most desirable, as no fires are permitted—it may be worth while to pave the whole internal area with flat tiles laid in cement, for these are not porous like common bricks, but when their quality is good are waterproof, or nearly so. A pavement of this kind in addition to the above precautions, will effectually prevent the damp from rising to the prejudice of the woodwork.

Fig. 296 represents a section through the foundation and floor of a powder magazine, in which the small areas or external spaces, *S S*, covered with Yorkshire stones the air holes, *a a a a*, beneath the floor, and the other particulars that have been described, are distinctly shown.

In this section both the external walls and the dwarf walls are supposed to be laid with cement for several courses beneath the woodwork. Indeed, it might be worth while, in so important a building as a magazine, to use cement instead of mortar for every part below that level.

It may appear almost needless to suggest that the water from the areas of buildings must be carried off by sufficient drains on a lower level.

Good slates being less affected by moisture than either bricks or tiles, courses of these laid in cement may be useful. Some authors on architecture have recommended a continued plate of copper, or of lead, to be laid all round near the base of the walls of a building, between two courses of bricks, in order to prevent damp from ascending, but I consider it preferable to use cement. Everyone may have observed that not only in roofs, but in external projections of brickwork, especially in old buildings, it has been common to lay a plate of lead over these projections, the upper part of which plate is usually introduced about an inch into one of the horizontal courses of brickwork above, either by laying it in at the time, or by clearing out the mortar after the brickwork is built.

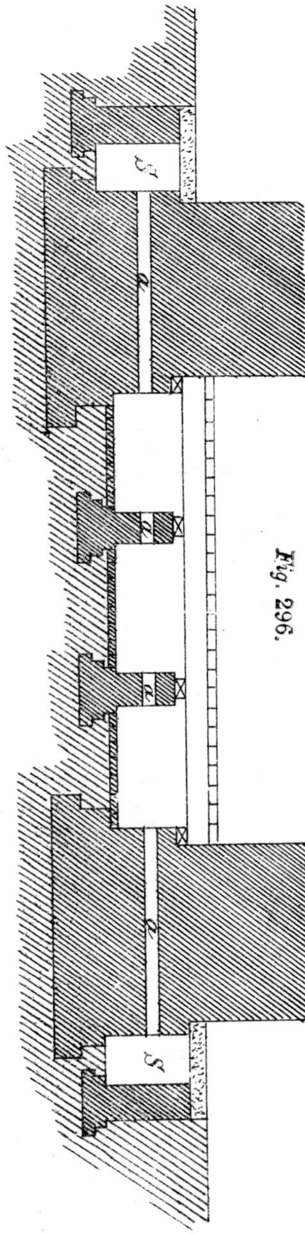

Fig. 296.

Of securing the interior of brick buildings against damp. The use of battening to prevent damp from penetrating through external walls.

In respect to the external walls of brick buildings, the thinner they are the less capable are they of keeping out the damp occasioned by beating rains in certain winds; and these walls not only admit damp, but also heat and cold, in a manner annoying to delicate constitutions. When brick walls are very thin there appears, therefore, no means of securing them against damp but to apply a coat of cement to the outside. When the use of cement is not judged expedient, it appears desirable that the external walls of brick buildings should be at least two bricks thick, in order to oppose a sufficient substance to prevent the penetration of damp, and, as was before implied, the thicker they are the less will they be affected by the humidity or temperature of the external atmosphere.

It has been found in this climate that stucco, or plastering, applied at once to the inside of an external wall of a new building, of the thickness usual in common dwelling-houses, either dries unequally in the first instance, or does not remain uniformly dry at all times, so that in some one part the damp will penetrate more than in others, and discolor the work; but where a vacant space is left with air between damp has less power of penetrating. Accordingly, experience has led to the battening of the inside of all external walls that are to be plastered, which, by separating the laths and plaster one inch or more from the face of the wall, produces the effect desired, of stopping the progress of damp. I believe there are few instances in this country of the stucco or plaster having been applied upon the brickwork itself, in which some imperfection may not be discovered by the eye of a nice observer. Usually the form of the several bricks below may be traced, although indistinctly, even when the coat of stucco applied over them is painted in oil colors.

There is another objection against applying stucco to the bricks themselves, even in internal walls not acted upon on either side by the weather. An efflorescence often takes place from bricks, which is caused by a portion of some salt in the earth with which they may originally have been formed. This is usually washed off from the external surface of new brick walls by the rain; but in inside work it would injure the appearance of the plastering.

Notwithstanding this reason, which may be urged in favour of battening even the internal walls of buildings, which is usually done about London, I have observed that a contrary practice prevails at Dover.*

Description of the hollow Brick Walls sometimes adopted as a security against Damp.

The principle of leaving a vacant space by means of battening, as a security against damp, has also been applied in another manner, by building the external walls themselves hollow. Mr. Nicholson, of

---

* In the internal walls of ordinary buildings, not battened, if one part should be damper than another, it has been recommended to nail on a thin plate of rolled lead, such as tea chests are made of, below the papering.

Rochester, has adopted this expedient in those two external walls of his dwelling-house in St. Margaret's which are the most exposed to rain. These walls are two bricks thick, but there is a vacant space of 4 inches in the centre, from the bottom upwards, and thus they may be said to be composed of two parallel walls, each one brick thick, and connected together chiefly by the sills, jambs, and soffits of doors and windows. In other parts heading bricks are occasionally thrown across to unite the two parts of the wall, of which some courses are shown in section in *Fig.* 297.

Fig. 297.

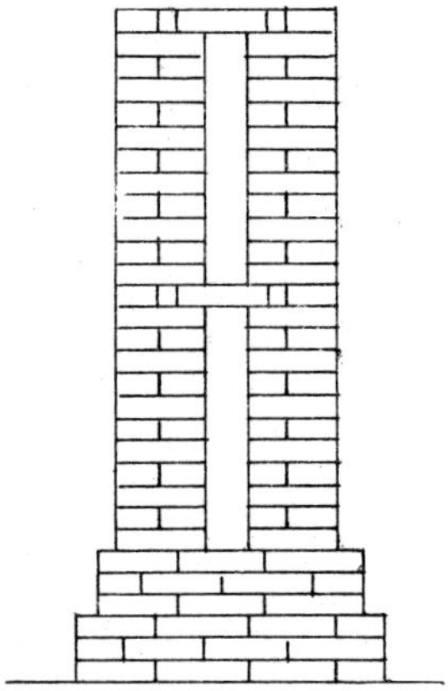

I have heard that the same system of hollow or double walls has been adopted with the same view in several of the Ordnance buildings erected in the Portsmouth District, and along the coast from thence to the eastward.

The system of hollow walls has also been applied to cottages, and is strongly recommended in a book published by Mr. Dearn, in 1821.* This author objects to the common mode of building cottages with brick walls one brick thick, and sometimes only half a brick thick, in which latter case they are brick nogged, in both cases being rendered inside, that is to say, plastered without laths or battening. Another method, no less common, of building cottages entirely with wood, lath, and plaster, like the inside partitions of more substantial houses, he considers equally objectionable; and, accordingly, he proposes to build cottages entirely of brick, making the walls one brick thick, with alternate courses of headers extending entirely across the wall, and of bricks on edge, laid as stretchers. Hence, there is in every alternate course a vacant space of about three inches in the centre of the wall, for a brick on edge is only about $2\frac{1}{2}$ inches thick, and in order to cause a complete vacuity throughout the whole wall, he leaves the centre of each joint of the heading courses dry, or without mortar, that being applied only to the sides of the wall, so that all the stretchers are entirely bedded in mortar, but the headers only partially. He states his opinion that nine-inch walls of this description are strong enough for the external walls of second rate, or third rate country houses, two stories high.

Fig. 298 is an elevation, and *Fig.* 299 is a section, of part of a wall thus constructed. The hollow parts are shaded in the section. The courses are of unequal height, because bricks on edge stand higher than bricks laid flat. This difference would not however offend the eye in buildings of the description to which he proposes to apply the above construction. The under courses for the foundation and footings of the wall he builds according to common English bond, as shown in the section.

It would be superfluous for me to enter farther into the system proposed by this author, which he explains in detail, describing in what manner the woodwork of doors, windows, and floors is to to be fixed or

* Entitled "Hints on an Improved Method of Building, &c.," by Thomas D. W. Dearn, architect.

Fig. 298.

Fig. 299.

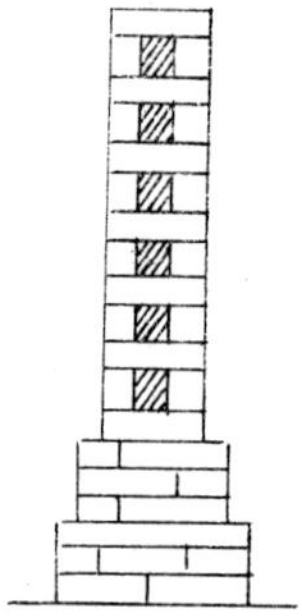

introduced, and stating also several varieties of construction, of which I have only explained one, that which can be carried into effect with common bricks, without any alteration by cutting, and without any deviation from the usual size.

He states that in such cottages nothing further is necessary than to whitewash the walls inside, which may also be done outside if thought proper, and that this construction, of which several specimens are to be seen in Milkhouse Street, in the parish of Cranbrook, in Kent, is the only one in which the walls will remain dry inside, without being battened and plastered in the usual manner.

After writing so far, I find, from Lieut.-Colonel Smart, now Commanding Engineer at Dover, that this system has not proved altogether satisfactory, as applied to the Ordnance Barracks at Shorncliffe, in that district, originally built for Horse Artillery, inasmuch as the wet actually penetrated across the vacant space between the two portions of the external wall, by means of the connecting headers, which occasioned much trouble. Indeed, I am informed that upon that coast the damp penetrates at times through the walls of buildings facing the sea, in a manner quite unparalleled in less exposed situations.

Upon the whole it appears to me to be scarcely worth while to construct the external walls of any buildings of importance in the manner now alluded to, even if the fact last stated did not tend to throw a doubt upon the efficacy of the principle itself, which, however, has been too often adopted in practice to admit of its being passed over entirely without notice in a treatise on brickwork.

Of securing Brick Walls against the effects of Rain penetrating vertically downwards.

Brick walls are of two kinds—namely, boundary walls, or walls of inclosure, such as dockyard walls, garden walls, &c., and the walls of buildings.

Brick walls of the former description are usually finished at top with a ridge to throw off the rain water, and it is usual to allow a projection of about a quarter of a brick on each side to prevent the wet from running straight down the face of the walls.

Figs. 300 and 301 are the sections of brick walls of inclosure thus finished with a ridge, which appears to me to be the best method of covering, or, as it is termed, of coping, a brick wall, when no other materials than common bricks are used for this purpose. In this case the triangular core beneath the ridge is formed with brickbats, or rubbish and mortar, and there may either be one projecting or gathering course, as shown in *Fig.* 300, or two such courses, as shown in *Fig.* 301.

Fig. 300.

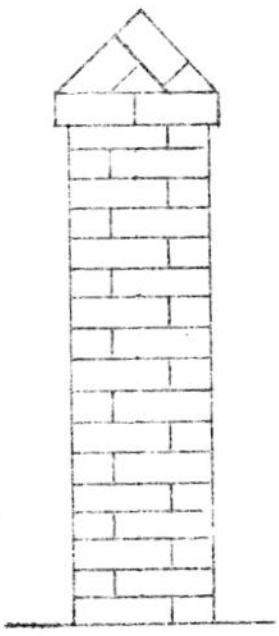

Fig. 301.

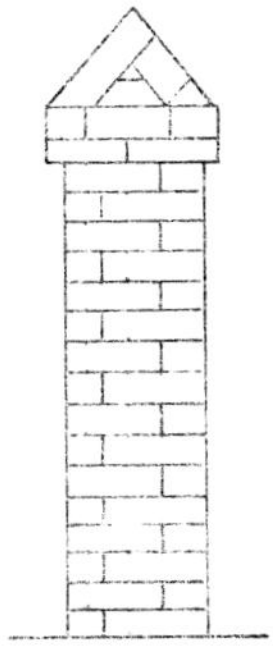

Another method is to use coping bricks, baked for the purpose, which is done in some parts of England, as represented in *Figs.* 302 and 303.

Common bricks being porous, it is evident they do not offer an advantageous covering for the summit of walls, unless they be laid in cement for one or two courses.

The handsomest coping for a brick wall, and one of the best, is to make use of long stones, which project two or three inches on each side beyond the face of the wall.

Fig. 302.

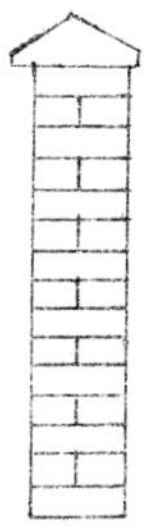

Fig. 303.

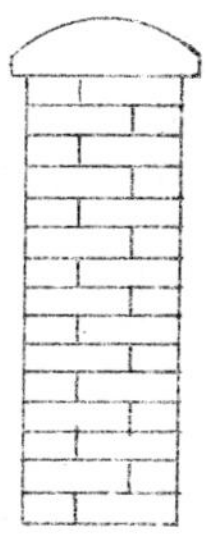

At top these stones are gently curved, as in *Fig.* 304, or gently ridged, as in *Fig.* 305, to turn off the rain water; and at bottom they are grooved in the projecting part, in a direction parallel to the face of the wall.

This groove, which is usually a semicircle, not exceeding one inch in diameter, in copings of moderate projection, is technically called *the throating*, and is necessary to prevent rain water from finding its way along the under surface of the stone, and thus descending by the face of the wall, which it would do without this expedient. The throating, on the contrary, causes rain to fall off in drops, and it therefore never is, or at least it never ought to be, omitted, in stone projections of any kind, whether large or small, such as cornices, window-sills, &c. The stones in a coping of this kind ought to be as long as possible, for the joints are the only part by which rain can penetrate, and therefore it will be useful to lay them all in cement. Indeed, a good and handsome imitation of a stone coping may be made entirely with cement, and at a less expense; but in walls intented to be ornamental, stone may obtain the preference.

Fig. 304.

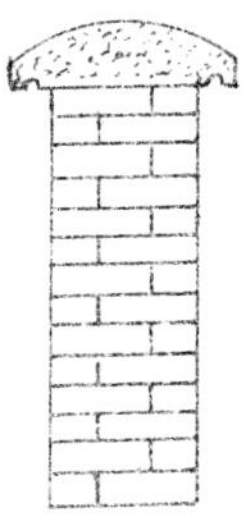

Fig. 305.

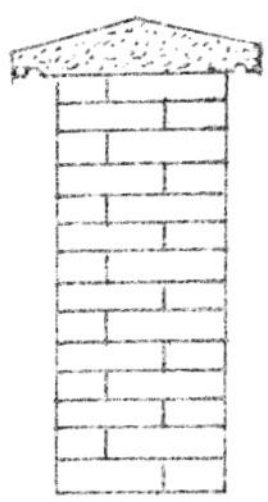

The brick walls at the summit of buildings, which in like manner require to be protected against rain entering from the top, are of two kinds, namely, those which are built with ridges or sloping sides, so as to follow the form of the roof; and those which are horizontal at top, or nearly so.

Of the former description are the party walls of houses, having roofs of the usual form, that is either with a single or double ridge, parallel to the front of the street or row of buildings, and also the extreme walls of a house or row of buildings, having a roof or roofs of uniform section, from one end to the other, and which walls must, therefore, necessarily follow the form of the roof. Walls of the latter description are called gables.

The second kind of walls which are horizontal at top or nearly so, are generally parapet walls of smaller thickness than the principal walls of the buildings of which they form the summit.

A building may either have parapet walls in front and rear only, with gables at the ends, or it may have parapet walls all round. In the latter case the roof is not of uniform section from one end to the other, but has slopes at the two ends, as well as in front and rear, and is technically termed a hip roof.

Indeed, the manner of finishing the walls of a building at top depends entirely on the form of the roof, which may either be with eaves or without.

A roof with eaves is that which projects entirely over the walls on both sides, and consequently requires no coping, or peculiar finishing, at the top of the brickwork.

A roof without eaves does not project so far as the walls of the building, which are, therefore, finished with the parapets that were before mentioned.

In London the party walls of buildings, which are required by the Building Act to be raised at least one foot higher than the roofs of the two houses which they separate, are generally coped with long stones laid along the slope of those walls, which, as was before stated, usually follow the form of the roof. Those stones are connected by iron cramps run in lead, which precaution is useful even in horizontal copings, and therefore seldom omitted in any. Sometimes, however, for the sake of economy, brick copings are used for the said party walls, in buildings of an inferior description. The usual mode of forming such copings is to lay whole bricks on edge along the slope which is previously formed by cutting the sides of the common brickwork, as shown in *Fig.* 306, and before the sloping bricks are laid, it is usual to introduce a line of flat tiles, as shown in the same fig., which is technically termed a tile creasing.

Fig. 306.

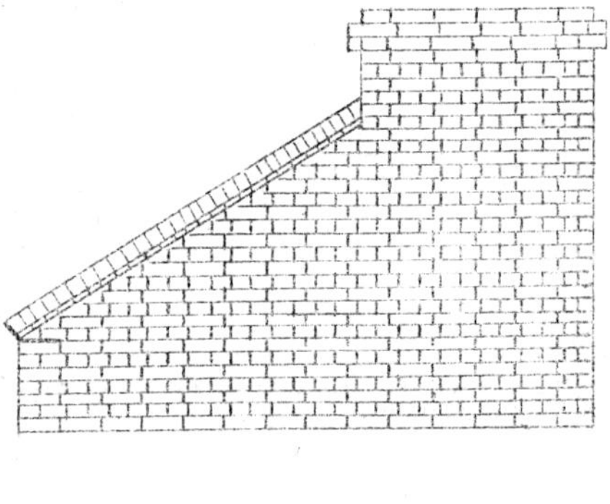

As bricks laid in the manner that has been

described are evidently not well bonded into the body of the wall which they cover, and would, if laid dry, have a tendency to slide, and even with mortar may settle differently, and as the party walls of houses are often constructed with two slopes on each side, forming an angle,* as shown in *Fig*. 307, instead of one continued slope, or line, which

Fig. 307.

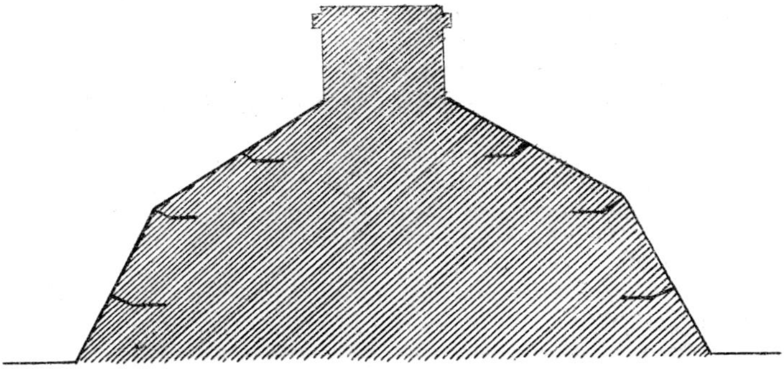

variety tends still further to injure the bond of a brick coping; it is usual to introduce hoop iron into some of the joints, in order to unite more effectually the horizontal courses of the body of the party wall with the sloping courses of the coping.

In consequence of the evident defects of a brick coping of the nature which has just been described, it has been recommended to build the bricks forming the sides of the ridge over which the coping is to be applied in steps of one or two courses each in height, and to toothe the lower ends of the coping bricks into those steps, in the manner shown in *Fig*. 308.

Fig. 308.

* This is done when the attic story forms a part of the roof, the latter being of the description called a kirb, or a Mansard roof

In this case it appears indispensable to lay the coping bricks partly as headers, otherwise there would not be depth enough to bear the cutting, without rendering the coping insignificantly thin. This last system may be recommended as the best for the parapets of brick towers, or slopes in fortification when not coped with stone, and it has accordingly been invariably adopted in finishing the sloping parts of the summits of the revetments on the western heights at Dover. But so far as houses are concerned, I conceive that neither of the brick copings that have just been described ought to be used for any public building or edifice of importance; for in such buildings, it is sounder economy to finish well in the first instance, than to adopt a cheaper construction, involving the necessity of frequent pointing and of continual repairs.

It now only remains to treat of the parapet walls of buildings. A coping of some kind is of course absolutely necessary for every wall of this description, and in order to prevent the rain water, which descends along the slope of the roof, from penetrating by the back of the parapet wall, a longitudinal gutter, usually of lead, is laid at the bottom of the slope, and close to the wall, and a leaden plate, called a flashing, is used to connect the side of the wall, and the gutter in such a manner as to prevent the water from insinuating itself between the two.

This is done by causing the flashing to enter horizontally about one inch deep into a joint of the brickwork, after which it is turned vertically downwards, and overlaps the upright side of the gutter, which rests against the wall. The opposite side of the gutter is turned up along the slope of the roof, and is covered at top on that side by the lowest course of slates, or of tiles, or other roofing material used.

Fig. 309 represents a brick parapet wall, nine inches thick, with a

Fig. 309.

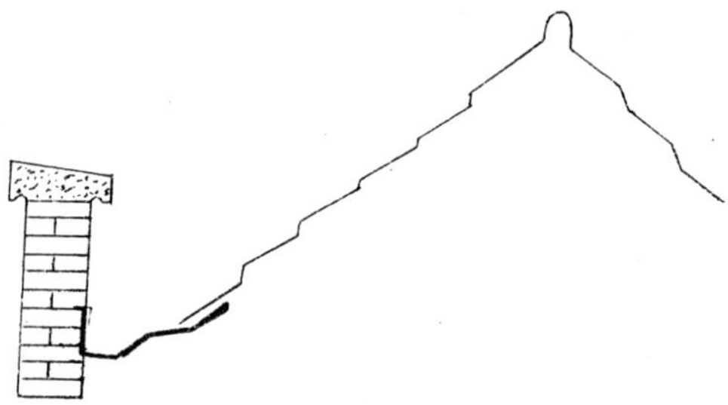

stone coping over it, and a leaden gutter and a flashing, connected in the manner that has just been described. The lead is marked by thick lines, and the joint between the flashing and that side of the gutter which it protects, is purposely left open for the sake of clearness. Part of the slope of the roof is shown in outline, but all other details are purposely omitted. It will be observed that the coping of the parapet wall has a superior slope, with a dip inwards, so as to throw all the rain water into the gutter. This slope is technically termed the weathering of the coping.

Every junction of a roof with a wall requires the greatest care to prevent water from insinuating itself by the joint, and therefore leaden flashings are also used to connect the sloping sides of roofs with the party walls which rise above them.* In this case the flashing is of course laid with a slope; but as it would not be convenient to groove the bricks for receiving the upper part of it in any other manner than by opening the joints, which are all horizontal, the lead is notched at top in the form of steps, in the manner shown in *Fig.* 310; one inch of each

Fig. 310.

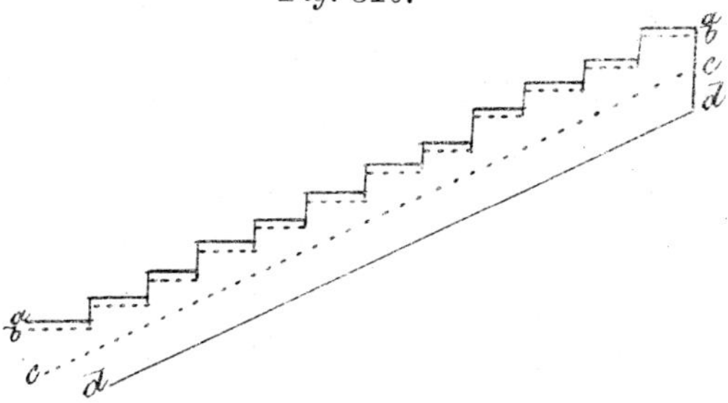

step, the upper part of it (*a b*) is then let into one joint of the successive courses of brickwork, whilst the lower part of the flashing (*c d*) being bent at right angles, is introduced along the slope of the roof, and is either covered by, or covers the extreme courses of, slates or tiles which rest against the wall, according to circumstances.

Fig. 311.

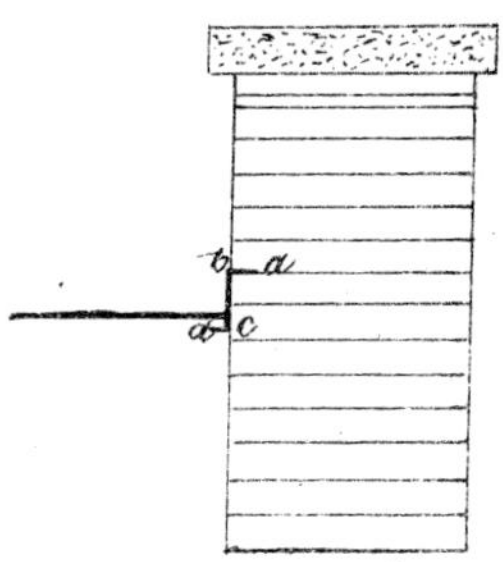

Fig. 310 is an elevation, *Fig.* 311 a section of the flashing, as placed along a slope, the former showing the lead cut, but not bent, the latter showing it bent according to rule, and inserted into the brickwork at top, and beneath the roof at bottom.

To enter further into this matter would be foreign to the subject of brickwork.

The tops of brick chimney shafts are also frequently coped with stone; indeed, the whole apertures of one or more flues are sometimes cut out of one solid slab of that material.

When stone is not used for this part of a building, it is common to lay a tile creasing beneath those salient courses of brickwork which are usual at the summit of a chimney shaft.

It should have been mentioned, in reference to the subject treated of in this article, that the brick copings of walls are most usually formed with bricks laid on edge; which, although it increases the number of joints, adds to the depth, and, therefore, is supposed to render the upper course less likely to be raised by the effects of frost. Mr. Howe is opinion that a coping of bricks, laid with offsets like the footings of walls, and diminishing half a brick in every course, is better than the common ridged copings described at the commencement of this article. The arrange-

* Leaden flashings are also used at the angles on the sides of the Mansard roof, mentioned in a former note, and of which the outline is given in *Fig.* 311.

ment he proposes appears good, provided that the offsets were sloped a little downwards on each side.

It is represented in *Fig.* 312, in which the wall is two bricks thick. The undermost course of the coping is $2\frac{1}{2}$ bricks thick; the next two bricks; the third $1\frac{1}{2}$ brick; the fourth, or uppermost course, is 1 brick thick.

Fig. 312.

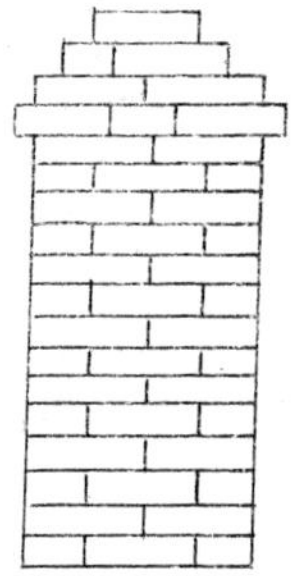

Of securing Brick Walls against the continued Action of Water. That this can be effected by cement alone. Experiments and Observations, tending to prove that common lime mortar is improper, even for the backing of Brick Walls exposed in front to the continued Action of Water.

In the former articles I treated of the preservation of brick walls against the effects of damp or of rain. It now remains to treat of their preservation against the continued action of water, which two cases must not be confounded, for their effects are widely different.

I might cut this matter very short by observing what I believe to be the fact, that the only method of securing brickwork against the continued action of water is to lay it in cement instead of mortar; but this opinion was formed under peculiar views of the subject, which I have not seen illustrated in any book, and which, therefore, it may be proper to explain.

In the first place, I tried a number of experiments upon lime stones and cement stones of various kinds, by burning small pieces of each not less than two hours in a smith's forge by a brisk fire, kept up by one man constantly blowing the bellows, using a crucible to prevent the coals from forming a concretion on the surface of the stone. When burnt, each piece of stone was well pounded in a mortar, and a small quantity of water being added to the dust, the compound was thoroughly mixed by a knife on a marble slab, and finally kneaded up into a stiff ball.* The several balls thus formed were in about 10 minutes afterwards immersed in bowls of water, and all the limes of this country of which I was able to procure specimens, excepting the *Aberthaw lime*, went to pieces sooner or later; whereas that lime, and all the cements, hardened in a few minutes, and when kept under water for months, showed no symtoms of going to pieces; the degree of hardness acquired by the several qualities being, however, very unequal. This circumstance, therefore, induced me to suspect that no mortar made of lime ought to be used in walls exposed to the constant action of water, but that cements only, including the Aberthaw lime, which more properly deserves that name, ought to be used in such situations.

---

* A friend, who is a chemist, has suggested that it would have been better to burn the stones in a slower fire and for a longer period, which I consider very probable, and I intend to try this method hereafter, using charcoal instead of coals, to save the crucibles; but in the meantime I must state the experiments as I first tried them.

Whilst employed in those experiments an idea occurred that the breach in the lower part of the external inclosure of Rochester Castle, which is so very large as to appear the work of violence, might in reality have been caused by the water of the Medway, which has access to it every tide, having decomposed the mortar, and also that the same cause may have produced a similar effect in the brick wall of Cockham Wood Fort, near Upnor, which was an old battery intended to defend the river, and said to have been used against the Dutch in the reign of Charles the Second. On examining those walls more minutely, I found that the mortar was perfectly hard at the upper part of each wall; at a lower level it was imperfect; but lower still, and were exposed to water at each tide, that little of it which remained was in the state of a soft paste.

To ascertain this matter further, I took a fragment of the old wall of Rochester Castle, composed of rubble masonry and mortar, and a fragment of a brick revetment that had been built within Chatham Lines, in the time of the American war, in both of which the mortar was very good, but evidently of common chalk lime, or at least of common lime. Having put these fragments into a large cask of water, changing the water once a week, at the end of three months the fragment of Rochester Castle had separated into several pieces, by the decomposition of the mortar; and the piece of brickwork seemed likely to undergo the same fate, for all the joints of the mortar were washed out to some depth in, and the surface of what remained was quite soft. In this state, the fragments, after being examined, were immediately put into the cask again, and left without changing the water for five months, when they were once more taken out of the cask. *

Here a curious and rather unexpected result presented itself. Instead of the mortar being further decomposed, that which remained had become hard on the surface, but still was soft below. No more stones and no bricks had separated, but a thin pellicle of lime, something like ice, had formed on the surface of the water; an effect which had also taken place in the former experiments, in the decomposition of the balls of new lime. This proved that the mortar of the several fragments of masonry and brickwork in the cask had undergone further decomposition at first, but that this effect had stopped entirely in consequence of the water not being changed, and in conformity with the chemical fact that a given quantity of water can only dissolve a very small proportion of lime, as is known in making lime water. This impediment to the decomposition of the mortar had not acted in the former experiment, for as the water was constantly changed, every new cask full of water decomposed a new portion of the mortar. Having since again changed the water regularly, the decomposition is again making rapid progress.

Hence it appears to me, that the hardest and oldest mortar, that has withstood the common action of the atmosphere for centuries, will go to pieces in a certain time, if exposed to the constant action of water, for although water in small quantities and not changed has very little effect, yet even the smallest stagnant pond to which the bottom of a wall can be exposed, has its water continually renovated by rains, land springs, or otherwise, and although such a pond would have much less effect in injuring a wall than a running stream, yet I believe that the final

* In the presence of Mr. George Rennie.

destruction of the mortar would no less certainly result from the former than from the latter, the period only being more distant.

Hence, in respect to brick walls, there is absolutely no precaution that will save them from the continued effects of water, excepting the use of cement instead of common mortar.

That cement is not liable to be destroyed by the continued action of water, so far as our present experience goes, was proved by the recent demolition of a singular wharf wall, constructed by Surveyor-General Bentham, in Sheerness Dockyard, which was removed to make way for the magnificent new works of Mr. Rennie. The lower part of the former wall was formed of brick masses, as they were termed, which were independent of each other in the first instance, but were afterwards surmounted and united by a continued wall at top, faced with stone. On removing the whole of this work, which was done after a coffer dam had been thrown out in front of it, which kept all dry, the brickwork, although exposed for many years to a great swell and strong tides, was compacted into one solid body by the cement, which was harder than the bricks themselves, and which was evidently the same kind of cement that is still found in that neighbourhood. On comparing this with the brick revetments in St. Mary's Creek, at the left of Chatham Lines, the difference was most striking. In the latter, although there is no swell and a very moderate tide, the external joints of the common mortar used there have been entirely washed out; and unless some measures of repair be adopted, the whole mortar will, in like manner, be entirely decomposed in process of time, and the partial fall of those revetments will be the consequence.

Although nothing is more common than the practice of facing wharf walls with cement, from 4 to 9 or 13 inches deep, and backing in the remainder of them with common mortar, yet, so far as my personal observation goes, there appears to be the strongest reason to conclude that this system is erroneous, and that common lime mortar should altogether be reprobated in aquatic works of any importance; for, on recently examining the wharf wall of the lower mast pond in Chatham Dockyard, which had given way in consequence of the timber land ties rotting, and which has, therefore, been partly pulled down, I observed that the cement which was used in front for about four inches deep, was perfectly strong and sound; that the mortar in rear of this, which evidently had been mixed with coal ashes for about two feet from the face of the wall, was in a state of pulp; but the mortar in the remainder of the wall, towards the rear, which was common lime mortar, was still softer; and that the whole, with the exception of the cement in front, added no more strength to the wall than building it with moist clay would have done; for those parts of the mortar which had more recently been dried by temporary exposure to the air, after removing the rubbish from the back and the water from the front of the wall, whilst the demolition was in progress, crumbled to pieces in the hands like sand.*

Moreover, on inspecting a portion of the old brick wharf wall of the same dockyard, towards the river, which had partly been pulled down to make way for the fine new granite dock, I observed that the same

* In rebuilding this wall, a singular arrangement has been adopted. The front part, which is 2 bricks thick, is built with cement, and has a straight vertical joint not bonded at all into the back part of the wall, which is built with mortar made of Dorking lime.

system of about four inches of cement in front, backed half way with coal ash mortar, and the remainder with common lime mortar, had prevailed; and that the same result had followed. The mortar was universally in a state of pulp, but the cement in front was harder than the bricks, and remained sound and good everywhere excepting in one part, where the backing mortar being acted upon more violently by a run of water escaping from the pumps, and sinking down in rear of the wall, had forced out the cement, and produced a curious concretion of the same nature in front of the wall, and had completely ruined even the cement itself in one or two joints.

Soon afterwards I had an opportunity of observing the same fact in the wharf wall of the inner basin of Dover Harbour, part of which, faced with cement and backed with common lime mortar, on the same principle, is now being pulled down, to be replaced with a new wall with a facing of Portland stone. The whole of the mortar of the old wharf wall was in the same state of pulp that has been described.

Hence, those three examples of wharf walls, built or faced with brick, which at this period may be seen and examined from the top downwards to the very foundation, as the water which acted upon them has been removed for the time, seem to prove that common lime mortar, and even coal ash mortar,* are quite unfit, not only for the front, but for the backing of wharf walls, for nothing more can be said in favour of the mortar that has been described than that it is better than none at all.

In examining those walls, the state of the bricks attracted also my attention, and I found that some few, of a very soft quality, had gone to pieces nearly,† but that the generality of them were exceedingly good and hard, and not at all injured by the moist state in which the body of the brickwork had remained for so many years; nor does it appear to me that bricks of a good quality are at all liable to deterioration from this cause.

Upon the whole, I would recommend that common mortar shall not be used at all in brickwork of any importance that is to be exposed either to the direct action of water, as in the external surface of a wharf, or of the pier of a bridge; or even to the indirect action of it, as in the back or centre of the same, for the moisture will assuredly find its way through, and reduce the mortar into a pulp, which, without absolutely destroying it, will rob it of all its strength.

Recently, I had occasion to see another proof of the propriety of this rule, in the pulling down of one of the piers of old London Bridge, for the temporary purpose of throwing two arches into one, during the erection of the new bridge. The centre of the pier, formed of rubble masonry and cement, was so hard that the workmen were obliged to make jumper holes to blast with gunpowder.‡ Had common lime mor-

* In the experiments on cements and limes before mentioned, in conducting which I at first employed Lieut. Renwick, an officer well acquainted with chemistry and geology, the idea occurred to him of mixing pounded chalk and coal ashes intimately together, and burning them afterwards. This produced a substance that immediately hardened under water, like the natural cement stones, and at first we were inclined to believe that it might lead to the discovery of a cheap and useful artificial cement. Unfortunately, it decomposed in water in process of time, although very slowly. Hence, I infer that coal ash mortar, which is useful for securing the joints of buildings against the weather, is incapable of resisting the continued action of water. We also tried the mixture of chalk, with various proportions of clay, &c.; all of which failed. But I have it in view to repeat those experiments again.

† This happens to place bricks, or badly burnt bricks, even in the open air.

‡ Mr. George Rennie is of opinion that Dutch tarras was the cement used in the construction of old London Bridge.

tar been used, I am confident that it might almost have been pulled to pieces by the fingers, as may be done in the body of the wharf walls of Chatham Dockyard, that have been described, and in the lower part of the exterior wall of Rochester Castle.

In concluding this article, I beg leave to suggest that it may often be very desirable for an Engineer Officer to ascertain by experiment what stones have the properties of cement. And I know no method so simple and easy as the mode described in the commencement of this article, which requires no nice chemical tests or analysis, and of which, as far as my experience goes, the result is certain and conclusive; for all those stones that had the highest character in this country, in consequence of having given satisfaction in buildings of importance, preserved their comparative character when burnt and made into balls, and put into water. The Sheerness cement was the best; next to it the Harwich cement; next followed the Aberthaw lime stone. Of other lime stones experimented upon, that which has the most reputation was the slowest in going to pieces when exposed to immersion. The importance of such experiments, and the inconveniency of neglecting them, may be judged of by the singular fact that until within the last thirty years, or thereabouts, the powerful cements found in the neighbourhood of Sheerness and Harwich were entirely unknown, and for all aquatic works of importance it was considered necessary either to import puzzuolana from Italy or tarras from Holland, at a great expense; and at the time when the celebrated Smeaton built the Eddystone Lighthouse, it was not even suspected that such excellent cements abounded on the English coast. Nothing, however, could have been better than the mixture of puzzuolana and Aberthaw lime which he adopted for that extraordinary and ingenious work.

The particulars which have just been stated would have been better placed in a former part of this book, in treating of cements, but when those articles were written the experiments alluded to had not been tried.

The same subject continued. Correction of an inaccuracy in the preceding Article, in consequence of further experience on the subject. That not only Aberthaw Lime, but the Blue Lyas, of Lyme Regis, and the Dorking Lime, may be considered as Cements, and used accordingly.

Subsequently to my having tried those experiments that have been recorded in the preceding article, and which induced me to doubt the accuracy of the prevailing opinion, that the blue Lyas lime, of Lyme Regis, and the Dorking lime, might justly be considered cements, or water limes, as well as the Aberthaw, to which I allowed that title, I had an unexpected opportunity of observing that all those limes might, with propriety, be used in aquatic constructions.

In consequence of the proposed enlargement of a small basin in the London docks, the end wall of that work is now being removed, with a view of extending the sides, in order to obtain greater length. On recently examining this wall, I found that the mortar was perfectly good and hard, although, with the exception of 14 inches in front, the whole had been made of Dorking lime. The facing mortar was said to have been composed either of Aberthaw, or of the blue Lyas lime of Lyme Regis, mixed with puzzoulana, and some preparation of iron, but in what state and in what proportions I was not able to learn at the time. In some parts I was informed that the backing mortar had been found moist, but that the moisture was of that nature which may be communicated to hard but porous bodies by water, without impairing their strength or durability.

This observation was to me most satisfactory; and such is, in my opinion, the only sure method of judging of the qualities of any lime as a water cement. Let any wharf wall, or other wall that has been exposed to the constant action of water, be opened at the proper depth, and if the mortar be found hard it is a sufficient proof that the lime of which it was made may be used with propriety for aquatic works. In respect to experiments on a small scale, with various limestones, such as were recommended in the preceding article, still they will have their use, for the limes which can stand the test there laid down may be used with confidence as cements. Those which fail need not, however, be condemned, as it appears that our experiments on a small scale yielded a less favourable impression of some limes than they deserved.

With regard to the Halling or grey chalk lime found in this neighbourhood, I have not yet seen it tried under water on a great scale, and, therefore, I am not prepared to give any opinion as to its qualities. But many persons hold it out as being equal to the Dorking lime, or nearly so. When burnt it resembles that lime in having the buff colour, which is usually one of the symptoms by which a water lime may be distinguished.

Of securing the Revetments of Fortresses against the effects of Wet.

When any revetment of a fortress has a wet ditch in front of it, or a ditch occasionally wet and dry, it must be considered as a wharf wall up to the known level of the water, and built accordingly with cement, or with the water limes which deserve that title, to the total exclusion of mortar made of chalk lime or of any other common lime.

Common limes are not, however, to be altogether reprobated for revetments not exposed to the action of water in mass, or to any very unusual degree of moisture; for experience seems to prove, provided that the mortar be allowed to set, and the wall to dry before the earth is applied in rear of it, that such mortar will not be much injured afterwards by the mere moisture of the earth of a common rampart; and provided also, that proper drains be made to prevent water in mass from collecting close to any part of the wall.

On recently pulling down the old brick revetment within Chatham Lines, that was before mentioned, which had been built with common chalk lime, the mortar of the back of the wall that had been in immediate contact with the earth was moist for about 4 inches in; but from thence to the front of the wall, the whole of the remaining mortar was perfectly hard, and as sound as the bricks themselves.

It is to be remarked that the revetment was of the leaning profile, having a slope of about $\frac{2}{11}$ths in front, and of $\frac{1}{11}$th in rear, so that the face and back of the wall were not parallel; but by this arrangement, the latter was of course rather less liable to injury than if it had been perpendicular in rear.

This revetment was probably more than 40 years old, and had a considerable external slope, as has just been mentioned, and yet it was not materially injured by the weather. In respect to such slopes, although they certainly appear to me to be objectionable, as I have stated in treating of revetments in my elementary fortification, yet when they cannot be avoided, it will be better upon the whole to build all the courses of bricks perpendicular to the slope, as was done in the old revetment alluded to, in preference to cutting them and laying the joints horizontally.

Indeed, in all external slopes of common bricks, oblique joints cannot well be avoided, unless facing bricks of the same height, but of a different form from common bricks, were burnt for the purpose, in order to admit of horizontal joints.

In respect to brick revetments, built with chalk or other common lime mortar, I beg to suggest, that it is improper to point the external joints with cement until the body of the wall shall have become completely dry, and not even then, unless it shall be positively ascertained that there is no possibility of water penetrating afterwards through the common brickwork, from the back of the wall towards the front; for in that case the moisture finding its way through the mortar, and meeting the hard, but thin joint of cement, through which it cannot penetrate and evaporate as through common mortar, will force it outwards in a mass, and break the outer surface of the bricks on each side, to which the cement strongly adheres.

In the massy revetments of fortresses, where moisture is not so likely to penetrate through the whole wall, this caution may be of less importance; but in thin retaining walls it should always be attended to. For I have observed in several thin walls in this neighbourhood, backed with earth, and built with common mortar, but pointed with cement, that a great number of the bricks have been defaced from this cause; at the same time that I have particularly remarked, that in similar walls, in which the common mortar alone has been used, no such effect has taken place.

It therefore appears to me, that the cement should either be applied to the whole mass of thin retaining walls, or to a sufficient thickness, not less than from 4 to 8 inches, to insure its resisting the moisture from the rear, but that it should never be used as a mere pointing or external coating to damp walls built with common mortar. To this method of using it, however, as applied to dry walls, there is no objection.

It seems probable, but this I have not ascertained positively, that the scaling away of the outer surfaces of bricks pointed with cement must have taken place at the period of frost, the action of which is known to be very injurious to brickwork in a moist state.

To conclude, and recapitulate; revetments exposed to the continued action of water must be built with cement or with water limes; but in those not exposed to that action, common lime mortar may be used, with a view to economy, and no farther precaution seems necessary to the preservation of the latter, after the mortar is once set, than a judicious system of drains, and that degree of attention on the part of the engineers in charge of them which all masonry requires.

In the wharf walls of the London Docks, and of St. Catherine's Docks, now in progress, a kind of puddling is applied to the back of the wall, composed of about equal parts in pure clay, and of small gravel rammed hard. This compound, which will stand perpendicularly of itself, not only diminishes the pressure of earth against those revetments, but assists to keep them dry in rear. Probably, the same substance might be used to advantage in backing the revetments of fortresses in all cases where the necessary materials are to be found on the spot; for it certainly would not be worth while to transport them from a distance.

Of the preservation of the Brickwork of Arches against damp. That pure cement should be used for the internal lining of cisterns, and for the external coating of casemates.

In arches or vaults built with common lime mortar, and exposed to the action of wet, from the circumstance of a mass of earth or rubbish liable to be penetrated with rain water resting upon the arches, the moisture by degrees decomposes the lime, which falls down from the joints in stalactites; and thus, in the course of years, the whole of the lime is dissolved, and the arch becomes little better than one built without any mortar at all.

The vaults in the basement stories of buildings being protected by the roof, are not liable to be ruined from this cause; but all others built with common mortar must necessarily suffer from it, unless they be covered at top with a coat of cement, which ought rather to be applied in the form of a ridge, as is usually done in casemates, than over the convex surface of the arch itself.

It may scarcely be necessary to observe that the mortar of arches built with any of the water limes, is not liable to decomposition from this cause, although water may penetrate through the work to a certain degree.

In pulling down Carlton Palace, and some other substantial old buildings in London, I have observed that a coat of puddled clay, about 8 inches thick, has been applied over the extrados of the coal vaults in front, which precaution appears to have been of great use, as the mortar of those vaults, although to all appearance made of common lime, was in good preservation and dry. The clay was quite moist, excepting where, by breaking down part of the vaults, a portion of it was exposed to the weather, where it had cracked in numerous fissures, as is usual with that substance.

I am not aware that puddling has yet been used for casemates, where I should not recommend this expedient in preference to cement, which we know will answer the purpose. I have known, for example, a range of casemates, having ridges covered with plain tiles, leading the moisture down to arched gutters in the valleys, accessible from above, by means of trap doors and shafts, which, notwitstanding these drains were rapidly going to ruin, owing to the moisture penetrating from the rubbish at top into the body of each arch, these casemates have since been effectually cured by removing the rubbish, and applying a coat of cement, after which they were immediately covered with rubbish again. Several years have elapsed since this was done, and yet no symptoms of damp have since appeared; and, in fact, they have recently been used as powder magazines.

For coating the ridges of casemates, and for all purposes where the total exclusion of water is the object, pure cement should be used, unmixed with lime or sand. The great military cisterns, or tanks, on the western heights of Dover, which were built of brick, in the form of two adjoining casemates, or bomb proof arches, in a large cavity, cut out of the chalk rock, were lined on this principle, with pure cement, and no leakage has taken place. The floor was formed with three courses of bricks, breaking joint, the uppermost of which was laid in cement. The intermediate pier, the abutment walls, as well as the two great arches themselves, were built with common mortar; but the whole was lined or coated with cement internally, the thickness nowhere being less than about $\frac{3}{8}$ or $\frac{1}{2}$ an inch, nor exceeding from $\frac{3}{4}$ to 1 inch. The cement was laid thickest on the floor, and lower part of the walls, and was applied in the usual style, by working from the top downwards.

Let it be understood that cement should always be applied on that side of a wall or arch by which the water must penetrate if it enters at all. For this reason, a casemate must be coated externally, whilst a tank or cistern must be lined internally, with that substance. To reverse this method of applying it would defeat the object in view.

I was informed by Mr. Knight, the very intelligent and zealous master bricklayer, who was employed in the Engineer Department at Dover under General Twiss, and afterwards under Colonel Ford,* whilst the principal works at that station were in progress, that he never experienced any difficulty in connecting the new cement with the old. The only thing necessary was to slope off and to score or roughen the surface of the cement of one day's work at the part where the new cement was to be applied on the succeeding day. He also stated that he has known a tank in that neighbourhood, lined with two coats of cement, in consequence of the first being of a bad quality, and not perfectly watertight—and that the second coat, which was of good cement, although applied long after the first, adhered to it ultimately, and remedied the inconvenience which was the cause of its being used.

In the neighbourhood of Dover it is not the custom to mix lime with cement; but for wharf walls, &c., in which the absolute exclusion of damp is of no importance, it is usual to employ fine sand in the proportion of from one-third to one-half of the whole compound, that is, one of sand to one or two measures of cement.

In some remarkably fine and large casemates within the inclosure of Dover Castle, I have been informed that no cement was used, and yet they are dry. This consequence has resulted from their peculiar situation, having been excavated at a prodigious depth from the surface, out of a solid mass of compact chalk rock. But it is not prudent, generally speaking, to attempt to build casemates without a coating of cement.

Of the thickness proper for external walls of brick buildings.

In private buildings, when the proprietor must suit his style of building to his means, or in those run up by speculators, it is usual to build very thin walls, for the sake of economy, as those persons are indifferent to the durability of the building, provided that it will last a certain time, the period of a lease, for instance, in leasehold property.

In the building act, which has been before quoted, buildings are divided into seven several classes, according to which they are called, first rate, second rate, third rate buildings &c., and in that act, the thickness

* Now Major General Ford, under whose direction the new works on the western heights were chiefly executed.

of the external, and also of the party walls for each several rate are fixed, but it ought to be understood that those thicknesses are the minimum dimensions allowed by law, not the dimensions proper, or suited to buildings of importance; and in like manner it ought to be understood by the student of practical architecture, that the precautions against fire, laid down in the same act, may be considered the minimum precautions that ought to be adopted in building. It is right to state these particulars, lest the building act, which is judiciously suited to the means of persons of limited income, should be supposed to afford rules worthy of being followed in the actual construction of public buildings, or even of private buildings of any importance.

Having made these observations, I shall take no further notice of the building act, the substance of which is given in all those useful annual publications, entitled Builder's Price Books.

In buildings of importance, it is not usual to make the external walls any where less than 1½ or 2 bricks thick; but for my part, I decidedly prefer the latter dimension for all public buildings, such as barracks, hospitals, &c.

Having decided on this dimension, which determines the thickness of the walls of the attic story, you may give an increase of half a brick more in thickness to the walls of each successive story, from thence downwards, by an offset, which is set off entirely on the inside of the wall until the level of the ground floor, on which level it is usual to divide the offset between the two sides of the building, allowing a projection of a quarter of a brick only each way.

It is to be observed that in very lofty buildings it is usual to make two successive stories of the same thickness, instead of always allowing an offset on the level of each floor; and in those buildings in particular, which are made two bricks thick in the attic story, it is more common to make the next story immediately below it also two bricks thick, than to give it the greater thickness of two bricks and a half.

In respect to the parapet walls at the summit of the external walls of brick buildings, these ought not to be less than one brick thick, and need scarcely exceed one and a-half brick in thickness, the greater dimension only being used when the wall of the attic story is two bricks thick or upwards.

Hence, in a building of four stories in height, including a basement below, and an attic story above the two principal floors, the thickness of the external walls may be according to the following scales:—

Parapet wall........	1 brick.........	1½ brick.
Attic story............	1½ ,,	......2 ,,
Chamber story......	2 ,,	......2 or 2½
Ground floor.........	2 or 2½ ,,	...2½ or 3
Basement	2½ or 3 ,,	...3 or 3½

In designing a building, it is best to decide upon the minimum thickness for the attic story, and from thence to work downwards, as in the foregoing example; but after having definitively fixed upon those dimensions, it is better to follow the more natural order of commencing from the foundation upwards. In describing them to others let it be understood that the thickness now under discussion apply to the walls of dwelling houses and of other buildings having apartments of moderate size, that is, not much exceeding 20 by 30 feet, and of a height not much exceediug 12 feet in the clear, which dimensions are seldom greatly exceeded either in soldiers' barrack rooms, or in the wards of hospitals, or in the drawing rooms of gentlemen's houses.

As a general rule, under ordinary circumstances, it appears to me that the thickness of the external walls of a public building, of the description now under consideration, may vary from three bricks thick in the basement to two bricks thick in the attic story.

When a building has its apartments of much greater size than the dimensions before specified, as is the case in churches, large public libraries, &c., the external walls ought generally to be made thicker than the proportions that have just been laid down; but the consideration of this matter will form the subject of another article.

Even in buildings of one story only, such as the United Service Library at this place, the cooking houses behind the soldiers' quarters or barracks, &c., it is not desirable to use walls of less than $1\frac{1}{2}$ brick in thickness, for although the first named building is only 1 brick thick, it was made so, not from an impression that this was the best dimension, but from necessity, having been built by voluntary subscription of the members, who were obliged to study economy.*

Of the thickness proper for the Internal Walls of Brick Buildings.

The party walls, which form the common boundary to two adjoining houses in a street or row, ought never to be made less than two bricks thick at top in any buildings of importance; and I believe it is not usual to vary their dimensions so much by offsets, as in the external walls. Indeed, I think it is not uncommon to make party walls two bricks thick from the bottom upwards; but, for my part, being an advocate for strength, I would prefer making them thicker in the basement and other stories, if there be more than one under the level of the grouud floor.

Parallel to the party walls in a street, or to the flank walls of an insulated building, are usually two passage walls, in a very large building having its staircase in the centre; and one passage wall in smaller houses, having the staircase at one extremity of the front, which is the most common arrangement in shops and dwelling houses.

When these passage walls have to support a stone staircase, inserted into the brickwork, or tailed into it, as it is technically termed, in the usual manner at one end only, they should not be less than two bricks thick, in order to admit of an insertion of nine inches into the wall, which I conceive is the least that ought to be allowed for that purpose, although I have observed, in some handsome new buildings in the Regent's Park, London, with ornamental cement fronts, that passage walls of nine inches only in thickness, have geometrical staircases, as the above are commonly termed, of not less than three feet six inches in width, tailing into them only from four to five inches, or half the thickness of the wall.

* Probably for the same reason, in a new brick house just erected in Brompton, in this neighbourhood, which is no less than four stories high, the walls are $1\frac{1}{2}$ brick thick in the basement story, but only 1 brick thick in the others.

In some new houses of three stories in height, erected on the Military Road in Chatham, the party walls to which the chimneys are attached are only half a brick thick in the two upper stories, without being brick nogged, according to the more usual custom, when so small a dimension is used. The external walls are of the same thickness as those of the building last described.

When the staircase is of wood, it is not absolutely necessary that the passage walls should be of the above thickness, so far as the mere service of carrying the steps is concerned.

Thus, in the soldier's quarters, in the Ordnance Barracks at this station, the two parallel passage walls are 2 bricks thick in the basement story, where the stair steps are of stone; but they are only one and a-half brick thick above that level, where a wooden staircase is used.

Having treated of party walls and passage walls, which are both transverse walls, it remains to treat of cross walls in a longitudinal direction, parallel to the front and back of a building.

In a double house, whenever those walls are central, they usually support part of a double ridge roof, or of an M roof, as is the case both in the soldier's quarters in the Ordnance Barracks, and in the new Naval Hospital at this station. In the hospital, this central wall is made of the uniform thickness of 3 bricks from the bottom upwards, which is thicker than the external walls, to which it is parallel, for the latter diminish from 3 bricks in the basement to 2 bricks in the attic story. This arrangement has probably been adopted by the architect, under the consideration that the central wall in this case supports as much of the weight of the floors and roof as the two other walls put together.

In the soldier's quarters in the Ordnance Barracks at this station, the arrangement is different, for the intermediate wall, which supports one half of the floors and roof, is precisely of the same thickness as the two external walls, which between them support the other half of that weight.

To conclude, I would recommend, in respect to the internal walls of buildings of importance, that the party walls should not be less than from $2\frac{1}{2}$ to 2 bricks thick; that passage walls to support stone staircases should be equal to the above in thickness; but that for wooden staircases, they may vary from 2 bricks to $1\frac{1}{2}$ brick thick; and that intermediate walls, intended to support floors and a roof of considerable weight, should be equal to the external walls, or at least to the party walls, in thickness. If no considerable weight rests upon them they may vary from 2 bricks to $1\frac{1}{2}$ brick thick.

In the generality of dwelling houses in London, the front and back walls, the two party walls, and generally one passage wall, are of brick; and in the basement story there is usually an intermediate wall one brick thick. But above this level the front rooms of the house are divided from the back rooms by partitions only, formed of woodwork, lath, and plaster.

Such partitions are not to be recommended; and indeed they are seldom used for substantial public buildings, such as soldier's barracks, hospitals, &c., in which strength and durability are of more importance than the mere saving of brickwork. They are, however, not unusual in the quarters of the principal officers, both in barracks and in hospital, which are sometimes assimulated to private houses in the mode of construction.

Practical examples of the Thickness of the Walls of Buildings actually constructed.

According to the rule laid down in this treatise, to illustrate each branch of my subject by practical examples of buildings actually executed, rather than by designs or ideas, I shall annex the thickness of the walls in some buildings of importance, stating also the heights of the several stories.

2 C

1st Example.

One of the new buildings erected by Mr. Baker, of Rochester, in Lancaster Place, Waterloo Bridge, London, of which the foundation was before described. The frontage of these buildings is generally 23 feet, and the depth from front to rear is 34 feet 9 inches in the clear, but in some of them, those dimensions are a little greater.

The thickness of the external walls, and heights of the several stories are as follows :—

Brick footings at the base	5 bricks thick,	3 feet high.
1st. In the cellar story...	3 ,,	,, 10 2 in. ,,
2nd. Next story............	3 ,,	,, 10 ,,
3rd. Story vaulted at top	3 ,,	,, 9 ,,
4th. Story (kitchen floor of the dwelling house)	$2\frac{1}{2}$,,	,, 9 ,,
5th. Ground floor.........	2 ,,	,, 14 6 in. ,,
6th. Drawing room floor	2 ,,	,, 15 ,,
7th. Chamber story......	$1\frac{1}{2}$,,	,, 10 6 in. ,,
8th. Attic story...........	$1\frac{1}{2}$,,	,, 7 6 in. ,,
From ceiling of attic story to coping of parapet wall	1 ,,	,, 4 6 in. ,,
	Total height......	93 2

In the above heights the dimension stated is taken from the level of the boarded floor of one story to that of the story above, excepting in the cellar floor, where one foot of rubbish and paving is included ; and in the attic story, which only includes the height to the ceiling.

It is to be observed that the present description differs from the former, in making the bottom of the walls half a brick thinner, but it is the most accurate of the two, having been taken from a correct plan, which Mr. Baker sent from London, at my request, whereas the former description was made out by him from memory, without reference to the drawings.

The party walls in those buildings are $2\frac{1}{2}$ bricks thick below, and 2 bricks thick above, whilst the passage walls are 2 bricks thick below, and $1\frac{1}{2}$ brick thick above. The difference takes place above the vaults, which are supported by those walls as piers. The arches are alternately semicircular and elliptical in pairs. The semicircular arches under the passages and staircases have a span of 7 feet 2 inches, whilst the elliptical arches under the apartments are in some of the houses 12 feet 6 inches wide, with a rise of about 3 feet 6 or 7 inches. In others they are about 9 inches wide. Those arches are all $1\frac{1}{2}$ brick thick, and the haunches are filled in externally up to the level of the bottom of the crown of each arch.

2nd Example.

The new house of Earl Manvers, in Portman Square, which is built of brick, with a handsome stone front and porch.

In this front the two wings project externally a little beyond the centre of the wall, and, therefore, two thicknesses are stated in bricks as usual, it being understood, however, that the average thickness of the stone facing, about 8 inches, is also included.

	Centre.	Wings.	Height.
Footings at the base......	4½ bricks	5½ bricks	not noticed.
1st. Basement story......	4 „	4½ „	12 feet 6 inches.
2nd. Ground floor......	3½ „	4 „	17 „ 0 „
3rd. First floor	3 „	3½ „	18 „ 0 „
4th. Second floor.........	2½ „	3 „	13 „ 6 „
5th. Attic story to ceiling.	2 „	2½ „	10 „ 0 „
Total height, not including the footings or parapet, nor the rise of the roof....			71 „ 0 „

In the back wall not faced with stone, the thickness at the attic story is only 1½ brick. In all other parts of that wall, and in the whole of the two party walls, which are independent of those of the adjoining houses, the thickness is 2 bricks throughout.

In all the above heights, excepting in the attic story, the thickness of the floor above is included, which is about two feet. The rise of the roof is six feet.

3rd Example.

The new Naval Hospital now being constructed at Chatham.

This hospital consists of three several distinct houses, with cooking houses between, and connected in front by a handsome stone arcade.

The external dimensions of each house are in plan, 70 by 54 feet, with a tower at the centre of the back end or front of each 14 feet wide, and projecting 7 feet. The principal apartments of each house are wards of 50 by 25 feet in the clear. The external walls are in the

1st. Basement story......	3	bricks thick	10 feet 6 inches high.
2nd. Ground floor.........	2½	„ „	15 „ 6 „ „
3rd. First floor............	2½	„ „	15 „ 6 „ „
4th. Attic story...........	2	„ „	12 „ 6 „ „
		Total......	64 „ 0 „ „

The intermediate wall running longitudinally, which makes the division between two wards in each story three bricks thick from the bottom upwards.

The cross walls at right angles to this for the entrance, staircases, &c., are 2½ bricks thick in the basement story, in all the others 2 bricks thick.

The tower at the back front is divided into two parts by an intermediate wall from the bottom upwards, being intended to contain water closets for the patients, and in consequence of the smallness of those spaces, its walls are only 1½ brick thick throughout.

In the cooking houses, and offices attached to this hospital, or within the same inclosure, and having only a basement and a ground floor, the walls of the former story are 2½ bricks thick, those of the latter are 2 bricks thick.

4th Example.

The new King's Mews Barracks, for His Majesty's Guards, recently erected in London.

1st. Basement story, including rubbish and pavement...	2½	bricks thick	8 feet	6 inches	high.
2nd. Ground floor..................	2	„	12	0	„
3rd. First floor......................	2	„	12	0	„
4th. Second floor....................	2	„	12	0	„
Parapet all of brick in rear........	1	„	4	6	„

In the several heights of those stories, the thickness of the floor is not included, which having girders of 15 inches in depth cannot be much less than 2 feet. The rise of the roof is 9 feet in 36.

5th Example.

The soldier's quarters in the Ordnance Barracks, Chatham. The two external walls and intermediate wall are in the

1st. Basement story	2½	bricks	thick	8 feet	11	inches high.
2nd. Ground floor.........	2	„	„	10 „	10½	„ „
3rd. First story............	2	„	„	10 „	10½	„ „

The passage walls are two bricks thick in the basement story, and one and a-half brick thick above that level.

The party walls are supposed to be of the same thickness as the external walls.

Those heights do not include the floors, which, being single, may be about 16 inches deep in all, including the boarding and ceiling.

6th Example.

In one of the Field Officers Quarters in the same barracks, having one story more than others of the same description, on account of the inequality in the original ground on which the barracks were built, the two external walls are in the

1st. Basement story.........	3	bricks thick	8 feet	8 inches	high.
2nd. Story..........................	2½	„	8 „	11½	„
3rd. Ground floor...........	2	„	10 „	10	„
4th. Chamber story.........	2	„	10 „	6	„
5th. Attic story...............	2	„	9 „	1	„

The passage wall is of the same thickness as the external walls, and the party walls are also supposed to be of the same thickness.

The intermediate wall, dividing the front from the back rooms, is, in the two lowest stories, 2 bricks thick, in the two next 1½ brick thick, but is not continued in the attic story, where there is a partition only.

7th Example.

Of a substantial old house in Hanover Square.

In consequence of the proposed erection of a new Oriental Club house in Hanover Square, the adjacent building was laid open, and the section of the walls exposed to view.

The walls had no footings. The building consisted of four stories, besides the attics, formed in a Mansard or kirb roof. The external walls in front was 3½ bricks thick in the basement story, 3 bricks in the ground floor, and 2½ bricks thick in the two next stories, as far as the parapet wall inclusive. The external walls in rear appeared to have been 4½ bricks thick in the basement story. The party walls were not exposed to view.

This substantial style of building seems better than that of more modern houses in every respect excepting in the want of footings, which I have since learned were often dispensed with in former times. This, however, was certainly a defect.

8th Example.

The new United Service Club house on the site of Carlton Palace.

The external walls are 3 bricks thick in the basement story, and $2\frac{1}{2}$ bricks thick from thence upwards. The principal internal walls are $2\frac{1}{2}$ bricks thick from the basment upwards, but in some places are recessed to a smaller dimension. Some of the internal walls of less importance, and inclosing small apartments, are of less thickness than the above.

There are pilasters $2\frac{1}{2}$ bricks wide, and projecting inwards 1 brick in the basment story, to assist in supporting groined arches, which are to be 1 brick thick.

9th Example.

Of lodging houses, recently erected at Dover, forming a street called the Marine Parade, and having a frontage of 21 feet, and a depth of 35 feet from out to out.

The houses are generally 4 stories high, including a basement, but not the attics, which are in the roof.

The external walls are 2 bricks thick in the basement, but $1\frac{1}{2}$ brick thick above.

The party wall, to which the chimneys of two houses are attached, is 2 bricks thick throughout.

Every fireplace has a projecting chimney breast, but a little above the level of the mantle all the flues, without exception, are led back from the chimney breast into the body of the party wall.

The other party wall, to which the staircases of two houses are attached, is one brick thick from the bottom upwards.

There is an intermediate wall of one brick thick from the basement to the floor of the attic story, where it terminates. Above that level are wooden partitions.

There is no passage wall between the staircase and the apartments. A wooden partition with lath and plaster is used in lieu of it.

The general heights of the apartments are 8 feet 6 inches in the basement, 9 feet 9 inches in the ground floor, 10 feet 6 inches in the first floor, 8 feet 6 inches in the second floor, and 10 feet 6 inches in the highest part of the attics. In those dimensions the thickness of the woodwork of the floors is not included, but being all single they may average about one foot each.

Some of the houses in the same row are one story higher, and rather larger in plan. In these the front wall are $2\frac{3}{4}$ bricks thick in the basement story, 2 bricks thick in the three next stories, and $1\frac{1}{2}$ brick thick in the attics, which are not entirely taken out of the roof like the others.

These may be considered rather a favourable specimen of buildings erected on speculation, for although not over abounding in strength, less substantial ones may very frequently be met with in other towns. But the above dimensions are too slight to be followed in permanent public buildings, which ought not to be erected at all without being well considered, and when finished ought to last for ages.

I consider it of great importance to any person whose business or duties are connected with practical architecture, to observe the dimensions of the parts of buildings actually executed, which may generally be measured in a very short time, not exceeding what other persons would employ in a more cursory view of the same buildings without measurement.

For this purpose I would recommend the student of practical architecture to provide himself with a cane of the description called a Dragon cane in London, which has only one natural joint in its whole length. Cut this cane to the length of 3 feet, so that the natural joint shall be exactly 1 foot from the end. Then divide the whole into lengths of 6 inches, and mark the remaining divisions by circular lines drawn by a hot wire. The extreme foot at bottom may be subdivided into 3 inches by finer lines, and afterwards into inches by dots. Thus, a cane having nothing unusual in its appearance, for those marks will resemble the natural joints, may be used to the same advantage as a regular 3 feet rod. This being carried in the hand, with a 2 feet folding ruler in 6-inch joints in the pocket, will answer all purposes of measurement that can be desired.* The former to be used for measuring the size of apartments, or the dimensions of massy pieces of masonry; the latter for measuring thin walls, scantlings of timber, and other more minute dimensions.

In observing the thickness of brick walls in finished buildings, it may be known by th sound, on knocking the wall with the knuckle, whether it is battened or not. In this case from 1½ to 2 inches must be deducted from the apparent thickness, on the side or sides plastered, and the remainder will give the nett thickness of the brickwork sufficiently near to ascertain its actual dimension in bricks and half bricks.

Further remarks on the Thickness proper for the Walls of Brick Buildings.

In the foregoing articles I stated, in general terms, the thickness proper for the walls of buildings under ordinary circumstances, and with apartments of moderate size.

It will scarcely be necessary to observe that the rule applicable to the above may require to be modified when any extraordinary circumstances occur.

For example, the thickness of walls may in some measure require to be regulated by the nature of the foundation, under considerations that were treated of in a former part of this work. (See page 22).

Moreover, if any of the walls of a building should act as the piers and abutments of arches, which is very common in the lower apartments, the thrust of those arches must be taken into consideration, and a greater thickness given to the said walls than might be absolutely necessary if no arches were used.

It was before stated, that the weight of the floors and roof should be considered in regulating the thickness of those walls which are to support them, and the size of the several apartments must also be taken into consideration, in reference to the same question.

---

* Colonel, afterwards General Eveleigh, under whose command I was first placed at Portsmouth on being transferred from the Royal Artillery into the Engineers, always carried a carpenter's rule in a regular rule pocket in his breeches.

For example, it must be evident, that as the several walls of a building are usually disposed at right angles, they must act as buttresses to each other, and the larger the apartments, or in other words, the greater the distance apart between such buttresses, the more substantial ought the walls to be, in order to possess equal stability with those of smaller apartments. In respect to height also, as the floors of buildings in some measure form the connection of opposite walls, and when deposited on contrary sides of a cross or party wall, do, as it were, buttress it against being forced either one way or the other, it follows that the walls of low apartments need not be made quite so substantial as very high ones, all other particulars being alike.

Under these considerations Mr. Smirke has judiciously made a remarkable difference between the thickness of the walls of the principal apartments of the New British Museum, and of the New Post Office, both designed by him about the same period.

The King's Library of the New British Museum, which is 300 feet long, and 40 feet 6 inches wide, for the space of 114 feet from each end, the centre of it being 43 feet wide, and of which the clear height is about 34 feet, has its walls $5\frac{1}{2}$ bricks thick in rear, and 6 bricks thick in front, including the thickness of stone with which this wall is faced. The great length of this apartment rather than its width is the cause of having made its walls so massy. Below it has a basement story vaulted; above is an apartment of equal size, with a moderate diminution of the thickness of the walls by offsets not exceeding one brick in all.

The two largest rooms of the New Post Office are divided from each other by an intermediate wall common to both. One of them, which is to the front of the building, is 103 feet 6 inches long, 32 feet 6 inches wide, and about 33 feet high. The other is 87 feet 4 inches long, 50 feet wide, and 28 feet high, to the lowest part of the panelled ceiling. The side walls of these two apartments are three bricks thick, being only about one half of the thickness of the former. The reason is not merely their inferiority of length, but also their being buttressed in rear by the walls of smaller apartments adjoining, whereas the great room of the New British Museum extends the whole width of the building, and, consequently, is not aided in that manner by other walls. In the basement story of the apartments last described the thickness is greater by half a brick, excepting in the front wall, where it is greater by one brick.

When strength is desired for a very large and lofty apartment, without increasing the general thickness of the walls, internal or external projections, or both, may be used, which will answer the purpose of buttresses, and may be rendered ornamental by giving to them the form of pilasters.

Breaks in the outline of a building, especially those which are so salient as to assume the form of towers, add to strength, by giving more stability than can be attached to two parallel lines of external walling. In the Naval Hospital at Deal, the passages, staircases, nurses apartments, &c., are formed in towers, of the description alluded to in the back front of the building.

Few cases can occur in which it can be absolutely necessary, for the sake of strength alone, to make the walls of the longest apartments more than 3 or 4 bricks thick. The latter dimension is seldom exceeded, even for the walls of churches, unless they have to carry an arched roof.

For example, in Waterloo Church, erected a few years ago, on the south side of the bridge of that name, and which is covered by a common roof, the side walls are 4 bricks thick in the body of the church, which is 103 feet long, not including the entrance, 62 feet 8 inches wide, and 35 feet high in the clear. In the basement story below, which is vaulted with rather flat segment groined arches, the walls are half a brick thicker.

To dwell further upon the various circumstances, according to which the thickness of the walls of buildings ought to be modified, would be superfluous, for all the considerations that ought to have their weight must necessarily suggest themselves to the mind of the architect whilst occupied in designing a building; and to decide upon them with propriety rather requires a sound judgement and the knowledge of what has previously been done by others, with accurate observations upon the success or failure of their designs, than upon any very nice theory or calculations.

Of the Thickness proper for Brick Walls of Inclosure.

Walls of inclosure are usually built with small buttresses or counterforts at intervals, which are generally on both sides, but sometimes on one side only.

Such walls should never be made less than one brick thick, which may be considered sufficient for heights not exceeding six feet. Above that dimension an increase of thickness should take place.

In common walls of inclosure, which seldom exceed from 10 to 13 feet in height, it is not usual to make the buttresses project more than half a brick; and in low walls, when there are buttresses on both sides, the whole of the above projection is divided between the two sides of a wall, allowing a quarter of a brick only on each side.

Buttresses of this description have the appearance of brick pilasters. Their width must, of course, be equivalent to some regular number of bricks and half bricks.

It is most common in the buttresses of walls built Flemish bond, according to the prevailing custom in this country, and not exceeding the heights above mentioned, to use buttresses of two and a-half bricks wide for rather low walls, and of four bricks wide for higher ones, those being the only pilasters that have a regular appearance when this kind of bond is used (see page 70). In English bond buttresses of any width may be used, but it is better to make that dimension an even number of bricks.

The central distances of buttresses apart is seldom less than 12 feet, even in low walls. In higher walls it seldom exceeds 18 feet or thereabouts.

Walls of inclosure are sometimes built with a projection or kind of plinth at bottom, which serves as a base for the buttresses; and the projection of the coping, by crowning the latter, gives to the spaces between the buttresses the appearance of panels in joinery. In this manner the inclosure wall of the Infantry Barracks at Chatham is built, which is two and a-half bricks thick at bottom, and at the extreme summit, as well as at each projecting buttress or pilaster. The general thickness of the wall in the recessed parts or panels is equal to two bricks. Its height is about ten feet, not including the brick coping.

Buttresses are sometimes made to taper from the bottom upwards. In this manner some of those of the inclosure wall of Chatham Dockyard are built, having a moderate projection at bottom, not exceeding about half a brick, which at top is sloped away to nothing. It is scarcely worth while to adopt this nicety, which must give much unnecessary trouble.

When inclosures walls exceed the abovementioned height of from 10 to 13 feet, they should diminish in thickness from the bottom upwards. But, in conformity with the opinion just stated, it appears to me that a regular slope in brickwork would not only give very great trouble in workmanship, but wonld also be liable to injure the bond. I would therefore recommend the following arrangement in preference :—

Fig. 313.

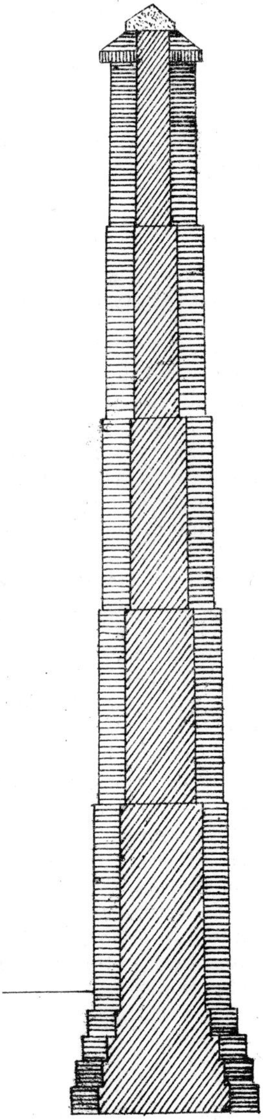

Let the minimum thickness at the top of the wall be $1\frac{1}{2}$ brick, and from thence downwards let it be increased at certain regular and equal intervals of 6 feet, or more, according to your judgment, or according to circumstances; and let the wall have a good base, attending carefully to the footings, in order to counteract the tendency which any long lines of walling have to lose their verticality if not properly secured at the bottom.

Thus, for example, *Fig.* 313 may represent a wall of inclosure 30 feet high, above the level of the ground, having 4 offsets in the footings, and 4 in the upper part of the wall. Thus, the thickness will be as follows :—

From $5\frac{1}{2}$ to 4 bricks in the footings, each step being in 3 courses at least of brickwork, and from $3\frac{1}{2}$ to $1\frac{1}{2}$ bricks thick in that part which appears above the surface.

Stating the same dimensions in feet, the thickness at the footings will vary from 4 feet $1\frac{1}{2}$ inch to 3 feet, and the thickness of the wall itself will vary from 2 feet $7\frac{1}{2}$ inches to 1 foot $1\frac{1}{2}$ inch, which does not appear to me to be too strong for a wall of that height.

In respect to the buttresses, there can be no doubt but that those which are square or oblong, with their length exceeding the width, give the greatest stability to the wall, but, by their great projection, such buttresses would take up too much room, and are not pleasing to the eye. It is, therefore, more usual to make the buttresses of inclosure walls oblong in plan, with the width greatly exceeding the length of projection.

Thus, for example, in reference to the above profile, the buttresses may be made to project only one brick each way beyond the body of the wall on both sides, thus adding 18 inches in all to the thickness in the buttressed parts. In regard to the width of buttresses, where the projection is so inconsiderable, it may be made about $\frac{1}{5}$th of the height of the wall at the ground level, and each buttress may have offsets of half a brick in elevation, corresponding with the offsets of the wall itself in section.

Fig. 314, which represents a part of the wall in elevation, explains

Fig. 314.

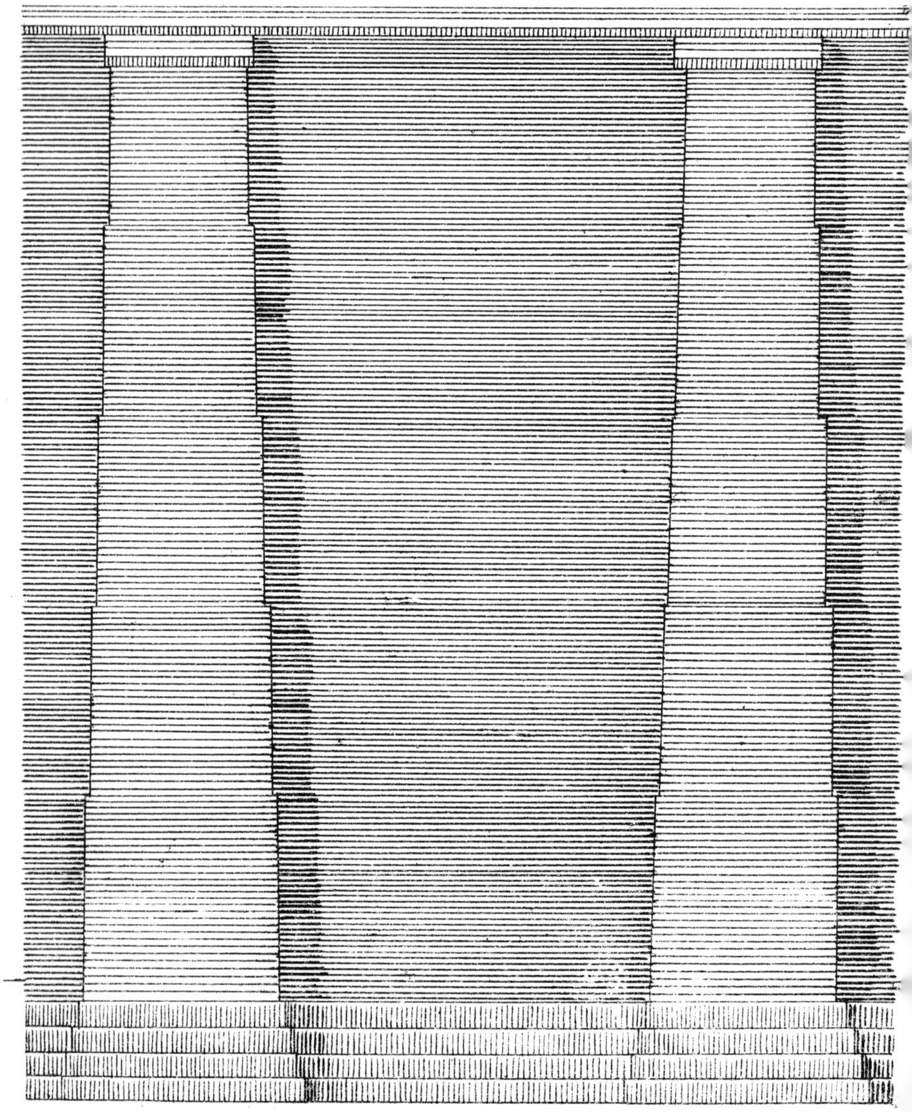

these particulars. At the bottom of the wall, or at the surface of the ground, the width of each buttress is 9 bricks, or 6 feet, being $\frac{1}{5}$th of the height. At the top it is diminished by four successive offsets to $6\frac{1}{2}$ bricks, or 4 feet $10\frac{1}{2}$ inches.

It appears to me more judicious to have buttresses on both sides of the wall than on one side only, for by dividing the whole offset into two parts, it makes the projection less. Moreover, which is of very great importance in a bad foundation, the wall is better balanced, as the centre of gravity falls on the centre of the foundation, which is not the case when the buttresses are all on one side.

Fig. 315.

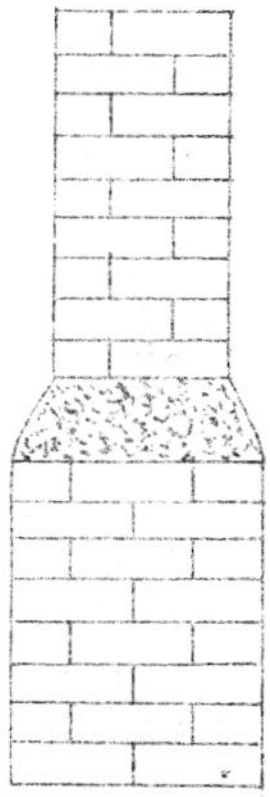

As brick offsets have not a pleasing appearance to the eye, and steps are not desirable in inclosure walls, it is worth while to insert courses of stone at all the offsets, and to round off the angles, as in *Fig.* 315. This may be done both in the main body of the wall and in the buttresses. In the thin parts of the wall those courses of stone may go entirely through; and, indeed, the same would be useful everywhere, provided that the expense of stone is not an objection. Those courses of stone may be made equal to two courses of brickwork in height.

In bad soil the buttresses may be connected with each other at the foundation by inverted arches, as has been judiciously done by Mr. Palmer, the present engineer at the London Docks, in the reconstruction of a part of the boundary wall. He has also connected these buttresses at a higher level by common arches, both extending entirely through the wall. Those arches do not appear above the surface, as the foundations, being bad, were cut very deep.

Having discussed the thickness proper for high walls of inclosure, we will now return to those of more moderate height.

Garden walls are sometimes made thinner than usual, without prejudice to their stability, by curving or indenting their outline; for it is evident that when a wall is built in curves alternately convex and concave, as in *Fig.* 316, or in angle alternately salient and re-entering, as in *Fig.* 317, each portion of wall serves as a buttress to the adjacent

Fig. 316.

Fig. 317.

portion, so that such a wall has more stability than a straight wall of the same height, and containing the same quantity of materials.

Specimens of curved garden walls, rather low, and apparently not more than half a brick in thickness, may be seen at Bexley Heath, in Kent; and a specimen of an indented garden wall only one brick thick, may be seen at Mr. Baker's house at Darland, near Luton, in this neighbourhood. This wall, which is covered with fruit trees, is quite as high as garden walls are usually made, and, therefore, if it had been built in a straight line, it ought to have been at least one and a-half brick thick, with the usual buttresses.

Practical examples of the Thickness of Walls of Inclosure.

In the preceding article, the thickness of the inclosure wall of the Infantry Barracks at Chatham was described, which is thicker than usual for a wall of that height, it being stated that the panels are 2 bricks thick, but that the wall measures 2½ bricks thick across the plinth or lower part, and also across the pilasters or buttresses, which appear on both sides of the wall, projecting a quarter of a brick on each side. The plinth is about 3 feet high. The buttresses are 2½ bricks wide, and are placed at central intervals of about 12 feet 6 inches apart, not being, however, everywhere regular.

Fig. 318.

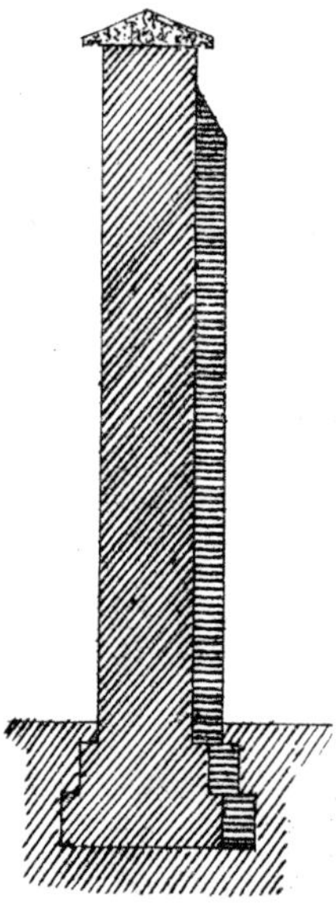

The boundry wall of the new Naval Hospital at Chatham is 1½ brick thick from the surface of the ground upwards, and 10 feet 6 inches high. It has internal buttresses 2½ bricks wide, and projecting half a brick from the surface of the wall, at central intervals of 13½ feet apart. The buttresses are sloped off for the three or four uppermost courses, so that they do not reach quite so high as the summit of the wall, which is crowned by a handsome stone coping.

The original inclosure wall of the London Docks was alluded to in the foregoing article, and as a considerable portion of it was blown down in a violent gale, it may be useful to describe it. It was planned by an architect who has succeeded better in his other works, which are handsome and creditable.*

It was founded on piles, of which three in width supported the whole section of the wall, including the buttresses.

The total height from the pile heads to the summit of the wall was 44 feet, of which about 12 feet were sunk below the level of the ground.

That part of the wall which was above the ground level, was 31 feet 8 inches high. It was 2½ bricks thick at the bottom, and from thence diminished to 1½ brick thick at the top.† It had internal buttresses 6 feet 6 inches wide at the bottom, projecting uniformly 2½ bricks from

* Failures, as I before remarked, are the most useful lessons in practical architecture, otherwise I would prefer avoiding any allusions to them.

† By offsets on the inside of the wall.

the face of the wall, and placed at central intervals of 18 feet. At the foundation, where the brickwork rested on the planking or piles, the wall and buttresses were 8 bricks thick, which, by successive offsets, diminished to 5 bricks at the level of the ground, as was before noticed.

Arches of 11 feet 6 inches in span connected the buttresses at the lower part of the wall, and these were not bonded with the panels or intermediate part of the wall, but were built separate, the panels being merely indented into the buttresses in the manner shown in plan in *Fig.* 319.

Fig. 319.

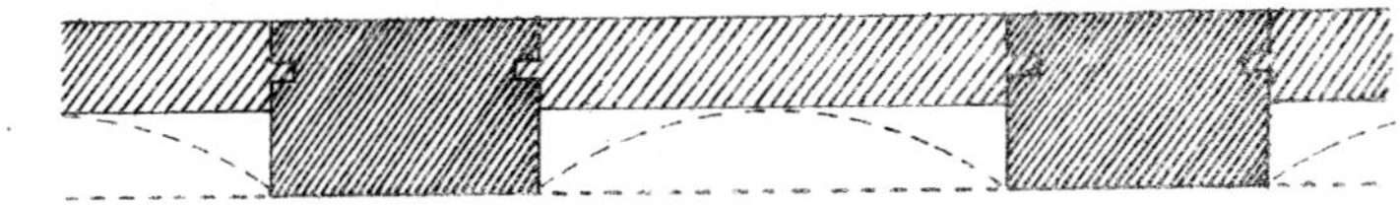

In the inside of the wall were several courses of bond timbers at certain intervals, the shrinking of which has injured the bond of the brickwork in some degree.

The downfall of part of this wall is generally supposed to have been caused by the buttresses being on one side only, which threw an unequal weight on the piles; and I conceive that the circumstance of the buttresses and panels not being built in one mass may also have had an unfavourable effect.

Mr. Palmer, the present engineer and architect of those docks, has since built a remarkable light and handsome wall in lieu of the above. The part above ground varies from $2\frac{1}{2}$ bricks at bottom to $1\frac{1}{2}$ brick at top, by a slope on each side, which is certainly much handsomer than offsets, although, all things considered, I should be inclined to prefer the latter, as stated in the preceding article. His buttresses appear on both sides of the wall, with an equal and regular projection of 1 brick each way from the bottom upwards, and they also slope regularly, not only in section but in elevation, and are placed at central intervals of 18 feet apart. Inverted as well as common arches are used to connect them below the surface. At the base of the wall Mr. Palmer has made use of the old piled foundation, but has added a substantial mass of grouting to aid it. I do not recollect the width of his buttresses, or the thickness of his wall at the foundation, but probably the latter does not exceed that of the former profile. No bond timbers are used.

Sometimes the walls of buildings may be made to answer as inclosure walls, which is usual in the mews or stables about London, and the same is done in the back of the soldiers' quarters in the Ordnance barracks at this station, and also at the cavalry barracks at Regent's Park. When there is no strong reason to the contrary, this arrangement is of course attended with considerable saving.

Of Brick Ornaments.

Formerly it was customary to have ornamental fronts of brickwork, which were prepared by cutting and rubbing the bricks, and when it was the fashion to build with red bricks, the ornamental parts were usually of a deeper red than the rest of the wall, and the bricks selected for this purpose bore a higher price, and were termed red rubbers.

In this manner, brick pilasters, with friezes, &c., were made, and it was also customary to rusticate brick piers, or the coins of buildings.

Specimens of ornaments of the description alluded to may be seen in most cities or old towns in this country, in which the houses are built with bricks.

Some of the storehouses in His Majesty's Dockyard at Deptford, are ornamented in this manner, with brick pilasters and a cornice.

In the present day, the practice of using brick ornaments of this description is almost obsolete, but there is one curious specimen to be seen on the east side of St. Martin's Lane, London, in a shop having an ornamental front of red brickwork, with fluted pilasters, very neatly and recently executed.

The only kind of brick ornament now frequently used, is to finish the wall of a building with brick dentils, by causing part of the headers of one course of bricks, the last but one, to project a quarter, or sometimes half, a brick from the face of the wall. Two courses of bricks, one above and one below these dentils, are also usually made to project a quarter of a brick, and over all a small stone cornice is sometimes added.

Fig. 320 represents this arrangement, which has been generally fol-

Fig. 320.

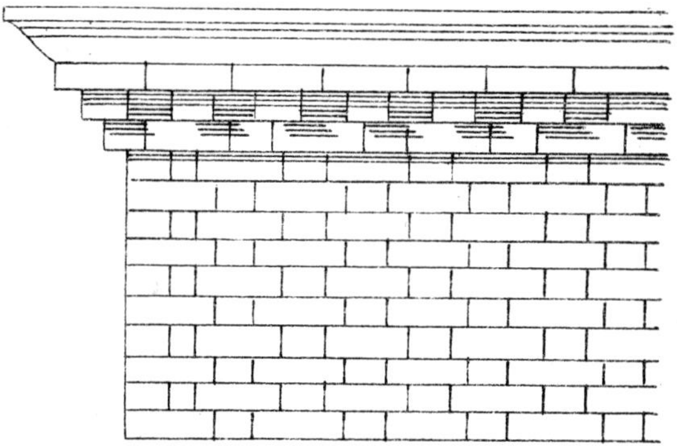

lowed in most of the low brick buildings in the Royal Arsenal at Woolwich. The Ordnance Hospital at this station has also brick dentils under a stone cornice, which, however, is larger and more ornamental than the one represented in the above figure.

Sometimes the brick dentils are placed at intervals apart equal to their own width, and form alternate headers in a heading course. Sometimes they are placed at intervals double of their width, and in this case they may be separated by stretchers, as in Flemish bond.

Another method of producing ornamental brickwork, without cutting or rubbing, is to set two projecting bricks upright on end, and with their edge on thin side to the front. These being included between projecting courses above and below them, assume the appearance of triglyphes in architecture.

Fig. 321 represents this arrangement, of which specimens may be

Fig. 321.

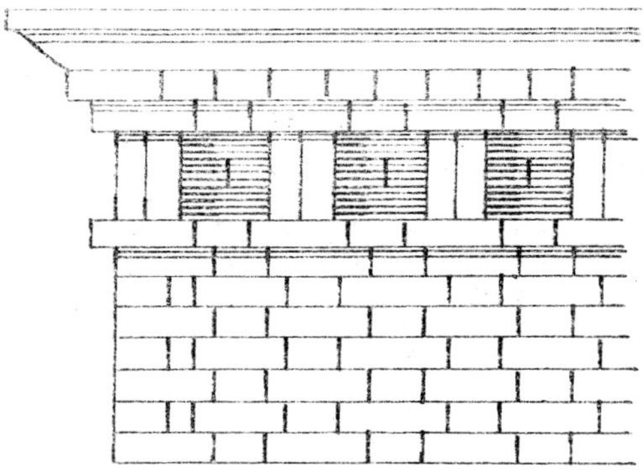

seen in a small school at Dartford, in Kent, and in some brick buildings on the south side of Waterloo Bridge, London. In the latter there is a small cornice of stone above, as shown in the figure.

It appears to me, that to attempt architectural ornaments in brick-work by cutting and rubbing the bricks, may be considered rather in a bad taste, as they never can have the beautiful effect that may be produced by the same thing executed in stone; but plain breaks, projections, and recesses, with occasional gathering courses, may relieve the uniformity of a blank brick wall in a manner pleasing to the eye, as has been done in the external wall of some of the mews attached to the new buildings on the south side of Regent's Park, and, if ornament be tolerated at all, the plain dentils and triglyphes that have just been described are the utmost that ought to be attempted.

Of preparing Brickwork for an Ornamental Front of Cement.

The use of cement, mastic, &c., has given quite a new character to the modern brick buildings about London, especially to those in Regent Street, and in Regent's Park, which have the appearance of ornamental stone buildings.

Brick columns are usually formed rough in houses built on speculation, by setting up a scaffolding pole of the desired height, as a core to the pillar, and working bricks or bats round it, all laid as headers, and allowing about an inch for the cement, which is to bring the column to the regular form and ornamental appearance afterwards proposed to be given to it.

In public buildings of importance about London, constructed by architects of reputation, I have observed, that a cast iron column of small diameter has sometimes been used for the core of a brick column, proposed to be coated in this manner. But, in respect to this point, much depends on the weight to be borne by the columns, which in some cases may be considerable, in others trifling.

After building the rough brick columns round a wooden or iron core in the manner that has been described, it is usual to lay a flat slab of stone over the top of each, to which the ornamental capitals of cement are afterwards attached.

Upon these slabs stout iron plates or bars are sometimes laid from column to column, with their ends turned up, over which a flat brick arch is built with cement to represent a stone lintel, as shown in *Fig.* 322, which is a sketch of the porch of a new building at Brighton pre-

Fig. 322.

viously to the application of the cement which was to give it the appearance of ornamental stonework.

The horizontal projections in the face of a building that is to be coated with cement are formed with gathering courses of bricks, when a moderate saliency of a few inches only is required, and in this case the lower courses of the bricks of the projection are laid as headers, sometimes on edge, and when the lower part of the projection is to be moulded, they usually have their lower ends dubbed off or rounded. But when a sudden and greater projection is to take place, Yorkshire stones, $2\frac{1}{3}$ or 3 inches thick, are used, which must either go through the wall or be so far bedded into it and loaded with such weight above as to render them perfectly stable.

In some of the new buildings in Regent's Park, London, the rusticated hollow joints which appear in the lower part of each front are done entirely with cement, the wall being previously a plain surface, but in the new United Service club house every sixth course of the original brickwork is sunk about an inch more than the face of the wall, in order to produce a similar effect of rusticated joints in the lower part of the building after it shall be coated with mastic as proposed.

It may appear scarcely necessary to observe that rough brick pilasters projecting from the face of a wall are invariably used to prepare for ornamental pilasters, when such are proposed.

In the new buildings at Regent's Park, London, it is common to use wooden beams, or bressummers as they are technically termed, to form the support of the brick architraves over the columns, and as they do not always fill out the whole space necessary to complete the form of the lintels over those columns, which are to resemble stone, they are often filled out with battens and laths, in the same manner as is practised for inside plastering.

It is obvious that relieving brick arches may be used over those bressummers, but I have seldom or never seen it done.

In buildings of any very great importance, I would prefer cast iron bressummers to wood for the above purpose, on account of the incombustibility and greater durability of the former material.

The bases and capitals of columns are usually made entirely of cement,

Fig. 323.

Fig. 324.

prepared in two or more pieces, and afterwards fixed and cemented together. I have observed that the upper part of a rough brick column is usually made square for three or four courses at the extreme top, immediately below the stone slab or cap, whenever the capital is to be of the Ionic, or Corinthian, or other highly decorated order. This probably affords greater facility in fixing the several pieces of which it is to be composed.

In reference to this subject, *Fig.* 323 represents a Doric archway.* One side is shown rough with the brick column, its stone cap and the projections above, partly consisting of gathering courses of bricks, and partly of Yorkshire stone, as before described. The bressummer immediately over the columns, which supports the upper part of the brickwork is in two pieces of 4 inch plank, each 12 inches deep, and bolted together at a clear interval of 6 inches apart. The other side of this archway is represented in the form which it will probably assume after being plastered with cement.

Fig. 324 is a section to explain these particulars more clearly.

Of the Mixture of Brickwork with Flints, or with small Stone Rubble in Building.

In some of those counties in England where chalk and, consequently, flints abound, it is common to build houses in a great measure with flints and mortar, but with all the coins of the buildings, and with the jambs and soffits of doors and windows of brickwork, intermixing also string courses of bricks thoughout the whole building on or about the level of the several floors. The fireplaces and chimneys are also built exclusively of brickwork.

In more ancient times the custom was to build with flints, but using stone for the coins, &c., instead of brickwork, of which there was specimens in many old churches in Kent.

Of the intermixture of flints and brickwork, which forms the subject

* This archway may now be seen, unfinished, in Regent's Park.

of this article, there are some few specimens in this neighbourhood ; but at Brighton it is the prevailing style of building.

The flints used there are boulders, generally of an oblong form, collected on the beach, of which the angles have been rounded off by the attritition of the waves. Their ends are presented to the surface of the wall, and they are laid in horizontal courses as regularly as may be when they are to appear in the external face of a wall. Coal ash mortar is generally used, the joints of which are necessarily thick from the nature of the materials which is to be cemented, and these joints are pointed so as to project about $\frac{3}{8}$ths of an inch in the centre. The brickwork at all the coins, and at the jambs of doors and windows, is usually in courses of threes, which extend alternately 2 bricks and $1\frac{1}{2}$ brick on the face of the wall, as shown in *Fig.* 325. The whole of the work thus combined has a neat appearance, and looks nearly as handsome as a common brick building.

Fig. 325.

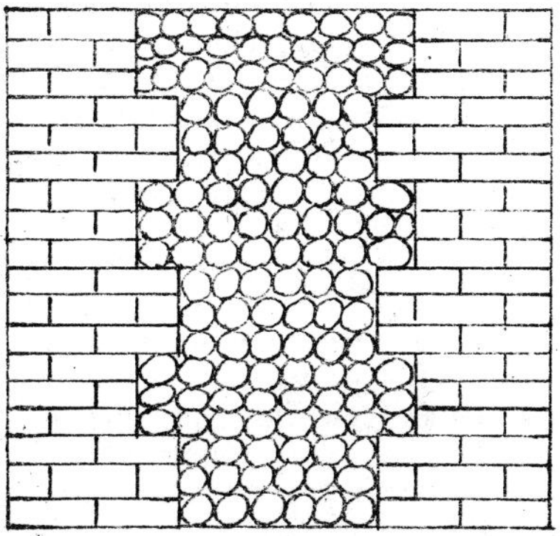

In the internal walls of houses thus built, the same attention is not paid to pointing the joints, or to the regularity of the courses. There the boulders are thrown in in a rougher manner, and, indeed, in every part of such a building, more depends on the cohesion of the mortar than on the bond of the other materials. In inside walls, in addition to the string courses of bricks, vertical courses or pilasters of brickwork are also introduced at intervals. In houses to be covered with cement, the whole of the walls are built equally rough, and sometimes bricks or bats are thrown in not laid courses, but diamond fashion, and without attention to bond.

It may be remarked that the common method of forming the coins with brickwork, before described, and illustrated in *Fig.* 325, is by no means good ; for although fair to the eye, there is not much bond in it, as you will find on analyzing it by means of your small models of bricks.

On considering the subject of brick coins interlaced with flint or other rubble work with attention, it will appear evident that single courses of brickwork only overlapping each other would not afford good bond, and that it is therefore proper to carry it up in portions of three courses each, as shown in the figure.

If the length on the face of the wall of one portion be made equal to the thickness of the wall, and the length the other way be made to exceed that thickness by three-quarters of a brick, good bond will be obtained by attending to the following rules :—

The several steps of three courses each are equal to each other, being rectangles resembling small piers of brickwork of $2\frac{1}{2}$ bricks by $2\frac{3}{4}$ bricks, of which the long and short sides alternately face different ways. You will, therefore, first build one course, which must show all headers on the long side, with a line of closers as usual near to the coin.

The next course will be a heading course on the short side. There let it have two closers at that end which appears in the face of the wall, and as it is a stretching course for the long side, let it finish on the reverse end with a header and closer, or with a three-quarter bat.

Fig. 326 shows the plan of the first and third courses, and *Fig.* 327 shows the plan of the second course of this part of the wall.

Fig. 326.

In the second portion of three courses of brickwork, in which the sides of the rectangle are reversed, the first, second, and third courses of the first portion, the bond being obtained by arranging the short sides of the second rectangle over the long side of the first rectangle, which is immediately below it, and *vice versâ*.

Fig. 327.

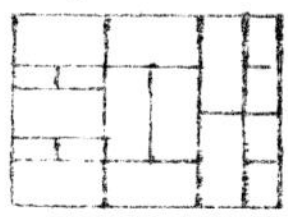

Using this arrangement, which I find on trial to be the only one by which good bond can be obtained, the elevation of one coin of the wall will be according to *Fig.* 328.

Fig. 328.

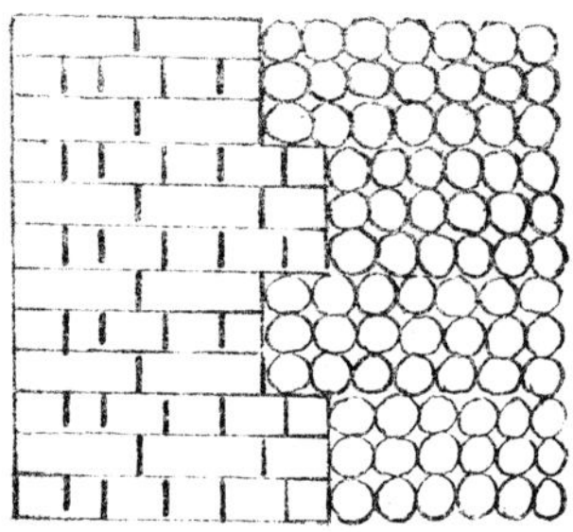

On the same principle good bond may be obtained for the brick coins of rubble walls of any other thickness.

In the brick pilasters in the centre of a rough flint or other rubble wall, it is also necessary, in order to obtain bond, to carry up the brickwork in portions, each consisting of three courses, and in building those portions each must be considered precisely as an independent brick pier of a square or oblong form. For example, if we suppose a rubble flint wall bonded by pilasters and string courses of brick to be equal to 2 bricks in thickness, the first three courses may be 2 bricks wide in the face of the wall, and consequently 2 bricks square, the next three courses may be $1\frac{1}{2}$ brick wide in the face of the wall, and consequently will form an oblong pier of $1\frac{1}{2}$ by 2 bricks square. These portions must have their closers running both transversely and longitudinally, and must be built with as much attention to bond as if there were no flints or rubble work on any side of them. To explain this arrangement here would be superfluous, it having been sufficiently illustrated in treating of square brick piers, as the same rules which apply to regular squares are no less suitable to oblong rectangles (See page 67).

At Dover, the fronts of buildings are usually built altogether with brick, the internal walls and back walls only being built with flint boulders bonded with brickwork in the manner first described.

In the inland parts of England, where field flints are used, they are cut transversely at one end, so as to present nearly a vertical surface in the face of the wall, and this kind of wall has a much darker appearance to the eye, than when boulders are used. This method also has rather a neat appearance when combined with brickwork in the front of a building.

When very small stone rubble is used in building, it is also customary to bond it together with brickwork at the coins, jambs, &c., in the same manner that has just been described, as applicable to flints. But if the stones to be used be a mixture of all sizes, it is more common to employ the largest for the coins, jambs, string courses, &c., and to square them for that purpose, leaving the smaller ones rough, and in this case bricks are not used.

In all cases when bricks are combined with any other material, the thickness of the wall is determined by a definite number of bricks and half bricks, in the same manner as if it were to be built exclusively of brickwork.

Of the mixture of Brickwork with Stone Ashler in Building.

In buildings of importance stone fronts are often used, which are backed with brickwork in one and the same wall.

The stone is built in regular courses, and in the architect's specification is, or ought always to be, made to bear fixed proportion to the dimensions of brickwork. For example, in his Majesty's New Palace, and in Lord Manver's new house, Portman Square, now in progress, the average thickness of stone is from 8 to 9 inches, being equal to one brick thick. In the New British Museum, the stone is from 8 or 9 to 13 or 14 inches thick, in order to correspond with from 1 brick to $1\frac{1}{2}$ brick.

The height of the courses of the plain part of a stone front should either be equal, or regulated in some other manner pleasing to the eye. Less than one foot is seldom allowed, the height being more usually 15 inches. In the New British Museum the heights of the several courses of stone are 18 inches.

It is always desirable that those heights should correspond with some regular number of courses of bricks, in order that the whole of the wall, at the finishing of every course of stone work should be exactly on the same horizontal level, for it is necessary every here and there to introduce bond stones, or bond courses projecting further into the wall than the general thickness of the stonework, which is usually a mere casing; and in buildings of importance it is usual to introduce here and there thorough bond stones or thorough stones, as they are termed, extending entirely across the wall, and sometimes thorough bond courses.

In stone pilasters projecting from the face of a wall having a stone front, the courses are usually equal to two or three courses in height, or even more, it being the custom to make such pilasters of much larger slabs than the rest of the wall. Door and window jambs are also generally made in much higher courses, or even sometimes in one piece, which is by no means unusual.

As the joints of the brickwork, being more numerous, are more liable to be compressed, and consequently that portion of the brickwork is liable to settle more than the stone front of the wall, it is prudent never to hurry the brickwork, but to carry it regularly all round the building by portions of a small number of courses at a time. Indeed, to the best of my observations, this caution must necessarily be followed in the neighbourhood of London, for it is seldom practicable to collect and prepare the stone for a large building so expeditiously, but that the bricklayers are obliged to wait for it.

Fig. 329 represents part of the wall of a stone front, backed with bricks in section. The stones are supposed to be 8 or 9 inches thick, and 15 inches high, but here and there some of 12 or 13 inches in thickness are intermixed with them, and thorough bond stones are used at intervals. The total thickness of the wall is equal to 3 bricks.

Fig. 329.

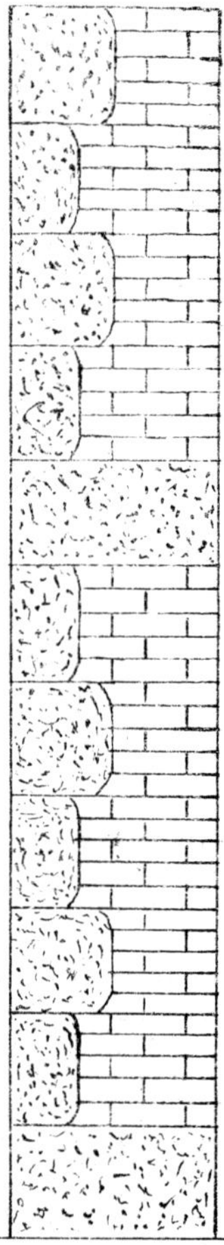

It is also not unusual to build part of a front of stone, and the remainder of brick, in which case the coins, the door and window jambs, and the arches or soffits over the same, as also pilasters, string courses, and cornices, are of stone, the remainder, that is, the plain parts or panels of the wall, being of brick. This arrangement is by no means handsome, when the dark red brick of former times was used, as may be observed in a part of Greenwich Hospital, but when malmstocks are used, or the pale Southamptom brick, it may produce a very good effect, as may be observed in the new storehouses in the Royal Arsenal at Woolwich, and in the new Parliamentary Mews near Westminster Abbey.

To enlarge further upon this subject would be superfluous, as it belongs more to stone masonry than to brickwork. I shall, however, conclude by observing, that an old brick building may be faced entirely with stone, by taking proper pains in the workmanship; and I should recommend cement for the joints of the stones, and metal holdfasts to connect them with the brickwork, as well as metal cramps from stone to stone.

Of this arrangement, of the perfect practicability of which I confess that I was not aware until I saw it done, there is a recent example in this neighbourhood, in the new facing of the square tower of Rochester Cathedral, which has been done with a casing of Bath stone set on edge, not exceeding from 4 to 5 inches thick, and laid in coal ash mortar, which casing, together with the addition of a flat roof and pinnacles at the angles of the said tower, has much improved the general appearance of that venerable edifice of which it forms a part.*

* The former tall square pryamidal roof of the same tower, which roof was itself a modern erection, was not unaptly compared to an extinguisher. The recent improvements have been carried into effect under the direction of Mr. Cottingham, the architect.

It will, perhaps, be scarcely necessary to mention, that this last mentioned arrangement ought never to be followed in new buildings, it being best to combine stone and brickwork into one body from the commencement, instead of making the former merely a casing to the latter. And under all circumstances in this and other northern climates, the greatest care must be taken to prevent the penetration of wet into the vertical joints between a stone front and its brick backing, which in a severe winter might ruin the building, in consequence of the effects of frost in those joints.

Of the connection of the Brickwork of Buildings with Staircases of Wood or Stone.

The prevailing custom in this country is to use geometrical staircases of wood or stone, both of which at first sight appear to derive their chief support from having all the steps inserted at one end only into a wall. This is actually the case in stone staircases, but in wooden ones, the steps themselves do not enter the wall, but are attached to a frame work technically termed the carraige of the staircase, which is partly supported by the wall. This, therefore, is a matter more belonging to carpentry and joinery than to brickwork, and as such, I forbear to enlarge upon it.

It was before stated that the steps of stone staircases of the above description ought to be inserted or tailed about 9 inches in a brick wall, although less is sometimes allowed. The grooves for this purpose are cut out of the brickwork after the wall is finished, and to give solidity to the steps, the butt end of each, which enters the wall, is made thicker than the projecting part, and is of a rectangular section, so that the groove cut into the wall is also rectangular, whereas the projecting part of each step of a stone staircase is cut away for lightness sake at the back, so as to assume very nearly a triangular section. Hence, in cutting into a brick wall for a stone staircase, there is a succession of rectangular grooves disposed step fashion, but on reaching the landing place a single oblong groove is of course cut, of much greater length, and corresponding with the extent of the said landing.

In order to save the trouble of cutting, it is not unusual, in preparing for landing places, to build that side of the wall into which the stones are afterwards to be inserted with dry bricks and sand for a couple of courses on height, and to the depth of 9 inches, or to whatever other depth may be required. Thus, the groove necessary for each stone is easily formed afterwards, by merely removing the dry bricks. This system, the best for landing places, is never used for the steps, as it would give more trouble, and weaken the wall more than the common method of cutting out the brickwork before described.

Of the execution of the Brickwork of Buildings, &c.

In those articles which treated of limes and cements, and of the various kinds of brick bond, the most important points relative to the execution of the brickwork were discussed, and these being understood, no further difficulty ought to arise, provided that you have bricklayers under you who understand their business, and are willing to do justice to their employer. But brickwork is run up so very quick that if the men be paid by taskwork, there is a great opportunity for unprincipled persons to do their work badly, or, as it is technically termed by artificers, to *scamp their work;* and this may often be done undetected, unless you have more foremen to stand over them than it is usual to employ. In order, therefore, to prevent this evil, some of the most eminent architects make it a rule in their specification,* that the *contractor shall not pay his bricklayers by measurement* but by the day, in order to remove from those men one of the chief temptations to do their work hastily, and therefore badly.

The too general custom amongst bricklayers is to neglect their vertical joints. They spread a bed of mortar over the top of each course of brickwork when finished. They then lay the outside bricks of the next course in mortar, which they apply, not generally, but to the external parts of those joints; and they are too apt to neglect the joints of the inside bricks in a similar manner. After a new course of bricks is thus laid, a horizontal bed of mortar is applied over the whole, which being too thick to sink down into such small crevices the consequence is, that the vertical joints are often laid dry, than which nothing can be more prejudicial to the stability of brickwork. It may easily be conceived that when the men are paid by taskwork, they may be tempted to neglect the vertical joints on purpose, which otherwise they can only do through carelessness; and it is not easy to discover any neglect in this point, without pulling part of a wall to pieces; for no sooner is any of those horizontal beds of mortar laid, than all the joints below it in the middle of the wall are completely hid from view. In the demolition of many old brick buildings in London, probably built by taskwork, I have observed the defect now alluded to of most of the vertical joints being left dry, or nearly so.

It therefore becomes the chief duty of persons superintending the execution of brickwork to see that all the joints are completely filled with mortar, or properly flushed, as it is technically termed. A certain method to ensure the object is the use of grouting, which is sometimes done in every course, but more frequently in every third course, or thereabouts, and which is most effectual in dry and warm weather. This operation is usually performed by laying a heap of mortar on the top of the wall in a concave form, so as to compose as it were a small basin of that material, and to pour water over it, after which it is mixed up with the water, and pushed backwards and forwards along the top of the brickwork by a tool resembling a flat hoe, until it is forced downwards into all the vertical joints. The only objection I have heard to grouting, as applied to ornamental brick fronts, is that it penetrates through every part of a wall, and therefore is not suitable for such fronts, which it disfigures by partially whitewashing them.

When a building is to have an ornamental back front, it is usual for an architect in his specification to direct that the courses shall be so regulated that they shall not exceed a certain height, which may vary according to the average thickness of the bricks used, for those made by different manufactures often vary a little in their dimensions. Generally in rough work, 4 courses of bricks will average one foot in height,

* The specification implies the written rules for the execution of a building drawn up by the architect, and accompanying the plans and other descriptive drawings according to which the work is to be executed.

but in ornamental brick fronts it is usual to direct that they shall rise only $11\frac{1}{2}$ inches, or thereabouts, including the joints of mortar; and the object of such an arrangement is to prevent any of these joints from exceeding $\frac{3}{8}$ths of an inch, and to make the whole of the brickwork have as regular an appearance as possible.

In carrying up brickwork it is usual, after a few courses are finished, and whilst the mortar is yet green, to draw a straight line along every joint by an iron instrument, which operation, technically termed striking the joints, compresses the mortar in the centre of each, and makes a small furrow in it. This is called jointing, and in this state the Ordnance Barracks at Chatham have been finished; and to me it appears superfluous to do more; but for private buildings, according to a practice before alluded to, it is very common to point all the joints in the manner that will be described in the next article. In preparing for that peculiar kind called tuck pointing, which is the most elaborate mode of finishing a brick front, it is considered necessary not only that uniformity in point of height shall be observed in regulating the several courses, but also that the whole of the corresponding upright joints, that is those of alternate courses, shall be exactly over each other in the same vertical line. For this purpose, after laying one course with care with bricks nearly of the same length, the bricklayer usually chooses a pattern brick, by which he tries all his bricks as he proceeds, and if any of them should vary in any sensible degree from that standard he rejects it and chooses another, and sometimes he is even obliged to cut them. In a former part of this work I have alluded with disapprobation to this unnecessarily troublesome and also expensive mode of proceeding, for in proportion to the time wasted by all these niceties, the workman must charge higher for the same quantity of brickwork executed.

In the construction of a building it is usual to lay not only chain timbers, wall plates, and other timbers that must enter into the wall, but also the joists of single floors, and the girders and binders of framed floors, as soon as the brickwork is carried up to the level at which they are to be placed. The framework of partitions is also fixed at an early period. In short, the whole skeleton of the carpentry of a building keeps pace with the brickwork; and the custom is for the bricklayers to prepare the bed, as it is termed, for every timber to rest upon, after which the timber itself is raised and placed by the carpenter.

Groined arches for the basement story of a building should never be executed, and, indeed, they are scarcely ever begun until the roof is covered in. The reason is, that the external walls of the basement story are seldom made thick enough, to act as sufficient abutments to such arches, without that aid which they are eventually to derive from the upper parts of those walls which belong to the stories above the basement. If, therefore, the groined arches were executed at the same time with the basement story, and before the building was raised to its full height, the thrust of those arches, not being met by the full weight of abutment that is eventually to counteract it, might force out the walls of the basement story, and ruin the building whilst in progress. It may also be remarked, that by waiting until the roof is finished, the mortar of the groined arches is of course completely secured againt the injurious effects of rain.

Wooden centring, which is made and placed by the carpenter, is necessary for all brick arches, with the exception of the arches of ovens.

These are built upon a solid mass of moist sand, moulded to the form which the internal aperture of the oven is to assume. After the brickwork is finished the sand is extracted through the mouth of the oven by a hoe.

There is a peculiarity in the finishing of brickwalls of inclosure which ought to have been noticed in a former part of this work. I allude to the circumstance of such walls being occasionally built of different heights in different parts, which is very common in the boundary walls that divide the ground in front of a row of houses facing towards a street. These walls are usually higher close to the houses than towards the other end. In this case, instead of making an abrupt break in the form of a step, it is usual to join the high and low portions by a concave and convex curve, resembling the letter S, so that if the coping consists of bricks on edge they are made to assume the archlike form, as in *Fig.* 330, which represents a wall of this description.

Fig. 330.

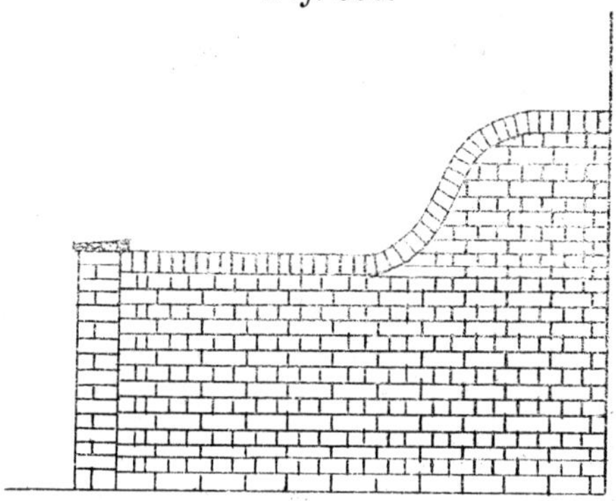

The arrangement now under consideration is technically known by the name of a brick ramp.

Of the Pointing of Brickwork.

There are two kinds of pointing in common use, one of which is called flat joint pointing, the other is called tuck pointing, which was formerly distinguished by the name of tuck and pat pointing.

In both the wall is prepared for pointing by scraping the common mortar out of the joints for a little way in; after which, in flat joint pointing, it is replaced by fine mortar, worked carefully into the joints with a small trowel made for the purpose, called a pointing trowel, and then the operation of jointing is performed, as before described.

In tuck pointing, after the common mortar is scraped or (as it is technically termed) raked out of all the joints, fine mortar is first put in, which is tinged of a brown, yellow, or bluish tint by a mixture of ochre, or of coal ashes and lampblack. This coloured mortar, which is called the stopping mortar, is well worked up, and a small groove is formed in the centre of it, by a pointing trowel. When this mortar has set, after which the joints are said to be stopped, fine white lime putty is worked in by a smaller pointing trowel; and whilst this is yet wet, the edges are cut off by a knife prepared for the purpose, called a pointing knife, which being applied to a straight edge, draws a couple of parallel lines at from a quarter to one-eighth of an inch apart, and scrapes off the superfluous putty at the same time. Joints thus pointed have the appearance of regular lines neatly formed with white paint over a darker ground, and in this process it is common to make the vertical joints a little narrower than the horizontal ones. The former joints are technically termed by the bricklayer "the perpens," being a curious contraction of perpendiculars.

Although, as I before stated, I prefer a plain front to all this minute regularity, yet it may be urged in favour of it, that the new pointing of a front that has been blackened by the smoke of the metropolis, gives it the appearance of a new building, as may now be observed in the Admiralty Office and the other buildings attached to it, which has recently been much improved in their appearance by this process.

Of Brick Drains and Sewers.

On the proper draining of a building, street, or city, much of the comfort of the inhabitants necessarily depends.

From every water pipe carrying off rain from the roof of a building, as well as from every sink, and from every water closet, a small drain leads into a greater one, the former of which is called a branch drain, whilst the latter is called the principal drain of the building. The latter likewise receives the rain water from all the areas, and often also from the ground in the immediate vicinity of the building, which for this purpose is levelled at a gentle slope, and has surface drains or gutters leading to iron gratings over the receiving drain, by which the water descends after a shower.

The branch drains are of two kinds, one flat at bottom and paved with perpendicular sides, and arched at top like the section of a small gallery; the other perfectly circular, and from thence termed a gun barrel drain, or simply a barrel drain.

Fig. 331.

The smallest size of the former kind of branch drain is 9 inches wide, 3 courses of bricks high up to the spring of the arch, and 4 inches thick at the sides and top, as shown in *Fig.* 331.

From the above, the size may increase to 12, 15, 18, or 24 inches in width, and from 4 to 8 courses of bricks in height, the custom being to make the width and height equal; and in this case it is usual in those larger kinds to make the sides and top one brick thick, as shown in *Fig.* 332.

Fig. 332.

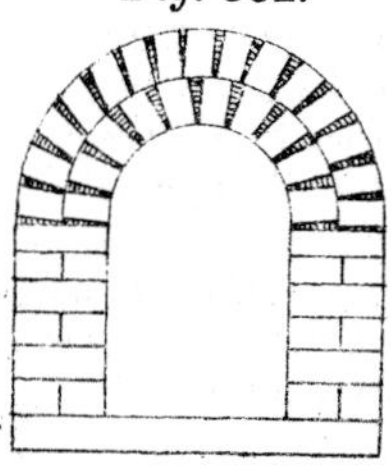

Of the barrel drain the smallest size used is 9 inches in diameter, which is always built with a 4-inch arch, as shown in *Fig.* 333.

This size is considered sufficient for carrying off clean water from the several pipes leading down from the roof of a building, and one 9-inch drain may consequently be used to every such water pipe. In this, and in other small cir-

Fig. 333.

cular drains, the joints must necessarily be very open at the extrados or external surface, for the reason before stated in treating of arches. Hence, in all the 9-inch drains in the Ordnance Barracks at this station, the bricks were ordered to be cut and gauged; but it scarcely appears worth while to go to that expense, which is by no means usual. Pieces of broken bricks, tiles, &c., or fragments of stone, ought, however, to be provided to put into the outside of the joints instead of using all cement or mortar.

In a drain leading from a water closet it is not considered desirable to use a less diameter than 12 inches in the clear; and, indeed, none of the branch drains in the new Naval Hospital at this station have been made less than the above diameter. The other sizes of barrel drains are 15, 18, 24, and 30-inch drains, the dimension stated always implying the clear diameter; and in these larger sizes it is most common to make the circular arch 1 brick thick, using, of course, in all cases two concentric 4-inch rings in preference to headers.

Fig. 334.

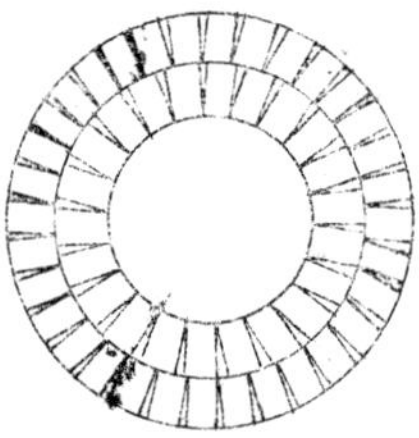

Fig. 334 is the section of an 18-inch barrel drain of the description alluded to.

The slope for a drain for clean water may be from $1\frac{1}{2}$ to 2 inches in 10 feet; the slope for a drain carrying down dirt or soil should be from 2 to 3 inches in 10 feet; but one cannot always command so great a slope as is desirable.

The barrel form is now most generally used for small drains, for which there seems to be good reasons, as the circle is the most compact and strongest form, and gives greater velocity, or, as it is termed, a sharper run to the water than a flat bottom drain of equal width.

Fig. 335.

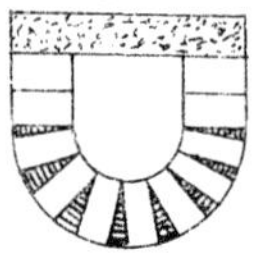

Another form of small drains have been sometimes used, in which a brick drain, semicircular at bottom and raised for a few courses above the inverted arch, have been covered with flat stones, as shown in *Fig.* 335.

One obvious advantage attends this arrangement, that the drains may be opened and examined by moving the stones, without materially deranging the work; whereas the small drain, constructed wholly of brickwork, cannot be thus examined without taking the upper part of the arch to pieces.

In all small drains, excepting those for clean water only, it is necessary to have cesspools of a square or circular form in plan, considerably larger and some feet deeper than the drain or drains to which they belong. These cesspools, being covered by a stone, or iron plate, or sometimes by a grating, receive the sand or earth washed down, and thus prevent the drains from being choked up, and when the cesspools themselves are nearly full, which must be ascertained by the persons in charge of the building, they must be cleared out by removing the stone or iron covering, which is replaced again as soon as this operation is done.

Besides these cesspools, which have been described, it is usual to form much smaller ones of a different description, usually termed stink-traps, to prevent offensive smells from passing along the whole extent of a drain. These are made by forming part of the bottom of a drain on a lower level, so that water must necessarily stand in it at all times, although the body of the drain may be occasionally dry. Into the little pool of water thus formed, a vertical stone or plate of iron is let down transversely from above, the under part of which enters lower than the surface of the water. This arrangement, without impeding the regular course of water in the drain, which finds its level after entering beneath the said stone or plate, does in a great measure prevent the passage of foul air.

The apparatus necessary for this purpose is sometimes made entirely of cast iron, as I have observed in the internal drains of the stables of the new Parlimentary Mews recently executed.

Care should be taken in arranging the small drains of a building, that nothing but the clean water drains shall have any possible communication with the tanks or cisterns for receiving rain water from the roofs, when such are used.

When the principal drain of a building is within a few feet of the surface, it may be a common barrel drain of a size merely suited to carry off with ease, and without obstruction, the greatest quantity of water that can be received from the branch drains and surface drains.

The above arrangement is not, however, always practicable. In every building having a basement story it is desirable that the bottom of the principal drain should be below the level of the area, which necessarily compels ycu to place it at a considerable depth below the surface. In this case, but more especially when the principal drain of a building passes under pavements, roads, walls, &c., which cannot conveniently be disturbed, it is proper to make it large enough for a man to go through and examine any part of it. Drains of this description are termed sewers or culverts, to distinguish them from the smaller kinds before described.

The smallest sized opening in which a man can well move by stooping is one of $2\frac{1}{2}$ feet in width, by $3\frac{1}{2}$ feet in height; but one of 3 by $4\frac{1}{2}$ feet is still more convenient.

Fig. 336.

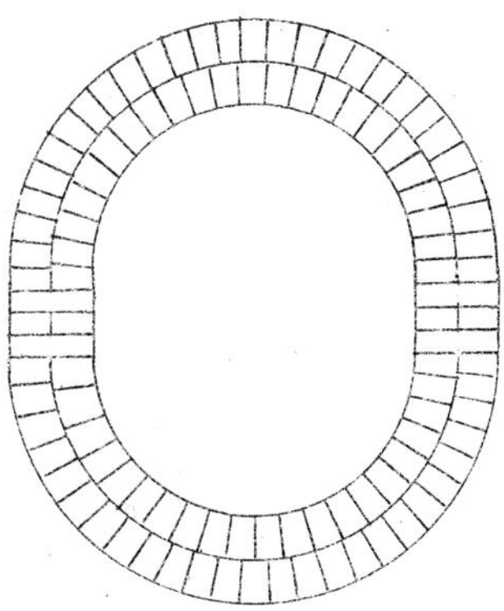

Such sewers may be elliptical in section, but the simplest form is to use semicircles at top and bottom, with a portion of the sides straight, as shown in *Fig.* 336.

I have selected the above section as the simplest, but others even more complex than the common ellipse, have been frequently chosen. For example, such sewers have sometimes been made semicircular, or in the form of an oblate ellipse at top, but in the form of an oblong ellipse or parabola at bottom, thus assuming a pearshaped section, as shown in *Fig.* 336.

From what has been already said, it may be inferred that the sections of sewers are in a great measure arbitrary. In London it is most common to make them with perpendicular sides and semicircular arches at top and flat at bottom, or nearly so, and even when counterarched, they often rest on a level foundation.

Fig. 337.

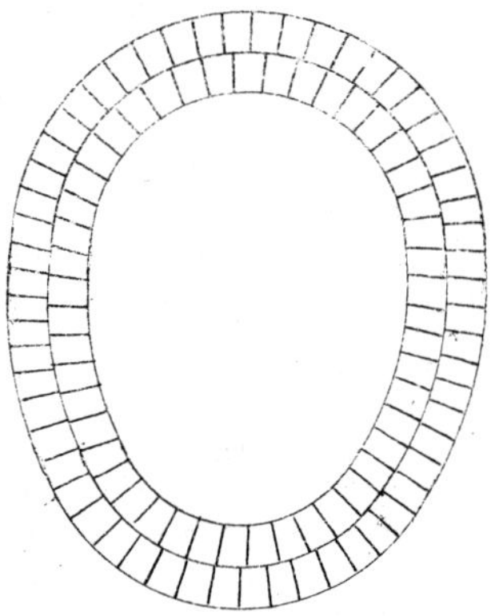

Fig. 337 is a section of the principal drain at the new King's Mews Barracks, near Charing Cross. It is 2 feet 6 inches wide by 4 feet 9 inches high in the clear. The sides are straight for nearly 3 feet in height, and are $1\frac{1}{2}$ brick thick. The arch at top is a segment nearly approaching to a semi-circle; at bottom a flat segment arch is used. These arches are one brick thick. The foundation is level, and small buttresses are added at intervals, as shown in the figure.

Fig. 338.

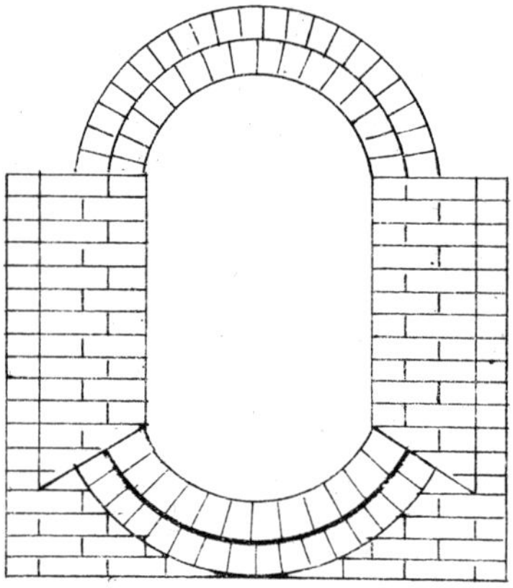

Into this sewer is conducted a pipe 3 inches in diameter, bored with numerous holes at the end, like the muzzle of a watering pot. Water can be turned into this pipe from the main pipe of the West London Water Company when required, for the purpose of cleansing the drain. A large cesspool, 7 feet deeper than the general level, and 7 feet wide, has been made just before the point where this drain enters one of the common sewers of that part of London. It may be observed that sewers or culverts do not require any great thickness of brickwork. Indeed, it appears to me, that the dimensions of one brick for the arch, and of $1\frac{1}{2}$ brick for the sides need scarcely be exceeded, even in sewers of rather greater magnitude than any yet described; but if the width and height of the aperture were increased to more than from about 5 by 7 feet, the culvert might then deserve the name of a tunnel, and as such it would be proper to use thicker brickwork than has been mentioned.

It would exceed the limits of this work to enter further into this subject of drains, which in most parts of England are comparatively easy, but which in low swampy countries, such as Holland, may be attended with considerable difficulties, and may therefore afford some exercise to the ingenuity of the engineer or architect who is called upon to design works or buildings on a great scale.

Drains are often constructed in common chalk mortar, excepting the cesspools, which in this case are generally built dry, because these being the only part of the work where water lodges permanently and acts in mass, the joints, if built with common lime mortar, would be ruined, in consequence of the decomposition of the lime, and the fall of the brickwork might follow.

In the drains of Fort Pitt, in this neighbourhood, in which common chalk mortar was used, the whole of the brickwork was laid on moist clay about 6 inches thick, and well beaten, and this system of puddling is no doubt good. But it appears to me that common lime, although very often used, is not a proper ingredient for the mortar of drains. I would therefore prefer laying the whole of the brickwork in cement, which is now generally done in the sewers of London, or, at least, I would recommend all the lower part, which is the most exposed to the action of water, to be laid in cement. The rest of the brickwork may in this case be built with a superior kind of mortar, such as that made from the Dorking or Blue lias lime, &c., &c.

I shall conclude by observing that all the drains should be well considered before a building is commenced, and they should be amongst the first things executed. The ground for the foundation should be well levelled according to the proper slope, and carefully rammed before the brickwork is begun. A drain should never be made to turn abruptly in any part of its course when it is possible to avoid it, and for the same reason, it is best that the branch drains should not enter the principal drains at right angles, but should be obliqued a little, so as to favour the new direction in which the water is to flow. It is also usual and proper to make the branch drains somewhat higher than the principal drains, to prevent the greater body of water in the latter, from flowing up into the former and choaking them.

Of the Tools and Implements used in the Execution of Brickwork.

1st. A large square, sometimes called a ground square, is used for laying out the angles of the foundation of a brick building.

2nd. Ground lines or tracing lines, for laying out the foundation of the kind commonly used by gardeners, which are attached to the pickets.

3rd. Line pins. Iron pins with large round heads for pushing into joints, to confine the ends of the common bricklayer's lines.

4th. Lines. These are cords much smaller than the ground lines, and consequently too light to be much affected by the wind; they are streched along both sides of a brick wall to regulate the heights of the courses whilst in progress.

5th. Tingles. Small pieces of lime tied round the bricklayer's line, and attached to temporary bricks placed at some central point, to keep the line straight.

6th. A plumb rule, being a wooden ruler or straight edge about 4 feet long with a plumb line and leaden bob fitted to it, to preserve the verticality of the sides of the wall in building.

7th. A large level to lay out the horizontal courses properly. Recently it has become usual to employ a substantial wooden straight edge, with a spirit level for this purpose, instead of the more cumbrous fitments of the common plumb line and bob.

8. A trowel. This is the bricklayer's principal tool, with which he spreads his mortar, lays the bricks, and cuts them when necessary, which in the style of building called Flemish bond is an operation continually going on. In cutting a brick, it is held loosely in the left hand, and struck with the edge of the trowel by the right hand.

9. A small wooden square for laying out the jambs of doors, chimneys, &c.

10. A wooden level. This is for laying out oblique lines, such as levelled window jambs, &c.

11. A hod. This is used by the bricklayer's labourer or assistant to bring bricks and mortar, and consists of an open box, having two sides about 16 inches long and 9 inches wide in the clear, and one end shaped as a quadrant of 9 inches radius. The hod is fixed upon a staff handle, and carried on a man's shoulder. At it holds 18 bricks, the weight carried in the hod approaches to 90lbs.

12. A gauge or wooden straight edge, about 16 or 18 inches long, having marks for regulating the heights of the courses in ornamental fronts.

13. A mortar board or plate of wood or iron about 3 feet square, to receive the mortar when deposited on a scaffold.

14. A larry, which is a flat iron hoe used in grouting brickwork.

15. Ladders as the work rises.

16. Scaffolding materials, consisting of poles, lashings and wedges, putlocks and planks. These will afterwards be described in a separate article.

The Tools and Implements used for making Gauged Arches are as follows.

1. A banker or bench, on which the bricklayers prepare their bricks.

2. A camberslip or ruler, curved in the proportion of about 1 inch in 6 feet. This marks the moderate curvature usually given to the intrados of flat arches over doors and windows. Sometimes the other side is also curved in the proportion of one half of the above, in order to make the still more moderate curvature sometimes, not always, given to the extrados of such arches. By this instrument the bricklayer describes his proposed curve, and draws his flat arch full size on

3. A large board, for which an old door is frequently used.

4. A lath is used for describing semicircular or segment arches, with a nail driven through one end for the centre, and a pencil at the other end of it.

5th. The instrument called a trammel for describing elliptical arches.

6th. A mould. A piece of wood 15 or 16 inches long, cut on each side to correspond with the radiating joints of the proposed arch. The form to be given to each arch brick is marked on the mould, after which it serves as a guide in cutting the several bricks. To save trouble, one face of a brick is always rubbed smooth and applied to one side of the mould, and thus the necessity of cutting two faces of the brick is obviated.

7th. A pair of steel compasses.

8th. Rules, such as by carpenters are termed straight edges.

9th. A small square with a brass blade.

10th. A small level with a brass blade for marking bricks that are to be cut obliquely, which is also necessary for the joints of groined arches, &c. The level is first applied to the brick, a line is then drawn close to it, across the brick by the trowel, and the brick is cut, and when the oblique side thus formed approaches to its proper state, it is rubbed smooth by another piece of brick, if it is to be exposed to view.

11. Templets, which are rectangular pieces for the purpose of marking the lengths of the several arched bricks. A long templet is used for marking those bricks which appear as stretchers; and a shorter one for marking those which appear as headers, in the face of the arch.

12. A small tin saw with a wooden back and handle, usually made in a rough manner by the bricklayer himself to commence the cutting of a brick, in order to prevent it from splintering.

13. A brick axe to complete the cutting begun by the saw. It has an edge at each end, like a very large chisel, with a round stem in the centre for grasping it. It is used by striking down over.

14. A chopping block.

15. A rub stone, to give the bricks a smooth surface, after being axed. This is a thin round stone fixed on the banker.

16. A float stone. This is a stone convex on one end,. to rub bricks to a concave form when necessary, as in niches, &c.

The Tools and Implements for Pointing Brickwork are as follows.

1. The jointing rule, a wooden straight edge, seldom exceeding 8 or 10 feet long and 4 inches broad.

2. The jointer. A small iron tool shaped like the letter S, with which the joints are compressed and marked in plain brick fronts.

3. The raker. A small iron tool, having a straight stem with two very short branches projecting from it in contrary directions, and pointed at the ends. By these points the mortar is raked or scraped out of the joints preparatory to pointing.

4. Pointing trowels, being smaller and finer than the common trowel.

5. Pointing knives, having an edge and a hook, the former to cut and mark the joints of putty, the latter to scrape off superfluous parts from those joints.

6. Hawks of different sizes, small boards from 5 to 10 or 11 inches square, with a handle for holding the mortar or putty in the left hand in the operation of pointing.

7. A watering pot or can.

8. A brush, such as is used in whitewashing, for damping the joints previously to stopping.

The Tools and Implements for Building Slopes in Brickwork, are as follows.

1. A battering plumb rule, of which one side is vertical, whilst the other is oblique, according to the required slope. A plumb line and bob hang near, and parallel to the vertical side, whilst the oblique side regulates the slope of the wall, technically termed its batter.

2. Profiles. These are made of substantial woodwork, to mark the whole height and form of the face of the wall. They are put up for regulating the form of sloping revetments of any importance, and more especially of such as are curved in front, like the docks and wharf walls

of some of our most eminent civil engineers; for to this last mentioned profile any kind of plumb rule is obviously inapplicable.*

The Tools and Implements for Cutting into Brickwork are as follows.

1. Brick chisels of different lengths, which need not be less than 1 foot nor more than 2 feet. They have a double arris or slope at the edge, which is of moderate width.
2. A light hand hammer for striking, similar to the miners boring hammer, or
3. A mason's wooden mallet, which may be used for the same purpose.

The above are necessary in all fine work, as in preparing the grooves for stone steps, &c., &c., but where the object is merely to destroy part of a brickwall, and there is room for a man to exert his full strength, the following tools will answer the purpose, viz. :—

4. A pickaxe, aided by
5. A crowbar.
6. A bricklayer's hammer is also sometimes used for cutting brickwork, having an edge at one end, and a flat broad head for driving at the other; but this is more properly a plasterer's tool.

Of the above tools and implements it will have been observed that many are common not only to the bricklayer but the stonemason. It may be remarked in concluding this article, that of all artificers the journeyman bricklayer requires the fewest tools, for he has seldom occasion to use any other than a trowel; whereas the carpenter and joiner, but especially the latter, are obliged to provide themselves with a great variety, which are not only bulky but expensive, it being the custom, and a very judicious one in all civil works in this country, that artificers shall find their own tools.

Of the Tools and Implements used in Making Mortar.

1. Water buckets.
2. A screen for sifting the lime and sand for common mortar.
3. A seive for finer purposes.
4. A rammer for pounding lime after being burned previously to using it, which is sometimes done in making mortar, but always in preparing for the grouted foundations recently adopted in this country, (See page 24).
5. A shovel for mixing the lime and sand, and compounding them into mortar.
6. A Bushel or other small measure for mixing the lime and sand in the exact proportions required, which, however, is seldom done in civil work in this country, as the expert bricklayer's labourer prefers judging of the proportions by his eye.

---

* A wooden coin on the principle of profiles has sometimes been used for vertical walls consisting of a flat block placed on the brickwork and kept down by a weight, to which is attached an upright gauge with marks for the courses. To these marks the line is fixed, instead of using the finished coin or temporary brick coin raised some courses higher than the rest of the wall, according to the more usual custom. Wooden coins are a modern refinement not generally introduced.

7. A cube yard measure, consisting of two parts, one of which fits on the other by dowels, each being a wooden box open at top and bottom, 3 feet square in the clear and 18 inches high. This is used in receiving lime and sand from the contractor who supplies them.

The following machines are sometimes used in making mortar, in preference to hand labour viz. :—

8. A mortar mill, working on a vertical axis turned by a horse. Two large vertical circular stones or iron wheels, move round in a trough, which is in the form of a circular ring, and grind the mortar, which is pushed from one to the other by scrapers. A hole on one side lets out the mortar when in a proper state.

9. A pugmill, similar to that used in brickwork. It consists of a vertical axis turning several knives, which are attached to it in a wooden vessel, having the form of an inverted conical frustum, and not exceeding four feet in height and three feet in extreme diameter. This machine also is worked by a horse, and the ingredients being introduced from the top, after having first been roughly mixed by a labourer with a shovel, are forced out through a small hole below, in the state of mortar, having been well mixed in descending by the action of the knives.

The pugmill has recently become more general than the mortar mill, from its being a less unwieldy machine.

Of Scaffolding.

After the bricklayer has built to the height of about 4 feet 6 or 9 inches above the ground level, it becomes impossible for him to raise the work higher without a scaffold; and in like manner, after building to the same height above the scaffold, it becomes necessary to raise it to this new level; and this process of shifting the scaffold at successive intervals of about $4\frac{1}{2}$ feet in height, goes on until the edifice is finished.

Scaffolding materials consist—

1st, of poles, which are usually 20 or 30 feet long, or even more, and from 6 to 9 inches in extreme diameter at the butt end, they being ufers or young pink trees.

2ndly, of putlogs, which are short poles about 6 feet long, and seldom more than 4 inches in diameter, but chopped square to prevent them from rolling. The ends are also square but cut still smaller, so as not to exceed $2\frac{1}{4}$ by $3\frac{1}{2}$ inches or thereabouts, in order that they may be less than the end of a brick.

3dly. Lashings and wooden wedges. The former of $1\frac{1}{2}$ inch rope and about 3 fathoms long.

4thly. Planks of the usual length of 12 or 14 feet, and $1\frac{3}{4}$ inch thick, which are generally hooped at the ends to prevent splitting.

With those materials the scaffolding for brickwork is put together, in the following manner:—

First, a line of upright scaffolding poles is erected on each side, parallel to the walls at the distance of about 5 feet, and at intervals of 8 or 10 feet apart. They are usually sunk about 2 feet into the ground at the butt end, and the earth rammed round them.

Secondly, a line of horizontal poles of the same description is lashed and wedged to those upright poles, in the position intended for the first scaffold. These horizontal poles, which are called ledgers, are continued all round the building, and where two meet it is usual to make the ends overlap, and to lash them not only to the upright poles but also to each other.

The ledgers and the wall combine in supporting the superstructure of the scaffolding, which is formed by the putlogs and planks.

The putlogs have a bearing of about 6 inches on the wall, and are laid in a position that ought to be the place of a heading brick. At the other end they rest on the ledgers. They are usually placed at intervals of 5 or 6 feet apart, excepting between doors and windows, where the piers are sometimes so narrow as to require them to be placed nearer, for they cannot, of course, be introduced where there is any opening without inserting an extra piece of timber across that opening as a bearer.

The planks are placed longitudinally over the putlogs, parallel to the wall, and it is common to use four or five planks alongside of each other, which forms a scaffold of about 3 or 4 feet in width. Care should be taken that the ends of planks shall never project any distance beyond the extreme putlogs on which they rest.

In high buildings, one tier of upright scaffolding poles is seldom sufficient. A second tier is therefore lashed to the first, having an overlap of not less than 10 or 12 feet, where the tops of those of the lower tier agree with the bottoms or butt ends of the upper tier; and, in this case, it is usual to introduce also diagonal scaffolding poles to connect the whole together, which extend longitudinally at an angle of 45 degrees, or thereabouts, along the line of upright poles and ledgers, and being lashed to both they stiffen the whole, on the same principle as the diagonal braces in field gates, or in frame work of carpentry.

In addition to those longitudinal braces, transverse struts are sometimes added to prevent the line of scaffolding from separating at top from the wall. These also consist of scaffolding poles, which are made to stand out at bottom at some distance from the line of uprights, and to those principal struts smaller struts, or braces, consisting of shorter poles, are occasionally added, which are fixed also in a transverse direction, and nearly at right angles to the former, to which they are connected about the middle height of the principal struts.

Of Machines necessary for Raising Bricklayers Materials, in the construction of a Building.

I consider it an axiom in practical mechanics never to employ a machine that is to be worked by men or horses, when the same object can conveniently be effected by the direct labour of the same number of men or horses, unaided by machinery. The same quantity of work is done without the machine in the same time, because it is well known that whatever you gain in power by a machine you loose in time, and it is done rather easier, because you get rid of the friction which attends all machines. It is also done cheaper, because you save the expense of providing them. Judgment, therefore, should be applied to the use of the mechanical powers, instead of considering every machine an advantage, which persons at first sight are apt to do. When absolutely necessary, use them, when they can be dispensed with, abstain from using them.

Generally speaking, machines are necessary in building for raising the masons and carpenters materials, for most of those are of a weight far exceeding the powers of a single man, or even of many men, and they cannot be subdivided into several portions without spoiling them. But with bricklayers materials the case is different. The single brick only weighs 5 or 6 lbs., and a mass of mortar may be subdivided into any number of portions you please. Hence, the bricklayer's labourer may carry just so many bricks, or so much mortar, as is suited to his strength. The robust men who follow this occupation about the metropolis seldom carry less than from 90 to 100 lbs. They take up those materials in hods, mount the ladders, and by means of the scaffolding deposit them on any part of the summit of the building where the bricklayers may require them. No arrangement is cheaper or more efficient than this, although it is not every labourer who is equal to the exertion.

Mechanical aids, however, are sometimes used, especially in very high building, the principal kinds of which I shall briefly describe.

1st. A large iron pulley of about 2 feet diameter, commonly termed a wheel, is erected on the top of a building, either by suspending it to the scaffolding, or from a pair of shears, which are two projecting spars lashed together at top, and secured by guys. Two square boxes are suspended by a rope, passing over this pulley, one of which ascends loaded with bricks or mortar, whilst the other is descending like the buckets in a well. Properly speaking, this is not a machine, for no mechanical power is obtained. Nothing is more common than this arrangement in high buildings, although I doubt the expediency of it. Two men are generally stationed below to fill the box, and haul upon the rope. One man above must unload it; and after all, as it is not convenient to move the shears and wheel continually from one part of the building to another, labourers with hods at top must take up the same materials, and carry them to the various parts, when the bricklayers require them.

2dly. I am informed that at the rebuilding of one of the great theatres in London, similar boxes of bricks or mortar were forced up by the weight of a man desending in the empty one, and holding by the rope. On reaching the bottom, which he did at the same time that the full box arrived at the top, he mounted by a ladder, in readiness to force up the next box in the same manner, as soon as it should be loaded. This method, which, however, I have never seen practised, appears to me to be preferable to the former one.

3dly. To the same apparatus of a wheel and shears, one box of greater size may be attached by a rope, the other end of which is fixed to the traces of a horse, after previously passing round a snatch block.* As soon as the box is filled with bricks or mortar, the horse is set in motion, who, walking straight forward in a path marked out for him, raises the full box to the top of the building. When emptied there, he is made to retrace his steps, and the box descends ready for a new cargo. This arrangement, in which no mechanical power is gained, appears to me to be the best method of using a horse for the purpose in question. I have seen it very recently employed at the new British Museum.

4thly. A drum may be used, which consists of a wooden cylinder of about 6 feet diameter, moving round on a vertical axis, turned by a horse, by the usual apparatus of a lever, &c. The horse in this movement de-

* I have never seen this apparatus used without the snatch block, but I am told it answers equally well; if so it is better without, for friction is saved.

scribes a circle of about 21 feet in diameter, or $10\frac{1}{2}$ feet radius. About the said cylinder half a dozen turns of a rope are taken, the ends of which, after first passing round a pair of snatch blocks to change their direction, are each led up to the top of the building, where they pass over a couple of wheels of the nature already described, and then boxes still larger than those before alluded to are attached to them.

The horse is made to turn the drum, first in one direction, afterwards in the contrary one, and thus one end of the rope is raised to the top of building with the full box attached to it, whilst the other box descends empty, after having been unloaded there. I have repeatedly seen this machine in use in London. Indeed, it is now used at the United Service Club House, in Pall Mall; but, on the principal before stated, I prefer the former method of employing a horse, as being much cheaper, much simpler, and equally efficient.

Of the Measurement of Brickwork.

The custom about London and in the south of England, is to measure brickwork by the rod, which is 272 superficial feet of walling,* supposed to be reduced to the assumed standard thickness of one brick and a half.

The general rule for measuring brickwork is, therefore, to find the superficial feet in any portion of a wall of uniform thickness, and to reduce it to the above standard of $1\frac{1}{2}$ brick thick by a simple proportion, and afterwards to divide the result dy 272, which will give the quantity in rods and superficial feet.

In measuring the brickwork of a building, the dimensions are taken on the spot by the two parties concerned, namely the contractor and the architect or engineer, who is to certify his accounts.† Every dimension is measured in feet and inches in their presence, with 10 or 5 feet rods, by a foreman or other competent person, and called out, and they enter it in ink, each in his own measuring book, which is a blank book ruled for the purpose, with alternate sheets of common paper and of blotting paper. During this process, if either party should object to a dimension stated, they examine the measurement over again, and the error, if any, is corrected on the spot. And to ensure perfect accuracy, the parties, before they leave the building, compare their books together, one of them reading over the whole of his dimensions whilst the other sees whether they correspond with his own; and in the event of any difference, it is rectified on the spot before they part. Thus, no mistake or misunderstanding can arise.

In these measuring books there are four columns for entering dimensions, &c., besides a wider space for remarks, called the waste.

In the first column, to the left of the book, the number is entered of any two or more parts of the building, which are exactly alike.

In the second column are entered the length and breath of any portion of a building in feet and inches.

In entering the dimensions of an arch in this column, the mean length

---

* A rod is an old English measure of $5\frac{1}{2}$ yards, or $16\frac{1}{2}$ feet. The superficial rod is therefore $272\frac{1}{4}$ superficial feet, but in practice the fraction is rejected.

† In civil life, the clerk of works employed under the architect usually represents him on this occasion, but in government works a surveyor of the Board of Works, or in the Ordnance Department an Engineer Officer must measure.

of the curve, supposed to be taken half way between the intrados and extrados, and the length or extent of the arch itself, are noted; and a small rough sketch of the section of the arch is also inserted.

The same is done in entering the dimensions of a drain.

The third column is reserved for the squaring or superficial content of each portion of the brickwork thus measured, which being a matter of calculation is done afterwards.

In the fourth column the thickness in bricks and half bricks is entered.

In the waste the portion of work measured is briefly described, and the addition of like quantities or other little calculations are sometimes performed on it, instead of using waste paper. Quantities thus added are technically said to be collected.

In measuring brickwork, walls are first measured solid, and the apertures which occur in them, such as doors and windows, are deducted afterwards. Hence, in stating the dimensions of those apertures the letters D D are inserted, opposite to them as an abbreviation for deduct.

In the second column, in which the dimensions of the several portions of a building are stated, a line is drawn under each portion to prevent confusion.

If any parts of the brickwork should be irregular, such as the spandrels of arches, &c., they are generally measured in cubic feet. In this case, the length, height, and thickness, are all inserted in feet and inches in the second column, and whenever three dimensions thus appear within the same space, cube measure is understood; and in the third column the cubing is inserted, instead of the squaring, which more generally occurs.

In squaring any two dimensions entered in the second column of the measuring book, they are multiplied together duodecimally, and the product generally yields feet, inches,* and twelfth parts of an inch, but no parts are inserted in the book, the rule being to reject all parts below 6, and to reckon 6 parts and upwards as an inch. Thus feet and inches only appear in the third column.

In cubing three dimensions by multiplying them together, the same rule is used.

The form of a measuring book is therefore as follows :—

	′ ″	′ ″		
2	10 0 3 9	75 0	3 B	
	75 7 5 0	377 11	51	South wall from under side of stone to top of stone plinth. D D Arch.
	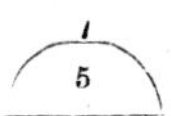	9 10	2	
	6 11 5 0 1 8	57 8		Add break.

These books are purchased ruled, but without any titles to the columns; and as the same form of measuring book applies also to carpenter's

* After two or three dimensions are squared or cubed, inches is an improper term, although generally used, for neither a superficial nor a cubic inch are implied, but the twelfth part of a superficial foot, or of a cubic foot.

work, &c., it is customary to write "bricklayer" at the head of the measurements of bricklayers's work, "carpenter" at the head of the measurements of carpenter's work, and so on.

When all the entries are made in the measuring book, including the squared or cubed dimensions, which are to be calculated afterwards, when the parties concerned return to their respective offices, the next thing necessary is to make out an abstract of the whole amount of brickwork executed.

Dr. Hutton, and some other writers, have suggested a rule for the abstracting of brickwork, which is to multiply the superficial feet or squaring of each portion by the number of half bricks in the thickness of the wall, and to divide the product by three.

Practical men, however, follow a different method, which after due trial I find to be much more advantageous than the above, as it saves a great deal of trouble and time.

In making up the abstract of their measurements from the measuring book, they use three columns, namely a 1 one brick column, a $1\frac{1}{2}$ brick column, and a cube column.

Having previously squared their dimensions in the measuring book, if the thickness of any portion of wall should be 1 brick, they set down the squared dimension in the one brick column; but if it should be $1\frac{1}{2}$ brick, they set it down in the $1\frac{1}{2}$ brick column.

If it should be $2\frac{1}{2}$ bricks thick, they set it down once in the 1 brick column, and once in the $1\frac{1}{2}$ brick column; thus entering the same squared dimensions twice, because 1 brick and $1\frac{1}{2}$ brick added together are obviously equal to $2\frac{1}{2}$ bricks.

On this principle they enter all other squared dimensions, either once or twice, in one or in both columns, as may be proper, but previously dividing or multiplying the squared dimension by some number, if necessary.

For example, if the wall be only half a brick thick they divide the squaring by 3, and put it down in the $1\frac{1}{2}$ brick column.

If the thickness of the wall be any multiple of $1\frac{1}{2}$, such as 3, $4\frac{1}{2}$, 6, $7\frac{1}{2}$, or 9 bricks, they enter it in the $1\frac{1}{2}$ brick column, previously multiplied by 2, 3, 4, or 5.

If the thickness of the wall be 2, 4, 5, 7, or 8 bricks, or any integer number of bricks not a multiple of $1\frac{1}{2}$, they enter the product of the squared dimension in the one brick column, previously multiplied by 2, 4, 5, 7, or 8.

If the thickness of the wall be $3\frac{1}{2}$, $5\frac{1}{2}$, or $8\frac{1}{2}$ bricks, or any number neither an integer nor a multiple of $1\frac{1}{2}$, the squared dimension is entered once in the $1\frac{1}{2}$ brick column as it stands, and its product multiplied by 2, 4, or 7, is entered in the 1 brick column, it being obvious that $3\frac{1}{2}$ is equal to $1\frac{1}{2}+2\times 1$, that $5\frac{1}{2}=1\frac{1}{2}+4\times 1$, and that $8\frac{1}{2}=1\frac{1}{2}+7\times 1$.

The cubed dimensions, after being calculated, are transferred at once from the third column of the measuring book to the cube column, without change.

When they are deductions in each of those three columns, which usually happens, the number of columns is doubled, there being an extra column of each class for the deductions.

All the quantities in each several column being summed up, and the deductions made, the whole brickwork executed is obtained in three amounts, namely—1st, the nett amount of the 1 brick column; 2dly, that of the $1\frac{1}{2}$ brick column; and 3dly, that of the cube column.

As $1\frac{1}{2}$ brick is the standard thickness, the amount of the 1 brick column must now be reduced to that thickness, by deducting from it one-third of itself. Thus, for example, 300 feet of walling, 1 brick thick, are equal to 200 feet of the standard thickness of $1\frac{1}{2}$ brick, because the proportional thicknesses of those walls are as 2 to 3.

In like manner the amount of the cube column must be reduced to the standard thickness of $1\frac{1}{2}$ brick, by deducting from it $\frac{1}{9}$th part of itself, because 12 inches are to $13\frac{1}{2}$ inches (the latter being equal to $1\frac{1}{2}$ brick) as 8 to 9.

Having thus reduced the amount of the three original columns of brickwork to the same standard of $1\frac{1}{2}$ brick thick, the sum of the whole must be taken, and divided by 272, as before stated, in order to obtain the quantity of rods and feet in the work executed.

This being done, the contractor sends in his account in the form of a bill, stating so many rods and feet of brickwork executed, at so much per rod.

If any of the quantities contained in the contractor's bill should be different from those made out by the architect or engineer, who has to certify his accounts, it will be evident, that as both parties start with the same dimensions, the error must be in the calculation on one side or the other, and may, therefore, easily be rectified on comparing the work together.

To give a more precise notion of an abstract, the following form is annexed, in making up which it is not the custom to rule the several columns, but to double or fold the paper, which is usually foolscap, a sufficient uumber of times, until the required spaces are obtainad. In summing up the whole brickwork, previously to reducing it into rods, odd inches are rejected, on the same principle before explained.

Abstract of the Bricklayer's Work at the—————*West Wing.*

Midsummer Quarter, 182—.

1 Brick. Add. ft. in.	1 Brick. DD. ft. in.	$1\frac{1}{2}$ Brick. Add. ft. in.	$1\frac{1}{2}$ Brick. DD. ft. in.	Cube. Add. ft. in.	Cube. DD. ft. in.
1889 7	107 2	1140 0	126 8	1142 9	312 9
303 4	16 8	270 0	30 0	608 0	312 9
860 10	21 2	165 0	6 4	480 10	1146 8
395 10	142 8	704 6	371 8	918 9	1329 0
622 11	25 8	232 6	556 3	1123 4	328 8
2255 0		370 9	90 0	23 4	
1858 4	313 4			312 9	3429 10
		2882 9	1180 11		
8158 10		DD. 1180 11		4609 9	
DD 313 4				DD 3429 10	
		1701 10			
7872 6				1179 11	
$\frac{1}{3}$ 2624 2		1701 10		$\frac{1}{9}$ 131 1	
		5248 4			
5248 4		1048 10		1048 10	

```
                    rods  ft
272 ) 7990  0 (  29   111
      544
      ----
      2559
      2448
      ----
       111
```

In measuring brickwork, it is the custom to measure all flues as if they were solid, the opening of each fire place being the only deduction made, and the contractor is therefore paid, with the above exception, for all the chimneys of a building as if they were solid brickwork, in consideration of the extra trouble which attends this part of the work.

For the same reason, small ovens are measured as solid brickwork, and the same rule applies to larger ones, excepting that in the latter the ash-hole is deducted.

The following kinds of brickwork bear a higher price per rod than common brickwork, viz. :—

1st. Brickwork laid in cement.

Those portions of a building which are thus executed, are entered separately in the abstract in double columns, in the same manner as common brickwork, but under the headings of "1 brick cement," "$1\frac{1}{2}$ brick cement," and "cube cement."

2dly. Brickwork circular in plan, such as the walls of towers and turrets, or bows in the front of a building. This is measured by taking the mean length of the curve, and the height, noting also the thickness in bricks and half bricks.

3dly. Sloping revetments, and more especially the curved leaning revetments used for dock or wharf walls by some eminent civil engineers, which must necessarily give more trouble to the workmen.

4thly. Rough arches, such as vaults, &c., not, however, including the small relieving arches over the soffits of doors and windows.

If, on the contrary, the agreement should be that rough arches are to be charged at the same rate as common brickwork, and the spandrels are filled in as high as the crown of the arch, they may first be measured as if they were all solid, from the level of the spring of the arch to the crown, and then the vacuity comprehended between that level and the intrados of the arch may be measured separately, and entered as a deduction, and in this case cube measure will of course be used.

5thly. Brick drains are also charged at a higher price per rod than common brickwork, on account of the trouble attending them ; and of the several sizes, although similar in section, the smallest is usually charged at the highest price per rod.

In measuring barrel drains by the rod, the length and the mean circumference are the two dimensions taken, whilst the thickness is noted in bricks and half bricks.*

6thly. Sometimes small drains are measured by the lineal foot, and paid at a certain price per foot, in proportion to their length. In estimating how much a drain ought to cost, when charged in this manner, you must first calculate how many feet of brickwork, reduced to the standard of $1\frac{1}{2}$ brick thick, there are in one foot in length of the given drain. Then by simple proportion, say as to 272 feet are to the proper price of such a drain, if paid for by the rod, so are the number of standard feet found by the aforesaid calculation to the proper price per foot of the given drain.

After having thus fixed the proper price per lineal foot of any brick drain, it is obvious that this method of measuring must save a good deal of trouble.

---

* This rule, like some others followed by practical measurers, is not strictly accurate, but very nearly so.

The same subject continued. Of Extra charges allowed for the nicer parts of Bricklayer's Work.

The rules that were stated in the preceding article chiefly applied to the measuring of plain brickwork, and in this manner the whole mass of brickwork of a building is measured and charged at the price or prices agreed upon.

Over and above the sums thus charged for plain brickwork, all the nicer and ornamental parts of the work are charged extra; and as the trouble and expense is not general, but applies to those portions of the work only which appear to the eye, these extras are measured separately, either by superficial measure, in which case they are entered in the second column of the measuring book in two dimensions, or by lineal measure in one dimension only. In abstracting them afterwards, the term superficial foot and running foot are usually abbreviated to the "foot super" and the "foot run".

The principal extras alluded to are as follows:—

1st. Extra facing, which includes the whole front of a building, excepting the side of four inch reveals and gauged arches, the former not being charged at all, and the latter being charged separately. Apertures are, of course, also excluded. This charge also implies not only superior workmanship, but also that a higher priced description of brick is used for the facing of a wall, and the extra price proper to be charged upon such bricks is usually calculated as if they extended $\frac{2}{3}$ds of a brick into the wall. The whole is, however, measured and valued by the superficial foot.

2nd. Extra labour and fine mortar for facing.

This implies that the same description of bricks are used throughout the whole mass of the walls, but that the best are selected or picked for the facing of the front, and that greater pains, and consequently more time are employed in the workmanship of this part, which renders it more expensive.

3rd. Gauged and rubbed arches, These are measured in superficial feet, according to the area of the figure contained between the intrados, the extrados, and the extremities of each.

4th. The points of groined arches. These require bricklayers of more than usual skill, and are very troublsome. They are measured by the lineal foot.

5th. Rough splays. These imply the oblique parts of window jambs and the splays or angular cuttings over the extrados, and on each side of the skew backs of rough arches, and for gables.*

6th. Rubbed splays. These imply the fitting of the brickwork over the extrados, and on each side of the skewbacks or gauged arches, and all other oblique work exposed to view in an ornamental front, which requires the bricks not only to be cut but rubbed. As this nicer fitment is known to extend no deeper than four inches, it is measured by lineal measure along the outline of all the curved or oblique lines to which it applies.

7th. Rough splays to angles. This implies that walls meet each other in oblique angles which may either be external or internal, salient or re-

* Extra labour for the coins of brick buildings has sometimes been claimed by bricklayers, but it is an improper and unreasonable charge, and has not been allowed to my knowledge for many years. It should always be resisted.

entering. In this case the bricks are cut only. These splays are charged by the lineal foot, measured, of course, according to the height.

8th. Rubbed splays to angles. Rubbing as well as cutting is necessary when those splays appear on the outside of an ornamental brick building, as in angular bows. These angles may of course be salient as well as re-entering, and are also measured like the former, by the lineal foot, but are paid at a higher rate.

9th. Bedding door frames or window frames. This is sometimes charged by lineal measure, by the running yard, sometimes by the piece, for those of equal size.

10th. Cutting chase, making good, and pargetting. This implies the cutting and plastering of a vertical groove, usually 4 inches deep, and 4 or sometimes 9 inches wide, cut into the side of one wall of a building, against which the end of another wall is to abut. The brickwork from the latter is to be let into this groove or chase, as it is technically termed, in the same manner as in carpentry one piece of wood is tongued into another. This is measured by the lineal foot.

11th. Pointing. This is charged by superficial measure, according to the style in which it is done, and applies to every part of a front pointed, excepting the gauged arches and cornices, or other ornamental work, if any, which are charged separately.

12th. Cornices or other mouldings set in putty. These are charged by superficial measure, not by taking the nett height, as it would appear in the elevation of a building, but by straining a line over all the projections, and into all the cavities, so as to embrace the whole outline of those mouldings, as they appear in profile.

13th. Brick dentils, as described in page 222. These are charged by the lineal foot.

14th. The excavations for foundations, drains, &c., are usually performed by the same person, who contracts for the brickwork of a building. This is measured and charged separately by the cubic yard.

The contractor for brickwork furnishes all his own scaffolding and machines, if he should think proper to use any, the expense of which he takes into consideration in giving in his tender for the execution of the work, but he makes no specific charge for those articles in his account.

It is customary to make no deduction for common bond timbers, or for other timbers let into the brickwork of a building, unless they be of very large scantling, and in that case it is usual to allow the bricklayer for every such large piece $\frac{2}{3}$ds of a course of brickwork of the same width. For example, if a piece of timber be 9 inches wide, and 36 feet long, an entry would be made of brickwork 36 feet long, 2 inches high, and one brick thick, in the measuring book, in favour of the contractor, as a compensation for his trouble in bedding the said piece of timber, whilst the remainder of the space occupied by it would be deducted.

The above are the principal extras which occur in brickwork, but many more might be added, which it would be superfluous in me to specify, for in all building contracts, care must be taken to define, regulate, and limit those particulars, as will afterwards be explained.

The same subject continued. Remarks on the Measurement of Brickwork. That the practice of measuring by the rod or yard ought to be abolished, and that all Brickwork ought to be measured by the foot.

Having from circumstances had very little experience in respect to measurements and estimates, I came almost entirely new to the subject, and such being the case, I could not help being forcibly struck by the inconvenience, or rather the absurdity, of the existing practice of measuring brickwork by the rod, an absolute fractional measure which tends only to give trouble without necessity.

In some parts of England they measure brickwork by the superficial yard of 1 brick thick, and it occured to me that a still simpler method would be to measure it by the square of 100 superficial feet, in which case, instead of the more complex division by 272, or even by 9, it would only be necessary to strike off two figures, to obtain the amount in squares and decimals.

On further consideration, however, and after making a reference to a friend of great skill and experience in the practice of building,* I have been convinced, that much the simplest mode of measuring brickwork would be to use the superficial foot reduced to the standard thickness of $1\frac{1}{2}$ brick in preference to any other mode.

If this system were adopted, the contractor tendering for the execution of brickwork would be required to name his terms at so much for the reduced foot, and he would be paid for that measure, instead of being paid by the rod. This would be a very great improvement, and would not occasion any change that would embarrass the practical man, who has been accustomed to the old system. For even now all abstracts of brickwork are first summed up in feet of the very same description that I propose, so that this new arrangement would merely be the means of saving a great deal of unnecessary trouble.

For example, if we suppose the foot of brickwork to cost one shilling, and that there are 1,574 feet to be charged in any part of a building, the price of the whole is $\frac{1574}{20}$ = £78 14*s*.

I need not point out, that if the same quantity of feet of brickwork is to be reduced into rods and feet by dividing by 272, how much more troublesome it will become to calculate the price of the rods and fractions thus obtained at £13 12*s*. a rod.

It may be remarked, that as in tendering for the rod of brickwork, pence are usually excluded, so if it were the fashion to tender by the foot, any fraction less than a farthing would be excluded; and in all cases try both ways, and you will find the new system of calculation now suggested to be the simplest.

Another beneficial effect will result from using no other measure than the foot. It will enable the Public Board Committee, or private gentleman, to whom a tender is offered for executing various descriptions of brickwork, to judge with more facility of the congruity of the prices, than when one part of the work is estimated by the rod, another by the yard, and another by the foot.

This remark struck me forcibly on looking at a contract for bricklayer's work in the Barrack Department at this station, in which the price of brickwork was to be at £3 9*s* 6*d*. per rod, provided that the government supplied their own bricks, whilst the contractor was to furnish labour, mortar, and scaffolding. In the same contract the price of raking the joints of an old wall, and pointing them in a plain manner, with coal ash mortar, was to be $2\frac{1}{4}$ per foot. Hence, the contractor's charge for building an entire new wall, with bricks supplied by govern-

* Mr. George Baker, of Montague Place, the son of Mr. Baker, of Rochester, and his partner in business as a builder.

ment, but he himself finding everything else that was necessary, would, according to this agreement, have been less than his charge for merely pointing a similar wall on two sides. Had all the prices in this contract been tendered for in feet, the incongruity of this arrangement could not possibly have escaped notice, as it actually did, until pointed out by a practical man, when the expediency of pointing the boundary wall of Chatham Barracks came under discussion.*

Of Contracts for the execution of a Building for a specific sum. Of the precautions necessary to insure the proper fulfilment of such Contracts.

In inviting persons to tender for the execution of a building or other work by contract, they are allowed access to the plans and other drawings prepared by the architect, together with the specification or detailed rules drawn up by him for the particulars of the work, which they usually copy; and from thence, after having measured and estimated every part of it, they are enabled to name the sum for which they are willing to contract for the execution of the whole building. This is the system usually followed in private works, whether executed at the expense of individuals or of companies; and in this case, although his employers will generally endeavour to impose some limit as to expense, the architect furnishes the designs and nothing more, without being obliged to enter into all the measurements, or to form a regular estimate of his own project.

This arrangement, although it saves a great deal of trouble to the architect, appears in some respects liable to objection. The task of estimating the whole expense of a large building in detail, which is not a very easy one, is thereby imposed upon a great number of competitors for the contract, only one of whom can obtain it, and only one of whom is, therefore, repaid for the time and trouble and expense bestowed on it, for some builders are obliged to hire a professed measurer to do it for them. Now, if only one out of the number of persons so employed should make an error in calculation, he may do so not only to his own prejudice, but to that of his competitors, for if the building be given to the lowest tender, such an error may lead to a person accepting the contract to his own loss, which otherwise he would not have done, and thereby throwing out others who were more competent to the proper execution of the work. Another evil results from this system. It prevents the Public Board Committee, or individual by whom the expenditure is to be defrayed or controlled, from being able to judge of the propriety of the various tenders; for none of those bodies or individuals can be expected to go over the whole estimate themselves. They can only judge by the comparative amount of the several tenders, which, by causing them to accept the lowest without reserve, whether a proper price or not, may lead them afterwards into the greatest embarrassment.

For example, if any such public body or private individual knew to a certainty that a proposed building could not be properly executed for less than £20,000, it would be the height of imprudence to accept the offer of any man, who undertook to contract for it at £10,000. Such a

* Mr. Brock, the present skilful and zealous Clerk of Works in the Royal Engineer Department at this station, who observed that it would be better to let the wall go to decay than to point it on these terms.

person must eventually either be incompetent, and his offer may be ascribed to ignorance, or he must hope to evade the proper execution of his contract by fraud and chicanery.

Of the impolicy of this system of accepting the lowest tender without reserve, which might be proved by a thousand examples, two very striking instances have occurred in the metropolis within the last twenty years ; one of a great public building, the other of a splendid new church, both of which, after having been contracted for in this manner, cost about twice as much as the original agreement, although neither the architects nor builders were changed; but the latter made a plea of deviations from the original plan, upon which they founded their claim to be paid by measurement for innumerable extras. It is remarkable, that in both these cases, no unforeseen difficulties occurred in regard to the foundation ; and neither were the area nor height of the building increased, nor the style of executing the work altered for a more expensive one, so that the inference drawn was, either that the original tenders were accepted, at far less than the fair value of the proposed work, or that the after charges were a gross imposition. It is superfluous to mention the names of the edifices alluded to, as all the facts are well known in the building world.

To prevent the risk of such very unpleasant excess of the expenditure of a building over the sum originally fixed upon, which under such circumstances as those that have been stated must in all cases be more or less discreditable to all the parties concerned, two precautions suggest themselves.

The first is to follow the rule now adopted by some of the most eminent British architects, who in drawing up the form of tender on behalf of their employers, always require each of the competitors for the contract not only to name the sum at which he is willing to execute the whole building, but also to specify the several prices at which he calculates, and according to which he would be equally willing to furnish the various materials, and execute the same work if paid by measurement, and they further require him, in the case of any alterations, to abide by those prices, which may either be in excess, or in diminution of the specific sum named, according as it shall appear on measurement that the alterations in question involve more or less work than those parts of the original design which are thus deviated from. To prevent all ambiguity on this head, a schedule of the various materials and descriptions of work necessary in the execution of a building, is usually annexed to the tender or form of contract, with blank spaces for prices opposite to each article, which the tenderer is to fill in, including also the rate he charges for his artificers and labourers, if employed by the day, which may sometimes be required.

The architect cannot be too careful in drawing up this schedule, which is termed the list of particulars, and he ought to leave no extras of importance unspecified that are likely to occur.

The second precaution alluded to is also generally followed by the same architects, who recommend their employers not to call for a general and indiscrimate competition by public advertisement, but to invite by letter only a limited number, such as ten or twelve of the most eminent builders, to send in their proposals. In this case the lowest tender may be accepted without hesitation, and without reserve, for there is no risk of disappointment in dealing with men whose skill has been repeatedly tried on a great scale, whose capital enables them to command all the means necessary without embarrassment or delay, and whose character for integrity is established.

If I were conducting a work on my own private account, I would adopt this system without hesitation, in preference to all others, on the same principle that any government would rather employ a general whose reputition is established than one who has not yet been tried. But strong reasons may also be urged, for not excluding general competition by public advertisement, which system has, therefore, been very often adopted both by public departments and by private companies.

When this course is resorted to, a rule ought to be established of never accepting any tender merely because it is the lowest. For if that mode of decision be adopted, which has been too often done, the certain consequence is, that in nine cases out of ten, the contract will fall into the hands of presumptuous speculators, without sufficient skill to conduct the work, or of unprincipled adventurers. A person who is without experience in such matters would naturally suppose that the conditions and penalities with which all contracts are fenced in, must completely guard against this evil; and that no man, who is either destitute of the skill or means for the proper execution of a contract would ever become a competitor, and that above all no one would be so rash or foolish as to engage to execute a work for less than what he knows to be the prime cost of the materials and workmanship. Experience, the only sure test of human measures, has, however, proved the very reverse of all this to be the case. It is well known that there are men who will offer to take any contract on any terms, even to their own apparent loss. Anxious to get into employment, they disregard the risk of failure, which has no terrors to men without capital and without established character; and in the meantime, if their offer be accepted, they obtain credit on the strength of their contract, which otherwise they could not command, and they take every possible advantage of their employers. But what is worse, this system when acted upon for a number of years in every public department, tends to demoralize all the tradesmen in its employment. Hence, it has been known, that persons really honest in private life, but who have accepted government contracts for less than the fair value of good materials, have either supplied bad, or have attempted frauds in respect to quantity, which in dealing with any private gentleman who agreed to give them fair prices, they would have disdained to do.

If, therefore, public competition be resorted to, I again repeat that no tender should, on any account, be accepted without reserve merely because it is the lowest. In the first place, the competency of the person offering to take the contract ought to be most strictly inquired into, and unless he be known to possess some skill and experience in the art of building, he should not be employed. Secondly, a correct estimate of the true value of the work should be obtained if possible, or at least a near approximation to it, and if the person whose tender is the lowest should prove to have offered to take the contract for a great deal less than the prime cost of the materials and workmanship, his offer should be peremptorily rejected.

For, under the most favourable of all suppositions, if those persons who tender for a public contract should generally be young men of great talent, but without capital, allow them fair prices, and they may succeed and eventually rise in the world by laudable industry. Curtail their prices to less than the fair and reasonable profits upon good materials and good workmanship, and they will be sure to fail, not only to their own detriment, but still more to that of their employers. For it is impossible to disguise the fact, that to grant a contract to any person, or to make any other agreement with him, on terms evidently disadvantageous to him, and leading as you suppose to your own gain and to his loss, is not only an act of great illiberality in point of principle, but in its results you will find it to be an act of folly, not of economy. According to the old English proverb, it will end in the biter being bit.

Of Contracts for the Execution of a Building by Measurement.

In government buildings, the system of contract described in the foregoing chapter is not always followed.

In the Ordnance Department, in Great Britain, the Commanding Engineer at each station is required to give in a detailed estimate of every new building proposed to be executed, as well as of all repairs for the current year; which being approved, the work is divided into the several heads of bricklayer's work, carpenter's work, &c. &c.; and those persons who are willing to contract for one or more of those branches of work, are provided with printed forms of tender, containing a specification of the manner in which every part of the work is to be executed, together with a schedule of particulars, with blanks for the prices thereof, including not only measured work, but all materials, and even day work, if such should occasionally be required, which, under particular circumstances, is sometimes the case. These tenders being filled in and signed by the several competitors, are sent sealed to the Engineer's offices, from whence they are forwarded to the Master General and Board for their decision.

This system appears to me to be much preferable to the former arrangement of contracting for a specific sum. If the Engineers estimate has been founded on just grounds, the probable expense of the whole building will be known beforehand with as much precision as can reasonably be expected. The competitors for the contract are saved the trouble and expense of making each his own separate estimate of a building, which only one of them can be called upon to execute, and the successful candidate, whoever he is, will not be liable to suffer from any error in calculation, but will be paid in exact proportion to his exertions.

Another advantage attends this system. The Public Board Committee, or individual by whom the expenditure is to be controuled, and of course the contracts approved, may easily ascertain the propriety of the terms contained in the several tenders by inquiry; whereas, according to an observation before made in the former system of tendering for a sum of money, it is quite impossible for them to judge between the several tenders, unless they were to go through the whole estimate themselves, which is out of the question.

When tenders are made to work by measurement, I again repeat, that the same rule laid down in the former article ought to be most strictly adhered to; which is no less sternly to reject the tender of him who inserts prices much inferior to the known value of the work in question, than of him who charges exorbitantly high. The honest tradesman ought in justice to have a fair and reasonable and moderate profit upon materials supplied and work performed.

In tendering to execute a whole building for a sum of money, a man may possibly make an error in calculation to his own prejudice, as was observed in the preceding article; but in tendering to work for prices, according to the system now under discussion, he who offers to accept a contract to his own apparent loss, must do it wilfully, and as it cannot possibly be his intention to fulfil his agreement faithfully, it is best to have nothing to do with him.

In contracts of the nature now under discussion, as well as in all others, it is desirable to leave no room for misunderstanding. Accordingly, the specification and schedule cannot be too carefully drawn up. They ought to embrace the known wants of the department. All extras likely to occur should be provided for, and no others should be allowed, notwithstanding that custom may be urged in favour of them; and if the peculiar mode of measuring any detail of bricklayer's, mason's, or carpenter's work, although usual, should be considered objectionable, as allowing more materials or labour than are reasonable, the precise mode approved in lieu of it should be clearly stated in the form of tender. Nothing of importance should be left undefined. At the same time, it would be idle to attempt to embrace in such a schedule every article without exception that may possibly be required, which would make it ridiculously complex, without adequate advantage, by including trifles of little value and of rare occurrence.

In the useful annual publications called Builders Price Books, the estimated prices of artificer's work of all descriptions for the current year are inserted, by a reference to which the nature of those schedules now under consideration may be understood; not that their contents can ever exactly agree in any one branch of work, but the one is sufficient to give a good notion of the other.

According to the present system of the department, all Ordnance contracts in Great Britain, including the affairs of barracks, are in force for three years, but determinable at the option of either party at the expiration of the first or second year, on giving three months previous notice. The materials and workmanship of the contractor are, according to custom, to be to the satisfaction of the Board of Ordnance, or of such person or persons as they may appoint to inspect the same; and in the case of neglect or failure, they are at liberty to employ another tradesman, and to charge the extra expense to the contractor; or in the event of gross neglect or misconduct, they may *immediately* annul the contract.

In the Ordnance Schedule for brickwork now in use, it appears to me that the articles are rather fewer than might with propriety be inserted. They are to be seen in all the Royal Engineers' Offices at the principal Ordnance stations, and are like all other forms, liable to vary from time to time, according to the views of the members of the board for the time being.

It may be remarked as not an uncommon practice, on the part of persons tendering for government contracts by prices, according to the system now allowed to, to speculate on what description of work is most likely to be required during the period for which the proposed contract is to last. This they ascertain, if resident on the spot, by observation or inquiry; and, accordingly, whilst they demand a reasonable price for those branches of work which are likely to be much wanted they will often tender at less than the prime cost of materials for those that are not likely to be required at all, or which they conceive will not be required in quantity. Hence, they endeavour to gain the preference by exhibiting a great number of prices, much lower than their competitors, if ignorant of this mode of speculation, are likely to offer. Such persons, however, are sometimes liable to suffer from an error in calculation, as happened at one great station in England, in which the contractor for carpenter's work was suddenly called upon to furnish two new drawbridges of massy oak timber, for which he had purposely inserted an inadequate price, not expecting any such timber to be used during his contract.

To guard against this kind of speculation, which ought to be repressed rather than encouraged, the prices of those articles which are known by the experience of the department to be the most important, and to be required to the greatest amount, ought chiefly, or rather almost exclusively, to be considered in deciding between the tenders of the various competitors for a contract of this description; and I believe that this is the rule now followed by the Board of Ordnance in such matters.

In the Engineer Department, works are not always executed by contract, sometimes they are performed by workmen in the pay of government, superintended by master artificers belonging to the department, under the direction of the Officers of the Royal Engineers. Sometimes part of the work is done one way, and part the other.

For the reason before stated, which often induces architects to enjoin contractors not to employ their workmen by taskwork, there are many branches of work, which, if executed not by contract, but under the Commanding Engineer himself, or by civil or military workmen in the pay of government, ought not to be given out by taskwork. The driving of piles, for instance, comes under this head; which, if badly executed, may eventually lead to the ruin of a fine and expensive building. But in those cases in which taskwork may be permitted, and those in which it should not, must be left to the judgment of the architect or engineer. Suffice it to say, that too much caution cannot be employed in this important point.

The same subject continued. Of the system of Contracts adopted by His Majesty's Board of Works in London.

The prices agreed upon in the Ordnance contracts, after being approved by the Master General and Board, are definitive for the whole period of the contract, provided that neither party find reason to take advantage of those articles which enable them to break off their agreement before the expiration of that period.

A different system is followed in the Department of the Board of Works in London, under whom most of the public buildings in and near the metropolis are constructed, excepting those which belong to the Ordnance and Navy.

The architects employed in this department are required to give an estimate of the probable expense of each of those buildings, for which they furnish designs; but when their projects are approved, the contractors for the execution of the work are always paid by measurement, both for materials and workmanship.

So far the system agrees with that of the Ordnance, but although similar in principle, it differs in respect to the details of the system of payment.

The Board of Works regulate their payments, not by definite prices for a given period of contract, but in reference to schedules of prices drawn up by themselves, and founded on the known market prices of the metropolis for materials and for workmanship for the time being, together with a certain estimated profit. These schedules are formed on such an amply remunerating scale, that the custom is for each competitor for a contract to state at what percentage less than the above schedules of prices he will be contented with; and if all the competitors are equally competent, which is usually the case, as the competition is more often select than public, they accept the lowest tender.

If the correctness of the data by which the Board of Works regulate their prices can be depended upon, which they do not usually alter oftener than once a quarter, this mode of payment, although fluctuating with the fluctuations of the markets, appears exceedingly equitable, and I believe it has been found to give general satisfication to all parties concerned.

I am informed that for very small buildings and repairs a profit of about $12\frac{1}{2}$ per cent.* is usually gained by persons contracting with this department, but that for great public buildings contractors are generally found who are contented with a much smallar profit.

I am also informed that recently a similar system has been adopted in the Ordnance department in Ireland. A schedule of prices of each kind of artificer's work has been drawn up for every great station or district, on a fair remunerating scale, and in just proportion, or nearly so, to the value of each branch of work; and in tendering to execute the same, the competitors are required to make an offer, in reference to the said schedule generally, not by particulars, that is to say, they may either offer to work for the same identical prices, or for a stated percentage more or less than the above, and the lowest tender is generally accepted, if no strong reason to the contrary should appear.

General Rules for forming the Estimate of the Brickwork and Foundations of a Building.

It was before stated, that in forming an estimate, the whole of the brickwork of a building, as well as the labour of excavating the foundations, must be measured and abstracted, and then nothing more remains than to fix the prices.

The price of excavation depends—First, upon the nature of the soil, whether easy or stiff. In the former a good labourer may dig and fill two cubic yards for one of the latter in the same time.

2ndly. On the distance to which the rubbish is to be wheeled or carted from the foundation.

3rdly. On the expense of the labour of men and horses for the time being, and also on the wear and tear of carts, and of tools and implements—namely, pickaxes, wheelbarrows, and planks.

The shovel or spade, as well as the clay spade and hod, when required, are found by the labourer himself in civil life, but when military working parties are employed, they are always supplied with every kind of tool necessary.

---

* Over and above the actual prime cost prices of materials.

The price of brickwork depends upon the following considerations :—

1st. Upon the number of bricks required, their prime cost, and the expense of conveying them to the spot by water, or in carts, or both.

2ndly. Upon the quantities of lime and sand required, the prime cost of each, and on the expense of conveying them to the spot.

3rdly. On the price of workmanship, including the bricklayers and their labourers, and the foremen superintending.

4thly. On the wear and tear of scaffolding, of machines when used, and of all implements not provided by the workmen themselves, such as large squares, levels, &c.

5thly. On the wear and tear of tools for military artificers, which article applies to the King's service only, for all artificers in civil life find their own, as was before mentioned.

6thly. A percentage for contingencies or unforeseen difficulties, usually ten per cent., is always added over and above the nett sum estimated.

7thly. When an architect is employed he is, of course, paid for his designs and trouble in superintending, and the clerk of works, who acts under his orders, and who is always on the spot, must also be included in the estimate. The latter is paid by the day so long as he is employed. In a very large building there are sometimes two clerks of works, which is the case at present in the new General Post Office. In civil life the architect is allowed a percentage on the whole expense incurred, usually at the rate of five per cent. for common buildings, but less is charged for the buildings of docks, such as warehouses, &c., in consequence of the greater simplicity of their construction, as well as their great number and uniformity.

In the Royal Engineer Department the officers and clerks of works are in constant pay, and it is not usual to include them in the estimates of any one particular building, but they are charged to the general expense of the station or colony where they may be employed.

In some departments of Government an architect also has a certain salary, usually a moderate one, in addition to which he is allowed to charge extra for his designs and drawings for new buildings only, and for the trouble of superintending them, but at a lower rate than usual, in consideration of his salary. In the Ordnance Department, designs furnished by the architect are sometimes executed by the Engineer Department, as was the case with the Ordnance Barracks at Chatham, which were planned by Mr. James Wyatt, and executed under the charge of Lieut.-Colonel D'Arcy, then the Commanding Engineer at the station. But, generally, it is more usual for the Engineer to design as well as to superintend, unless the building should be much more decorated than the barracks that have been mentioned.

Having stated general rules for forming an estimate, I shall recapitulate, and explain, if necessary, and make some observations.

In respect to excavations, it has been found that an able-bodied man, inured to labour, will, without over exertion, dig one cubic yard of earth in an hour, in easy soil, or one cubic yard in two hours, in hard stiff soil. Also, that he will wheel the same quantity of earth 20 yards up an inclined plane, or 30 yards on level ground, in the same time. Hence, the following prices were allowed as task work to military labourers in excavating for the foundations and ditches of the towers in the Eastern District, viz. :—

		Easy soil.	Hard soil.
Digging and throwing out per cubic yard		1d.	2d.
For digging, filling, and wheeling	20 yards......	2	3
" "	50 "	3	4
" "	80 "	4	5
" "	110 "	5	6
" "	140 "	6	7

The price of military labourers of the Line is fixed by the regulations of the Army at 1d. per hour, or at 10d. per diem, for the day of 10 hours. But if the above scale be correct, as I believe it to be, the same proportion may be easily applied to civil labourers with equal accuracy.

For example, if you pay your labourers 2 shillings a day, which is about 2½ times as much as the working pay of the soldiers, you have only to multiply the above prices by 2½ in fixing the rates at which you may contract for their labour by taskwork.

It is to be observed that the above prices applied to very large excavations, in which the men had ample room to work. In excavating for the foundations of common walls, which are narrow trenches, and in which a man works with constraint, and in which also time is lost by the necessity of levelling the bottom, double the price was allowed for this kind of excavation,—that is to say, 2d. per yard was allowed for it in easy soil, and 4d. in hard stiff soil for the diggers, whilst the prices for wheeling remained the same as before.

It is also to be remarked, that if the system of throwing earth from one level to a higher one be adopted, which is sometimes used in deep cutting, instead of wheeling it up an inclined plane, the price per yard for throwing it to the height of 5 or 6 feet is the same as for wheeling the 20 yards up the plane; but in this case the first throwing performed by the party of men who excavated was not allowed as extra work. Moreover, the men were not allowed in any case to charge extra for time lost in placing and moving their wheeling planks, &c.

I am informed that by this task the military working parties usually earned about a day and a-half, which appears a fair proportion for labourers work, as 1¼ day of great exertion would scarcely be enough, and 1¾ or 2 days would be too much.*

Estimate of Materials for One Rod of Brickwork, according to the Builder's Price Book.

4500 bricks, including waste.†
1½ hundred of lime equal to 37½ striked bushels.
3 loads of sand equal to 66 striked bushels.

Estimate of labour.

It is computed that a good bricklayer, without over exertion, and having bricks and mortar continually supplied to his hand, will build one standard rod of plain brickwork of Flemish bond, in which the outside wall only is worked fair, according to custom, in 5 days, of 10 hours each; but in boundary walls, in which both sides must be worked fair, it will require from 5½ to 6 days. English bond, which dispenses

* This information was furnished by Mr. Robert Howe, who was overseer in the Engineer Department in the Eastern district at the time those tasks were fixed.

† In the brickwork of wells and cesspools, which are often built dry, more bricks must necessarily be employed, but mortar is of course saved.

with the continual cutting of bricks, may be executed in less time, and consequently cheaper.

On first commencing a building, one labourer may supply each bricklayer, provided that the bricks and mortar are close to the work; otherwise, more than this proportion of labourers may be necessary. As the walls rise the number of labourers must be increased, so that eventually each bricklayer may require two labourers to assist him, or even more than that proportion. But when bricklayer's work is let by the principal contractor to two or three working bricklayers, as sub-contractors—which will sometimes be done, although certainly an improper practice, and which the engineer or architect ought not to permit—in this case, the bricklayers, being personally interested, select and pay their own workmen, and, in respect to labourers in particular, they employ only men of great vigour to assist them, whom they pay at a higher rate, to incite them to extraordinary exertion.

In day work it is of course most expedient to employ none but able-bodied men, but even in task work, such as foundations, the same rule should be followed, for the most robust and expert labourers will, at the same price, get through the work much quicker; and all unnecessary delays in building are prejudicial, and generally attended afterwards with extra expense. On the other hand, all hurry in the execution of the brickwork itself is to be reprobated, as it may cause partial settlements that would not otherwise take place.

In a former article I mentioned the number of extras usually charged in bricklayers' work, and stated that they should be well defined and limited in the schedule of particulars. Whenever it is practicable, it is best to curtail them as much as possible. I find that this has very judiciously been done in the present contract for bricklayers' work in the new British Museum. There are only three prices, all by the rod—1st, for the brickwork generally, as executed in Dorking lime mortar; 2nd, for the brickwork of the vaults, which bears a higher price; and 3rdly, for brickwork laid in cement, which bears a still higher price than either. Thus, the whole is measured and paid in the simplest possible manner.

It was also mentioned that architects in civil life seldom or never make out a regular and detailed estimate of their own designs, excepting, I believe, those employed under the commissioners for the erection of new churches, who, from the regulations laid down by those commissioners, have found it absolutely necessary to enter into this matter in detail, and with more than usual attention.

Yet, as not only his employers, but the architect himself, must be desirous of knowing what the building is likely to cost, I am informed that the custom is for an architect to make a rough estimate, on the following grounds:—

First, the number of cubic feet contained under the roof of the proposed building, which is found by taking the area or ground covered by it, and the heights, as shown in a transverse section.

Secondly, the number of floors.

Thirdly, the style in which the building is ornamented outside and finished inside.

From those particulars combined, and knowing by his own experience, or by information, the expense that has been actually incurred in erecting former buildings of a similar description, he is enabled to make a rough estimate of the probable expense of any new design sufficiently near for his own satisfaction, if not for that of his employers.

It has been suggested that it would be an improvement in the mode of tendering to execute a building for a specific sum, if the competitors were relieved from the trouble and expense of each measuring the whole work separately, by the employment of an intermediate person, or of two persons, of known skill in the art of measurement, who should give in a statement of all the quantities of each branch of artificers' work contained in the architect's design, and who should be paid by the whole of the competitors jointly; the latter, however, being at liberty, for their own security or satisfaction, to take out their own quantities if they doubted the competency of the professed measurers thus employed. It appears to me, however, that this is only a half measure, and that it would be better in all cases to call upon the architect himself to furnish the competitors for a contract with those quantities. To this arrangement the architects of the present day would naturally object, unless they were paid extra, for they would be obliged to employ and pay professed measurers or surveyors in addition to the clerks and draughtsmen, by whose assistance they get through their present duties; and on the other hand, in the case of such extra allowance being granted, which certainly would be no more than reasonable, it may be objected that the expense of estimating, which now falls entirely on the contractor, would fall on his employer. To this I reply, that a builder who is relieved from the trouble and expense of measuring the quantities alluded to would be able to take his contract on cheaper terms, in proportion to the money or time saved; for the time of men of business is of value, even when they measure and estimate for themselves, instead of having another person.

Whichever mode be adopted, a certain benefit would result from the various tenders being more in accordance with each other than is usually the case at present, for it has been remarked that the very great difference which often occurs in the sums demanded for the execution of a whole building by various competitors, generally arises from a difference in the calculated quantities of work, proceeding from errors in the measurement of one or more of those persons, and not from any great difference in the prices according to which they estimate.

Supplementary Remarks, correcting an omission on the subject of Chimneys. The conclusion of Brickwork.

In treating of chimneys, I enlarged more upon double stacks than on single ones. The latter are, however, no less common, or rather more so.

In London and in most other cities a row of houses is built with one door and two windows in the front of each. When the doors of all the buildings in the row alternate regularly with the windows, the stacks of chimneys are always single. When the doors of each pair of adjacent houses are next to each other, the stack is always double, and the chimneys are placed back to back.

In the new buildings at Dover, in which the two doors of every pair of houses are next to each other, all the fireplaces are placed back to back; but yet the stack of chimneys are not, properly speaking, double, for, at the height of two or three feet above the mantle-piece, each flue is carried obliquely into the centre of a two brick party wall, which is thus completely honeycombed towards the top.

Some of those houses being 5 stories high, and all of them having 2 rooms in depth, there are no less than 20 flues inserted into one of those party walls, 10 of which belong to one house, and 10 to the neighbouring one; and those flues being 1 by $1\frac{1}{2}$ brick square in the clear, with 4 inch withs, the 20 occupy a space no less than about 30 feet in length at the extreme summit of the wall, where each is usually crowned by a chimney pot. Thus the chimneys, as they appear at the roof, occupy nearly the whole depth of the building from front to rear, which is not the case in any of the buildings about London, or in this neighbourhood.

It will be remarked that the Dover method of chimney building renders the party wall much weaker than the other, in which a much more solid wall is buttressed by the chimney breasts, which contain the fireplaces and flues.

In the extras of brickwork I also omitted to mention Mr. Hiort's patent chimneys, which would be charged over and above the usual price allowed for common flues, in proportion to their dimensions, as they require chimney bricks baked on purpose, which are more expensive than common bricks. The present prices for the execution of those chimneys by the lineal foot, as well as for the various materials composing them, may be ascertained by referring to the inventor's treatise on the subject.

Having now finished that part of practical architecture which comprehends the brickwork of buildings, and the matters more specially connected with it, I shall not now enter into any other branches of workmanship, such as plastering, tiling, paving, &c., merely because they are generally executed by the bricklayer in the country. The more natural order of studying practical architecture is, first to consider the brickwork; secondly, the carpenter's work; and thirdly, the finishing of the building, which involves plastering, joinery, the covering of the roof, &c. At the conclusion of the first of those branches I stop for the present.

INDEX.